Yeast Strain Selection

Bioprocess Technology

Series Editor

W. Courtney McGregor

Xoma Corporation
Berkeley, California

Yeast Strain Selection

edited by

Chandra J. Panchal
VetroGen Corporation
London, Ontario, Canada

CRC Press
Taylor & Francis Group
Boca Raton London New York

CRC Press is an imprint of the
Taylor & Francis Group, an **informa** business

CRC Press
Taylor & Francis Group
6000 Broken Sound Parkway NW, Suite 300
Boca Raton, FL 33487-2742

First issued in paperback 2019

ISBN-13: 978-0-367-40297-6

Visit the Taylor & Francis Web site at
http://www.taylorandfrancis.com

and the CRC Press Web site at
http://www.crcpress.com

Series Introduction

The revolutionary developments in recombinant DNA and hybridoma technologies that began in the mid-1970s have helped to spawn several hundred new business enterprises. Not all these companies are aimed at producing gene products or cell products, as such. Many are supportive in nature; that is, they provide contract research, processing equipment, and various other services in support of companies that actually produce cell products. With time, some small companies will probably drop out or be absorbed by larger, more established firms. Others will mature and manufacture their own product lines. As this evolution takes place, an explosive synergism among the various industries and the universities will result in the conversion of laboratory science into industrial processing. Such a movement, necessarily profit driven, will result in many benefits to humanity.

New bioprocessing techniques will be developed and more conventional ones will be revised because of the influence of the new biotechnology. As bioprocess technology evolves, there will be a need to provide substantive documentation of the developments for those who follow the field. It is expected that the technologies will continue to develop rapidly, just as the life sciences have developed rapidly over the past 20 years. No single book will cover all of these developments adequately. Indeed, some books will be in need of replacement or revision every few years. Therefore, our continuing series in this rapidly moving field will document the growth of bioprocess technology as it happens.

The numerous cell products already in the marketplace, and the others expected to arrive, in most cases come from three types of bioreactors: (a) classical fermentation; (b) cell culture technology; and (c) enzyme bioreactors. Common to the production of all cell products or cell product analogs will be bioprocess control, downstream processing (separation and purification), and bioproduct finishing and formulation. These major branches of bioprocess technology will be represented by cornerstone books, even though they may not appear first. Other subbranches will appear, and over time, the bioprocess technology "tree" will take shape and continue growing by natural selection.

W. Courtney McGregor

Preface

Although aware of yeasts and their usage in traditional biotechnology industries, such as bread making and wine and beer production, we microbiologists continue to be fascinated by the intricacies of these rather unique organisms, which have contributed immensely to human civilization (as well as suffering). The developments in molecular biology and genetic engineering have quite naturally lent themselves to a more thorough investigation of yeasts at the molecular level.

And what have we found? We have seen that yeasts, particularly *Saccharomyces cerevisiae* and some of the *Kluyveromyces* species, are readily amenable to genetic manipulation and possess many properties that make them ideal candidates for commercial exploitation as well as academic research. It has been learned that a whole host of foreign proteins can be quite readily produced in and secreted by yeast strains. It has also been revealed that some yeasts possess viruslike particles, and lately we have been exposed to the striking similarities between yeast transposable elements (Ty) and human retroviruses. We have recently, however, also been privy to information showing major differences among yeast strains, particularly between "laboratory" yeasts and "industrial" yeasts. Thanks to the much-refined techniques of DNA karyotyping, some of these differences have been revealed clearly and some of the similarities among the species and genera have been highlighted. An increasing amount of attention is being paid to usage of yeasts for commercial production of novel biologicals; it seems pertinent that greater emphasis be placed on selection of appropriate yeast strains for the desired processes.

A look at how this is accomplished in the traditional bread, wine, and beer industries would help one assess the criteria used in these particular enterprises to select appropriate yeasts—a decision that could arguably be the single most important one in some of these industries. Chapters addressing these issues reveal that some aspects of classical yeast genetics as well as some aspects of yeast selection in nature play important roles. The criteria, however, are radically different from those used for selecting yeasts for the production of enzymes, amino acids, or heterologous pharmaceutical proteins. In the latter case, emphasis has been placed on stability of plasmid vector systems or the inability of host yeast strains to degrade foreign proteins. Should this be the case, or should some lessons be learned from a long and successful history of yeast usage in industry, and these be adapted for the more recent and novel needs in biotechnology? It is hoped that some of the answers will flow from the chapters of this book.

The editor wishes to thank the publisher for initiating the project and the production staff, particularly Elaine Grohman, for continued support and help in completing the book. Sincere appreciation and gratitude go to all the authors for their time, efforts, and dedication to completing the chapters and above all for their patience.

Chandra J. Panchal

Contents

Contributors

Carl A. Bilinski Labatt Brewing Co. Ltd., London, Ontario, Canada

Douglas A. Campbell Miles Inc., Elkhart, Indiana

Gregory Paul Casey Anheuser-Busch Companies, St. Louis, Missouri

Rathin C. Das* Miles Inc., Elkhart, Indiana

Malcolm A. J. Finkelman Genex Corporation, Gaithersburg, Maryland

Ronald D. Klein The Upjohn Company, Kalamazoo, Michigan

Cletus P. Kurtzman Northern Regional Research Center, Agricultural Research Service, U.S. Department of Agriculture, Peoria, Illinois

*Present affiliation: Miles Research Center, West Haven, Connecticut.

Marc-André Lachance Department of Plant Sciences, University of Western Ontario, London, Ontario, Canada

Nelson Marmiroli* Institute of Genetics, University of Parma, Parma, Italy

Tilak W. Nagodawithana Universal Foods Corporation, Milwaukee, Wisconsin

Chandra J. Panchal VetroGen Corporation, London, Ontario, Canada

Ronald Ernest Subden Department of Microbiology, University of Guelph, Guelph, Ontario, Canada

Flavio Cesar Almeida Tavares Escola Superior de Agricultura, Instituto de Genetica, University of São Paulo, Piracicaba, Brazil

Nayan B. Trivedi Universal Foods Corporation, Milwaukee, Wisconsin

Phillip G. Zaworski The Upjohn Company, Kalamazoo, Michigan

Present affiliation: Division of Genetics, Department of Biology, University of Lecce, Lecce, Italy.

Yeast Strain Selection

1
Culture Collections as Sources of Strains for Industrial Uses

Cletus P. Kurtzman
Northern Regional Research Center
Agricultural Research Service
U.S. Department of Agriculture
Peoria, Illinois

I. INTRODUCTION

The rapid advances taking place in biotechnology have introduced large numbers of scientists and engineers to the need for locating yeasts and other microorganisms that have a diversity of metabolic functions. Questions frequently raised by these researchers concern general sources of cultures, location of strains with particular properties, requirements for handling cultures, preservation methods, and even taxonomic relationships. This chapter will focus on these questions and emphasize the importance of culture collections as a primary source of unique strains and of information on these strains. The organisms maintained in culture collections have originated from a wide variety of substrates and from many parts of the world. Consequently, the culture collection offers access to strains of demonstrated industrial significance as well as to taxa of varied genotype whose properties are yet to be exploited.

II. MAJOR COLLECTIONS

Many of the world's major yeast collections are listed in Table 1. The reasons for which these collections were initially established vary, and as a consequence, this has influenced the types of germplasm that predominate. Most maintain

1

Table 1. Major Yeast Culture Collections

USA
   Agricultural Research Service Culture Collection (NRRL), Northern Regional Research Center, 1815 N. University St., Peoria, IL 61604

   American Type Culture Collection (ATCC), 12301 Parklawn Dr., Rockville, MD 20852

   Culture Collection, Department of Food Science and Technology (UCD), University of California, Davis, CA 95616

   Yeast Genetic Stock Center (YGSC), Department of Biophysics and Medical Physics, University of California, Berkeley, CA 94720

The Netherlands
   Centraalbureau voor Schimmelcultures (CBS), Yeast Division, Julianalaan 67A, 2628 BC Delft

Japan
   Institute for Fermentation (IFO), 17-85 Juso-honmachi 2-chome, Yodogawa-ku, Osaka 532

   Japan Collection of Microorganisms (JCM), RIKEN, Hirosawa, Wako-shi, Saitama 351-01

United Kingdom
   National Collection of Yeast Cultures (NCYC), AFRC Food Research Institute, Colney Lane, Norwich NR4 7UA

Federal Republic of Germany
   Deutsche Sammlung von Mikroorganismen (DSM), Grisebachstrasse 8, D-3400 Göttingen

Czechoslovakia
   Czechoslovak Collection of Yeasts (CCY), Institute of Chemistry, Slovak Academy of Sciences, Dubravska cesta 9, 84238 Bratislava

   Research Institute for Viticulture and Enology (RIVE), Matuskova 25, 83311 Bratislava

Peoples Republic of China
   Center for Collection of General Microbiological Cultures (CCGMC), Institute of Microbiology, Academia Sinica, P.O. Box 2714, Beijing

USSR
   Department of Type Cultures of Microorganisms, Institute of Biochemistry and Physiology of Microorganisms, USSR Academy of Sciences, Pushino, Moscow region, 142292

representative strains of the major taxa but also have numerous isolates in their area of particular interest. For example, the National Collection of Yeast Cultures has a preponderance of brewing strains whereas the ARS Culture Collection has focused on agricultural and general industrial isolates and the Yeast Genetics Stock Center has predominantly genetic selections of *Saccharomyces cerevisiae*.

Cultures may be obtained from any of the collections by written and telex requests, and some accept telephone orders. Nearly all charge for their cultures and other services, but these fees almost never cover the total cost of operating the collection, and the deficit is made up from public funds. Holdings are generally listed in catalogs which indicate the sources of the strains and sometimes their applications. A few collections, such as the ARS Culture Collection, do not issue a catalog, but curators provide lists of specific groups of cultures on request (1).

III. SOURCES OF STRAINS

Culture collections obtain strains from a variety of sources: (a) requests to researchers, (b) unsolicited contributions from researchers, (c) isolates sent for identification, and (d) strains collected by the curators during their research. Culture collection curators generally actively scan the scientific literature for new taxa and interesting strains of known species. This is the main means whereby curators keep their collections current, and their own research interests may prompt the collection to acquire large holdings in particular areas.

New taxa and strains of currently recognized taxa from varied and unusual habitats provide the biotechnologist with novel germ plasm. It is generally beyond the capability of any one scientist or small group of scientists to do extensive worldwide collecting over an extended period of time. However, the holdings of a culture collection offer the researcher the advantages of access to a worldwide collection program staffed by hundreds of scientists. It is for this reason that thoughtful selection of strains from culture collections may provide the necessary germ plasm for new products and processes. Sometimes it is helpful to communicate needs for particular cultures directly to the curators since not all strain data can be placed in a catalog.

IV. GENETIC, ECOLOGICAL, AND PHYSIOLOGICAL PROPERTIES OF STRAINS

The sexual properties of many of the strains in culture collections have been tested, and some knowledge of their genetics is usually available. These comparisons may have been done by the original investigator before the strain was accessioned, or they result from the systematic examination of strains by the curators. Genetic makeup as a reflection of nucleic acid base sequences will be discussed in Section VI.

The identification of yeasts is based in part on their fermentation and assimilation responses to a large variety of carbon compounds as well as to their utilization of certain nitrogen compounds. Comparisons may also include growth at various temperatures, vitamin requirements, and production of proteases, lipases, and amylases. Compounds commonly used in fermentation tests include D-glucose, D-galactose, sucrose, maltose, lactose, raffinose, trehalose, D-xylose, and sometimes melibiose and soluble starch. The wide variety of compounds employed in assimilation tests is listed in Table 2.

Responses to growth and fermentation tests reflect some of the metabolic activities of strains. Ability to assimilate starch demonstrates the presence of extracellular amylase, a property that might be useful industrially if the particular enzyme can function under desired reaction conditions. Additional enzymatic activity can sometimes be predicted on the basis of other test results. The fermentation of D-xylose is such an example. Earlier, yeasts were not believed to be able to ferment this compound, and it was not included among fermentation tests. However, by looking at yeasts that ferment hexoses and assimilate pentoses, it was possible to predict that species such as *Pachysolen tannophilus* might ferment D-xylose. This prediction proved true and serves as an example of the types of pathways that can be inferred from standard tests (2, 3).

Assimilation and fermentation test results are summarized in taxonomic treatises such as *The Yeasts, A Taxonomic Study* (4) and *Yeasts: Characteristics and Identification* (5). The reactions listed represent the consensus for all known strains of a species. Consequently, interesting combinations of variable reactions that might be found in a single strain are usually not given in these works but are frequently available from culture collection records.

Table 2. Compounds Typically Used in Assimilation and Fermentation Tests for the Identification of Yeasts

---

Hexoses: D-glucose, D-galactose, L-rhamnose, L-sorbose
Pentoses: D-xylose, D-ribose, L-arabinose, D-arabinose
Disaccharides: sucrose, maltose, cellobiose, trehalose, lactose, melibiose
Trisaccharides: raffinose, melezitose
Polysaccharides: soluble starch, inulin
Alcohols: erythritol, ribitol (adonitol), D-mannitol, inositol, methanol, ethanol,
                glycerol, galactitol (dulcitol), D-glucitol (sorbitol)
Organic acids: succinic acid, citric acid, DL-lactic acid, malic acid, gluconic
                acid, glucuronic acid, 2-ketogluconate, 5-ketogluconate
Glycosides: α-methyl-D-glucoside, arbutin, salicin
Other compounds: glucono-δ-lactone, D-glucosamine hydrochloride, decane,
                hexadecane

The recent effort by many culture collections to computerize strain data has made it possible to quickly retrieve cultures possessing particular combinations of physiological and genetic characteristics, something previously done only with difficulty. If isolation data are also available, physiological properties not examined on standard tests can sometimes be inferred from the strain's habitat. Such strategies will be discussed in a later chapter.

The maintenance of certain yeast strains in culture collections for up to 80 years allows comparisons of the properties of these early cultures with those of recent isolates that have been subjected to selective pressures imposed by present-day society. Is the antibiotic resistance seen in current clinical isolates to be found in strains from an earlier age? Is the distribution of plasmids among contemporary brewing strains different from that of 50 years ago? An opportunity to understand the genetics of such problems may be provided in the germ plasm maintained by culture collections.

V. PRESERVATIONS OF STRAINS

The impact of preservation methods on strain selection needs to be considered from the perspective of stability of various properties. Will a preservation method maintain the particular trait once it is located, and just as important, can the trait be found during a survey if strains have been preserved in a particular way?

Serial transfer on agar slants is one of the most common methods for culture maintenance. Experience in many laboratories has shown that certain strains can have a rather short lifetime on agar. Notable examples are cultures of *Brettanomyces*, which die within a week or so because of acetic acid production. Not so obvious is the selection of variants that are more competitive under laboratory growth conditions. Many changes go unnoticed, but two examples may be found in sporogenous diploid cultures of the heterothallic species *Yarrowia lipolytica* and *Issatchenkia terricola*. In the author's laboratory, a culture of the first species changed from diploid to become exclusively one mating type after several serial transfers over a 2-month period. In the other case, the type strain of *I. terricola* was described as a sporogenous diploid, but when it was received after an unknown number of serial transfers, it too represented a single mating type (6).

Many culture collections now preserve their strains as dried preparations to preclude selection during repeated transfer. It is common in genetics laboratories to air-dry a drop of the culture on a small square of filter paper and freeze the preparation in a sterile foil wrap. Quantitative survival data are generally unavailable for this method, but genetic stability appears good. Several Japanese culture collections use the process of L-drying in which in vacuo drying occurs while the cells are suspended in a liquid (7). Lyophilization (freeze-drying) is the most commonly used drying method for microbial cell preservation. Cell survival is frequently low following both L-drying and lyophilization, but cell death appears random and the surviving cells frequently remain viable for decades (7, 8). Kirsop (9) has reported some changes in sugar fermentation and vitamin requirements following lyophilization of brewing yeasts, but the rate of change was much less than experienced for cultures maintained by serial transfer. Little is known concerning the stability of plasmid-bearing and recombinant strains following lyophilization. Experience in the ARS Culture Collection, as well as in other culture collections, suggests that changes in these types of cultures may not be a significant problem.

Cultures maintained by liquid nitrogen freezing show high cell survival, suggesting that this procedure is the most reliable method of preservation (10, 11). Reports of genetic change are nearly nonexistent, and the life span of liquid nitrogen frozen cells is expected to exceed that of lyophilized preparations (12).

VI. TAXONOMY AS A GUIDE TO STRAIN SELECTION

A. State of Contemporary Taxonomy

The value of taxonomy as a guide to strain selection for industrial uses lies in its potential as an indicator of genetic relatedness. If we assume that different species are indeed reproductively isolated, we can exclude the possibility of strain improvement by interspecific hybridization using conventional mating techniques. This, however, does not preclude interspecific crosses made by such techniques as protoplast fusion. It is generally assumed that protoplast fusion products are more stable if constructed from closely related species. If this is true, a knowledge of species relationships becomes quite important.

Contemporary taxonomy relies heavily on phenotypic characters for the definition of species and genera. Prominent among these characters are the ability to ferment and to assimilate various carbon compounds such as sugars and sugar alcohols. Growth on different nitrogen sources and growth in the absence of vitamins are also important tests. Morphological features such as the presence or absence of hyphae and pseudohyphae have been used to define both species and genera. Ascospore shape and ornamentation are other criteria used to separate taxa (4).

Knowing the genetic basis of morphological and physiological features is central to determining their value for recognizing biologically distinct species and genera. Early genetic studies, especially those by Winge and Roberts (13) and Lindegren and Lindegren (14), showed that a single gene could confer on a yeast the ability to ferment or assimilate a sugar, and this raised the possibility that species might be inadvertently and artificially established simply because of the functionality of one or a few genes. Following these findings was the discovery by Wickerham and Burton (15) that ascospore shape in *Pichia ohmeri* might be either hat-shaped or spheroidal, depending on which complementary mating types were paired. While these studies cast some doubt on the validity of physiological tests and ascospore shape as a means to separate species and genera, there were insufficient comparisons to justify discarding criteria that generally seemed to offer reliable resolution of species. Furthermore, the difficulty of doing definitive genetic studies, especially among homothallic species, has hampered the genetic evaluation of many traditional taxonomic criteria.

Comparisons of nucleic acid relatedness have provided a more convenient genetic means for evaluating commonly used phenotypic criteria. The methods employed and their limits of resolution are discussed in the following sections.

B. Nuclear DNA Relatedness

The first of the molecular criteria used to define species was the guanine + cytosine (G+C) content of the nuclear DNA. The taxonomic uses of G+C values are mainly exclusionary because the 600 or so known yeast species range in G+C content from approximately 28–70 mol%, and overlap between unrelated species is inevitable. The G+C values are usually determined from buoyant density in cesium chloride gradients generated by ultracentrifugation (16) or from thermal denaturation (17). When determined by buoyant density, a difference in G+C content of 1.0–1.5 mol% or greater indicates strains to be different species (18–20) while this difference is 2.0–2.5 mol% if determined from thermal denaturation (21).

Examination of G+C values for a variety of taxa illustrates the interesting fact that the G+C contents of ascomycetous yeasts is about 28–50%, whereas that of basidiomycetous yeasts is approximately 50–70% (22, 23). Except for the narrow range of 48–52%, where some overlap occurs, the taxonomic class of anamorphs (imperfects) can be reliably determined from their base composition. The range of G+C contents among species within a genus is quite often 10% or less as found in *Debaryomyces*, *Kluyveromyces*, and many other genera. By contrast, species in *Pichia* differ by as much as 22%, and it seems likely for this and other reasons that the genus may be phylogenetically heterogeneous.

Although G+C values have been helpful for separation of taxa, their uses are clearly limited. Considerably greater resolution is offered from comparisons of DNA complementarity as measured from reassociation reactions. The methodology has been described in detail elsewhere (18, 19, 24). The results from these comparisons are expressed as percent relatedness, but there has been uncertainty in deciding the point at which strains are considered separate species. Several workers have suggested that relatedness above 70–80% demonstrates conspecificity (18, 25).

Kurtzman and co-workers (19, 20) examined this question through comparisons of heterothallic yeasts in which DNA complementarity can be correlated with fertility. In the first of these studies, *Pichia amylophila* and *P. mississippiensis* were shown to mate well, but ascospores were not viable. The two species showed only about 25% DNA relatedness, thus correlating lack of fertility with low DNA complementarity. However, the relationship between *Issatchenkia scutulata* var. *scutulata* and *I. scutulata* var. *exigua* was somewhat different. These two taxa also showed only about 25% DNA relatedness, but genetic crosses gave 3–6% viable ascospores. Furthermore, crosses between these F_1 progeny gave 17% viability among F_2 progeny. Reciprocal crosses among the F_1 progeny were fertile, as were backcrosses to the parents. From this, it appears that all stocks had essentially homologous chromosomes and that progeny were neither amphidiploids nor aneuploids. Clearly then, the lower limit of DNA-DNA homology values suggesting species delimination is not yet well defined, but base sequence divergence, as estimated from whole genome comparisons, may be as great as 75% before genetic exchange can no longer occur. Data in Table 3 summarize the preceding findings and provide further comparisons of other taxa. In these studies, progressively less DNA relatedness parallels decreasing mating competence and fertility. These data also indicate that resolution goes no further than the detection

Table 3. Correlation of Mating Reaction and DNA Complementarity Among Closely Related Heterothallic Ascomycetous and Basidiomycetous Yeasts

Species	Mating reaction	% DNA relatedness and reference
Filobasidiella neoformans var. *neoformans* X *Filobasidiella neoformans* var. *bacillispora*	Fair conjugation 0–30% basidiospore viability (F_1progeny; F_2 not determined)	55–63 (65)
Issatchenkia scutulata var. *scutulata* X *Issatchenkia scutulata* var. *exigua*	Good conjugation Ascospores viable: F_1 = 5%; F_2 = 17%	21–26 (20)
Pichia amylophila X *Pichia mississippiensis*	Good conjugation Ascospores not viable	20–27 (19)
Pichia(Hansenula) bimundalis X *Pichia (Hansenula) americana*	Poor conjugation Ascospores not produced	21(42)
Pichia (Hansenula) alni X *Pichia canadensis (Hansenula wingei)*	Poor conjugation Ascospores not produced	6 (41, 66)
Issatchenkia orientalis X *Issatchenkia occidentalis*	Infrequent conjugation Ascospores not produced	3–8 (20)

of sibling species. Beyond this, all species, regardless of extent of kinship, show essentially less than 10% complementarity.

However, despite the strong correlation between DNA relatedness and fertility exhibited in Table 3, exceptions may occur, and whole-genome DNA complementarity should therefore be regarded as a strong but not infallible indicator of biological relatedness. Exceptions to this trend would include chromosomal changes that affect fertility between strains such as inversions, translocations, autopolyploidy, and aphidiploidy (allopolyploidy). Most of these changes would not be detectable in whole-genome DNA comparisons. However, amphidiploidy is recognizable if parental species have been included in the study or should be suspected if genome sizes vary between strains showing substantial relatedness. For example, *Saccharomyces carlsbergensis* shows relatively high DNA homology with both *S. cerevisiae* and *S. bayanus* even though these latter two species demonstrate little relatedness (Tables 4, 5). A comparison of genome sizes suggests *S. carlsbergensis* to be a partial amphidiploid that arose as a natural hybrid of *S. cerevisiae* and *S. bayanus* (26). In this case, we would predict *S. carlsbergensis* to be infertile with the proposed parents despite relatively high DNA relatedness.

DNA relatedness studies have had an enormous impact on our assessment of the criteria used to define species and genera among the yeasts (Table 6). Yarrow and Meyer (27) combined the genera *Candida* and *Torulopsis* when it was found that presence or absence of pseudohyphae might occur in a single species. Similarly, the 75% relatedness detected between *Pichia lindneri* and *Hansenula minuta* prompted Kurtzman (28) to propose that the two genera, which are separated on ability to assimilate nitrate, be combined. The DNA comparisons have also complemented earlier genetic studies in their demonstration that differences in fermentation and carbon assimilation are common among strains of a species.

C. Mitochondrial DNA Relatedness

Comparisons of mitochondrial DNA (mtDNA) provide another means for examining taxonomic relationships. Relatedness has been examined through comparisons of fragment patterns generated with restriction endonucleases. The small size of the mitochondrial genome makes such comparisons practical, whereas the larger nuclear genome could present more fragments than might be reasonably handled. McArthur and Clark-Walker (29) used mtDNA restriction patterns to correlate perfect-imperfect relationships between the yeast genera *Dekkera* and *Brettanomyces*. They found identical restriction patterns for the pair *D. bruxellensis/B. lambicus*, as well as for the imperfects *B. abstinens/B. custersii* and *B. anomalus/B. clausenii*. This strongly suggests the pairs to be conspecific. Size differences in mtDNA among the other species assigned to these genera prevented an unambiguous assessment of their relationships. It appears that

Table 4. Extent of DNA Reassociation Beween *Saccharomyces cerevisiae* NRRL Y-12632[Ta]and Other Species of *Saccharomyces* sensu stricto

Species	NRRL No.	% DNA reassociation[b]
S. aceti	Y-12617[T]	96
S. beticus	Y-12625[T]	99
S. capensis	YB-4237[T]	94
	Y-12629	99
S. cerevisiae	Y-2034	91
S. chevalieri	Y-12633[T]	96
S. cordubensis	Y-12636[T]	100
S. coreanus	Y-12637[T]	94
S. diastaticus	Y-2416	92
S. gaditensis	Y-12644[T]	95
S. hienipiensis	Y-6677[T]	95
S. hispalensis	Y-11846[T]	94
S. italicus	Y-12649[T]	96
S. norbensis	Y-12656[T]	90
S. oleaceus	Y-12657[T]	94
S. oleaginosus	Y-6679[T]	89
	Y-12659	86
S. prostoserdovii	Y-6680	93
	Y-12660	101
S. carlsbergensis	Y-12693[T]	57
S. abuliensis	Y-11845[T]	15
S. bayanus	Y-12624[T]	5
S. globosus	Y-12645[T]	3
	Y-12646	8
S. heterogenicus	Y-1354[T]	20
S. inusitatus	Y-12648[T]	24
S. kluyveri	Y-12651[T]	-3
S. uvarum	Y-12663[T]	14
Debaryomyces melissophilus	Y-7585[T]	6

[a]T = Type strain.
[b]Data from Vaughan Martini and Kurtzman (26).

Table 5. Extent of DNA Reassociation Between *Saccharomyces bayanus* NRRL Y-12624[Ta] and *Saccharomyces* sensu stricto Species Showing Low DNA Relatedness with *S. cerevisiae*

Species	NRRL No.	% DNA reassociation[b]
S. abuliensis	Y-11845[T]	91
S. globosus	Y-12645[T]	104
	Y-12646	87
S. heterogenicus	Y-1354[T]	86
S. inusitatus	Y-12648[T]	94
S. uvarum	Y-12663[T]	98
S. carlsbergensis	Y-12693[T]	72
S. kluyveri	Y-12651[T]	24
Debaryomyces melissophilus	Y-7585[T]	1

[a]T = Type strain.
[b]Data from Vaughan Martini and Kurtzman (26).

Table 6. DNA Relatedness Between Yeast Species Differing in Traditional Taxonomic Characteristics

Species	Characteristic/+ or /−	% DNA relatedness and reference
Candida slooffii	Pseudohyphae/+	80 (67)
Torulopsis pintolopesii	/−	
Hansenula wingei	True hyphae/+	78 (41)
Hansenula canadensis	/−	
Debaryomyces formicarius	Glucose ferm/+	96 (18)
Debaryomyces vanriji	/−	
Schwanniomyces castellii	Lactose assim/+	97 (18)
Schwanniomyces occidentalis	/−	
Hansenula minuta	Nitrate assim/+	75 (28)
Pichia lindneri	/−	
Sterigmatomyces halophilus	Nitrate assim/+	100 (68)
Sterigmatomyces indicus	/−	

the taxonomic resolution from mtDNA restriction patterns is no greater than afforded by whole nuclear DNA reassociations, but the technique might allow detection of subspecies from specialized habitats.

D. Ribosomal DNA Relatedness

The rather narrow resolution provided by whole-genome DNA reassociation and mtDNA restriction analysis does not allow verification of species assignments within genera or an understanding of intergeneric relationships. The DNA coding for ribosomal RNA (rRNA) contains some of the most highly conserved sequences known, and it offers a means for assessing affinities above the species level. One of the commonest methods for measuring rRNA complementarity consists of immobilizing single-stranded DNA onto nitrocellulose filters and incubating the filters in a buffer solution with radiolabeled rRNA (30–32). A modification of this method was developed by Baharaeen and co-workers (33) in which complementary DNA was synthesized on 25S rRNA fragments and then allowed to hybridize with rRNA in solution. As we will discuss, another method that holds promise for phylogenetic studies is the comparison of sequences from rRNA subunits.

Bicknell and Douglas (30) were among the first to employ rRNA comparisons to assess phylogenetic relatedness among fungi. Their study focused on the genus *Saccharomyces*, some species of which have now been assigned to the genera *Zygosaccharomyces*, *Kluyveromyces*, and *Torulaspora*. In general, the more highly related species clustered into groups corresponding to present generic assignments. Similar results were obtained for this same group of species by Adoutte-Panvier and co-workers (34) through electrophoretic and immunochemical comparisons of ribosomal proteins. Bicknell and Douglas (30) also showed that 25S rRNA comparisons cannot resolve the phylogeny of closely related species because of the highly conserved nature of the sequences.

Walker and Doolittle (35, 36) used 5S rRNA sequences to investigate relatedness among the Basidiomycetes, including several yeast taxa. Their work identified two distinct clusters that correlated with the presence or absence of septal dolipores rather than with the traditional separation of these species into the classes Heterobasidiomycetae and Homobasidiomycetae. Further, the data suggest that capped dolipores evolved from capless dolipores which may have evolved from single septal pores. In continuing work, Walker's (37) analysis of 5S rRNA sequences from several ascomycete genera suggested much greater divergence than previously suspected. Blanz and Gottschalk (38) have provided comparisons of 5S rRNA from certain of the lower basidiomycetes.

Because 5S rRNA is such a small molecule, only about 119 nucleotides are available for comparison (38). Sequences of the larger rRNA molecules potentially offer much grater resolution, but sequencing the whole 18S or 26S rRNA molecule is at present an enormous task, especially considering the large number of species that most taxonomists need to compare. Another approach is to partially sequence the larger molecules (39). For example, with specific primers one may select a portion of the molecule and perhaps detect up to 350 nucleotides on a single sequencing gel. Additional primers may be used depending on the degree of resolution needed. There also exists the possibility that not all of the rRNA sequences are highly conserved, and with appropriate selection of primers, variable regions will offer resolution of relatively close relationships.

E. Other Methodologies

Examples have been given in which species defined by traditional methods did not coincide with those determined from comparisons of nucleic acids. Taxonomic conclusions drawn from other methodologies, such as numerical analysis or from proton magnetic resonance (PMR) spectra and serology of cell wall mannans, have sometimes been at odds with data from nucleic acids studies. For example, using numerical analysis, Campbell (40) proposed reduction of *Hansenula wingei, H. canadensis,* and both varieties of *H. bimundalis* to a single species. Fuson and co-workers (41) showed from studies of DNA complementarity that *H. canadensis* and *H. wingei* are conspecific, whereas this latter species and *H. bimundalis* show little relatedness. Kurtzman (42) demonstrated only 21% DNA complementarity between the two varieties of *H. bimundalis.* This disparity of the two methods probably results from the relatively limited amount of phenotypic data available for numerical analysis, whereas DNA studies have the whole genome as a data base. This also appears to be the case for separations made using PMR, since Spencer and Gorin (43) found the spectra of cell wall mannans from *H. beckii* and *H. canadensis* to be quite similar, but different from the pattern shared by *H. wingei* and the varieties of *H. bimundalis.* In further comparisons of this group, Tsuchiya and co-workers (44) reported the cell surface antigens of *H. wingei, H. canadensis,* and *H. beckii* to be indistinguishable. The reason for this discrepancy is not at all clear, but Ballou (45) has shown for *Saccharomyces* mannans that single gene changes can impart significant differences in immunological reaction. The difference between the two main PMR spectral types of mannan in *S. cerevisiae* was also found to be controlled by a single gene (46).

Differentiation of many ascosporogenous genera is based on ascospore shape and surface structure. An opportunity to test the soundness of this convention in the genera *Schwanniomyces, Saccharomyces, Debaryomyces,* and *Pichia* arose through a comparison of the DNA data of Price and co-workers (18) with the scanning electron microscopy studies by Kurtzman and Kreger–van Rij (47), Kurtzman and Smiley (48, 49), and Kurtzman and co-workers (50, 51). All four species of *Schwanniomyces* had high DNA relatedness (18) and similar spore architecture (50, 52). The least related, *S. persoonii* at 80% complementarity, showed fewer and less pronounced spore protuberances than the other three species, all four of which are now regarded as conspecific. Ascospore topography for many species of *Torulaspora* was essentially identical and also frequently indistinguishable from the majority of species in *Debaryomyces*. Although many species within their respective genera were found to be conspecific by DNA homology, others proved distinct despite spore similarity. The few species with recognizably different spores showed little DNA relatedness. Thus, species with unlike spores can be expected to show little DNA complementarity, but no prediction can be made concerning species with similar spores. A few exceptions exist among other groups. Ascospores of *Pichia ohmeri* may be either spheroidal or hat-shaped, depending on mating type (15, 53), and spore shape in *Yarrowia lipolytica* is mating type dependent (54). Spore shapes among *Kluyveromyces* spp. with high degrees of DNA complementarity may be either spheroidal or kidney-shaped (55).

Kreger–van Rij and Veenhuis (56) demonstrated by transmission electron microscopy a basic difference in hyphal septa among ascomycetous yeasts. Three classes of septum ultrastructure were observed: a single central pore (micropore), plasmodesmata, and a dolipore configuration. All mycelial species of *Hansenula* and *Pichia* have the micropore, whereas plasmodesmata are common to all *Saccharomycopsis* species except *S. lipolytica* (= *Yarrowia lipolytica*). Dolipores were found in species of *Ambrosiozyma*. Future comparisons of these species by molecular techniques that detect distant relatedness might show ultrastructural differences in septa to be of evolutionary significance, as now appears to be the case among the lower basidiomycetes (35, 36).

One issue that needs attention concern the means for detecting phylogenetic relationships of species within genera. As we have seen, whole-genome DNA complementarity is too specific, whereas rRNA comparisons may be too broad in resolution to detect closely related species (30). However, the DNA coding for certain enzymes such as glutamine synthetase and superoxide dismutase, although less conserved than that coding for rRNA, is still less divergent than is apparent from comparisons of whole-genome DNA (57, 58). By determining changes in the amino acid sequence of these enzymes through such techniques as quantitative microcomplement fixation, an estimate of intermediate relatedness may be found. Immunological studies of protein similarity are very sparse for yeasts, and the resultant information is highly dependent on the kinds of proteins studied. Lachance and Phaff (59) used exo-β-glucanases from species of *Kluyveromyces* but found that the enzyme was poorly conserved and that immunological distances between most species were too great to be reliable for determining relationships in the genus.

Electrophoresis of allozymes presents another means for estimating molecular diversity and, as with immunological studies, resolution depends on the extent of sequence conservation. The proportion of point mutations that are electrophoretically detectable is estimated at approximately 0.27, because of the redundancy in the genetic code and the large proportion of amino acids that are electrically neutral (60–62). The nearly universal occurrence of extensive protein polymorphisms in natural populations has led some workers to believe that variation would prove physiologically important and therefore under selective control; others regard it as without phenotypic effect and thus selectively neutral (61). Regardless, data derived by this technique do provide insight into evolutionary processes and taxonomy. Baptist andd Kurtzman (63) utilized comparative enzyme patterns to separate sexually active strains of *Cryptococcus laurentii* var. *laurentii*from nonreactive strains and from the varieties *magnus* and *flavescens*. Yamazaki and Komagata (64), utilizing similar techniques, examined species relationships in *Rhodotorula* and its teleomorph *Rhodosporidium*. Holzschu (62) studied the evolutionary relationships among some 400 strains of various cactophilic species of *Pichia*. His study of the banding patterns in starch gels of 14 metabolic enzymes allowed a determination of the genetic distances among the various yeast populations.

Our examples show that taxonomy is still in a state of flux but that many problems have been resolved through molecular comparisons. These new findings should result in a better understanding of yeast genetics and phylogeny, which will have a direct impact on strain selection for industrial purposes.

REFERENCES

1. Kurtzman, C. P., The ARS Culture Collection: Present status and new directions, *Enzyme Microb. Technol.*, 8:328–333 (1986).
2. Slininger, P. J., R. J. Bothast, J. R. Van Cauwenberge, and C. P. Kurtzman, Conversion of D-xylose to ethanol by the yeast *Pachysolen tannophilus, Biotechnol. Bioeng.*, 24:371–384 (1982).
3. Schneider, H., P. Y. Wang, Y. K. Chan, and R. Maleszka, Conversion of D-xylose into ethanol by the yeast *Pachysolen tannophilus, Biotechnol. Lett.*, 3:89–92 (1981).
4. Kreger–van Rij, N. J. W., *The Yeasts, A Taxonomic Study*. Elsevier, Amsterdam (1984).
5. Barnett, J. A., R. W. Payne, and D. Yarrow, *Yeasts, Characteristics and Identification*. Cambridge University Press, Cambridge, U.K. (1983).
6. Kurtzman, C. P., and M. J. Smiley, Heterothallism in *Pichia kudriavzevii* and *Pichia terricola, Antonie van Leeuwenhoek*, 42:355–363 (1976).
7. Mikata, K., S. Yamauchi, and I. Banno, Preservation of yeast cultures by L-drying: Viabilities of 1710 yeasts after drying and storage, *IFO Res. Commun.*, 11:25–46 (1983).
8. Wickerham, L. J., and A. A. Andreason, The lyophil process, its use in the preservation of yeasts, *Wallerstein Lab. Commun.*, 5:165–169 (1942).

9. Kirsop, B., Maintenance of yeasts by freeze-drying, *J. Inst. Brewing*, 61:466–471(1955).
10. Hwang, S. W., Effects of ultra-low temperature on the viability of selected fungus strains, *Mycologia*, 52:527–529 (1960).
11. Goos, R. D., E. E. Davis, and W. Butterfield, Effect of warming rates on the viability of frozen fungous spores, *Mycologia*, 59:58–66 (1967).
12. Mazur, P., Survival of fungi after freezing and desiccation. In *The Fungi, An Advanced Treatise*, Vol. 3, *The Fungal Population*, G. C. Ainsworth and A. S. Sussman (Eds.). Academic Press, New York, pp. 325–394 (1968).
13. Winge, O., and C. Roberts, Inheritance of enzymatic characters in yeast and the phenomenon of long-term adaptation, *C. R. Trav. Lab. Carlsberg*, 24:263–315 (1949).
14. Lindegren, C. C., and G. Lindegren, Unusual gene-controlled combinations of carbohydrate fermentations in yeast hybrids, *Proc. Natl. Acad. Sci. U.S.A.*, 35:23–27 (1949).
15. Wickerham, L. J., and K. A. Burton, A clarification of the relationship of *Candidaguilliermondii* to other yeasts by a study of their mating types, *J. Bacteriol.*, 68:594–597 (1954).
16. Schildkraut, C. L., J. Marmur, and P. Doty, Determination of the base composition of deoxyribonucleic acid from its buoyant density in CsCl, *J. Mol. Biol.*, 4:430–433 (1962).
17. Marmur, J., and P. Doty, Determination of the base composition of DNA from its thermal denaturation temperature, *J. Mol. Biol.*, 5:109–118 (1962).
18. Price, C. W., G. B. Fuson, and H. J. Phaff, Genome comparison in yeast systematics: Delimitation of species within the genera *Schwanniomyces*, *Saccharomyces*, *Debaryomyces*, and *Pichia*, *Microbiol. Rev.*, 42:161–193 (1978).
19. Kurtzman, C. P., M. J. Smiley, C. J. Johnson, L. J. Wickerham, and G. B. Fuson, Two new and closely related heterothallic species, *Pichia amylophilia* and *Pichia mississippiensis*: Characterization by hybridization and deoxyribonucleic acid reassociation, *Int. J. Syst. Bacteriol.*, 30:208–216 (1980).
20. Kurtzman, C. P., M. J. Smiley, and C. J. Johnson, Emendation of the genus *Issatchenkia* Kudriavzev and comparison of species by deoxyribonucleic acid reassociation, mating reaction, and ascospore ultrastructure, *Int. J. Syst. Bacteriol.*, 30:503–513 (1980).
21. Meyer, S. A., M. T. Smith, and F. P. Simione Jr., Systematics of *Hanseniaspora* Zikes and *Kloeckera* Janke, *Antonie van Leeuwenhoek*, 44:79–96 (1978).
22. Nakase, T., and K. Komagata, Taxonomic significance of base composition of yeast DNA, *J. Gen. Appl. Microbiol.*, 14:345–357 (1968).
23. Kurtzman, C. P., H. J. Phaff, and S. A. Meyer, Nucleic acid relatedness among yeasts, In *Yeast Genetics, Fundamental and Applied Aspects*, J. F. T. Spencer, D. M. Spencer, and A. R. W. Smith (Eds.). Springer-Verlag, New York, pp. 139–166 (1983).
24. Meyer, S. A., and H. J. Phaff, DNA base composition and DNA-DNA homology studies as tools in yeast systematics. In *Yeasts, Models in Science and Technics*, A. Kochová-Kratochvilová and E. Minarik (Eds.). Publishing House Slovak Academy of Sciences, Bratislava, Czechoslovakia, pp. 375–386 (1972).
25. Martini, A., and H. J. Phaff, The optical determination of DNA-DNA homologies in yeasts, *Ann. Micro.*, 23:59–68 (1973).
26. Vaughan Martini, A., and C. P. Kurtzman, Deoxyribonucleic acid relatedness among

species of the genus *Saccharomyces* sensu stricto, *Int. J. Syst. Bacteriol.*,35:508–511 (1985).

27. Yarrow, D., and S. A. Meyer, Proposal for amendment of the diagnosis of the genus *Candida* Berkhout nom. cons, *Int. J. Syst. Bacteriol.*, 28:611–615 (1978).

28. Kurtzman, C. P., Synonomy of the yeast genera *Hansenula* and *Pichia* demonstrated through comparisons of deoxyribonucleic acid relatedness, *Antonie van Leeuwenhoek*, 50:290–217 (1984).

29. McArthur, C. R., and G. D. Clark-Walker, Mitochondrial DNA size diversity in the *Dekkera/Brettanomyces* yeasts, *Curr. Genet.*, 7:29–35 (1983).

30. Bicknell, J. N., and H. C. Douglas, Nucleic acid homologies among species of *Saccharomyces*, *J. Bacteriol.*, 101:505–512 (1970).

31. Kennell, D. E., Principles and practices of nucleic acid hybridization, *Prog. Nucleic Acid Res. Mol. Biol.*, 11:259–301 (1971).

32. Johnson, J. L., Genetic characterization, In *Manual of Methods for GeneralBacteriology*, P. Gerhardt (Ed.). American Society for Microbiology, Washington, pp. 450–472 (1981).

33. Baharaeen, S., U. Melcher, and H. S. Vishniac, Complementary DNA-25S ribosomal RNA hybridization: An improved method for phylogenetic studies, *Can. J. Microbiol.*, 29:546–551 (1983).

34. Adoutte-Panvier, A., J. E. Davies, L. R. Gritz, and B. S. Littlewood, Studies of ribosomal proteins of yeast species and their hybrids: Gel electrophoresis and immunochemical crossreactions, *Mol. Gen. Genet.*, 179:273–282 (1980).

35. Walker, W. F., and W. F. Doolittle, Redividing the basidiomycetes on the basis of 5S rRNA sequences, *Nature*, 299:723–724 (1982).

36. Walker, W. F., and W. F. Doolittle, 5S rRNA sequences from eight basidiomycetes and fungi imperfecti, *Nucleic Acids Res.*, 11:7625–7630 (1983).

37. Walker, W. F., 5S Ribosomal RNA sequences from ascomycetes and evolutionary implications, *Syst. Appl. Microbiol.*, 6:48–53 (1985).

38. Blanz, P. A., and M. Gottschalk, Systematic position of *Septobasidium*, *Graphiola* and other basidiomycetes as deduced on the basis of the 5S ribosomal RNA nucleotide sequences, *Syst. Appl. Microbiol.*, 8:121–127 (1986).

39. Weisburg, W. G., C. R. Woese, M. E. Dobson, and E. Weiss, A common origin of rickettsiae and certain plant pathogens, *Science*, 230:556–558, (1985).

40. Campbell, I., Numerical analysis of *Hansenula*, *Pichia*, and related yeast genera, *J. Gen. Microbiol.*, 77:427–441 (1973).

41. Fuson, G. B., C. W. Price, and H. J. Phaff, Deoxyribonucleic acid sequence relatedness among some members of the yeast genera *Hansenula*, *Int. J. Syst. Bacteriol.*, 29:64–69 (1979).

42. Kurtzman, C. P., Resolution of varietal relationships within the species *Hansenula anomala*, *Hansenula bimundalis*, and *Pichia nakazawae* through comparisons of DNA relatedness, *Mycotaxon*, 19:271–279 (1984).

43. Spencer, J. F. T., and P. A. J. Gorin, Systematics of the genera *Hansenula* and *Pichia*: Proton magnetic resonance spectra of their mannans as an aid in classification, *Can. J. Microbiol.*, 15:375–382 (1969).

44. Tsuchiya, T., Y. Fukazawa, M. Taguchi, T. Nakase, and T. Shinoda, Serological aspects of yeast classification, *Mycopathol. Mycol. Appl.*, 53:77–91 (1974).

45. Ballou, C. E., Some aspects of the structure, immunochemistry, and genetic control of yeast mannans, *Adv. Enzymol.*, 40:239–270 (1974).
46. Spencer, J. F. T., P. A. J. Gorin, and G. H. Rank, The genetic control of the two types of mannan produced by *Saccharomyces cerevisiae*, *Can. J. Microbiol.*, 17:1451–1454 (1971).
47. Kurtzman, C. P., and N. J. W. Kreger–van Rij, Ultrastructure of ascospores from *Debaryomyces melissophilus*, a new taxonomic combination, *Mycologia*, 68:422–425 (1976).
48. Kurtzman, C. P., and M. J. Smiley, A taxonomic re-evaluation of the round-spored species of *Pichia*, In *Proceedings of the Fourth International Symposium on Yeasts*, Vienna, Austria, Part I, H. Klaushofer and U. B. Sleytr (Eds.). Vienna: Hochschulerschaft an der Hochschule fur Bodenkultur, pp. 231–232 (1974).
49. Kurtzman, C. P., and M. J. Smiley, Taxonomy of *Pichia carsonii* and its synonyms *Pichia vini* and *P. vini* var. *melibiosi*: Comparison by DNA reassociation, *Mycologia*, 71:658–662 (1979).
50. Kurtzman, C. P., M. J. Smiley, and F. L. Baker, Scanning electron microscopy of ascospores of *Schwanniomyces*, *J. Bacteriol.*, 112:1380–1382 (1972).
51. Kurtzman, C. P., M. J. Smiley, and F. L. Baker, Scanning electron microscopy of ascospores of *Debaryomyces* and *Saccharomyces*, *Mycopathol. Mycol. Appl.*, 55:29–34 (1975).
52. Kreger–van Rij, N. J. W., Electron microscopy of sporulation in *Schwanniomycesalluvius*, *Antonie van Leeuwenhoek*, 43:55–64 (1977).
53. Fuson, G. B., C. W. Price, and H. J. Phaff, Deoxyribonucleic acid base sequence relatedness among strains of *Pichia ohmeri* that produce dimorphic ascospores, *Int. J. Syst. Bacteriol.*, 30:217–219 (1980).
54. Wickerham, L. J., C. P. Kurtzman, and A. I. Herman, Sexuality in *Candida lipolytica*, In *Recent Trends in Yeast Research*, Vol. I, D. G. Ahearn (Ed.). Georgia State University, Atlanta, pp. 81–92 (1969).
55. Presley, H. L., and H. J. Phaff, Personal communication (1981).
56. Kreger–van Rij, N. J. W., and M. Veenhuis, Electron microscopy of septa in ascomycetous yeasts, *Antonie van Leeuwenhoek*, 39:481–490 (1973).
57. Baumann, L., and P. Baumann, Studies of relationship among terrestrial *Pseudomonas*, *Alcaligenes*, and enterobacteria by an immunological comparison of glutamine synthetase, *Arch. Microbiol.*, 119:25–30 (1978).
58. Baumann, L., S. S. Bang, and P. Baumann, Study of relationship among species of *Vibrio*, *Photobacterium*, and terrestrial enterobacteria by an immunological comparison of glutamine synthetase and superoxide dismutase, *Curr. Microbiol.*, 4:133–138 (1980).
59. Lachance, M. A., and H. J. Phaff, Comparative study of molecular size and structure of exo-β-glucanases from *Kluyveromyces* and other yeast genera: Evolutionary and taxonomic implications, *Int. J. Syst. Bacteriol.*, 29:70–78 (1979).
60. Baptist, J. N., C. R. Shaw, and M. Mandel, Comparative zone electrophoresis of enzymes of *Pseudomonas solanacearum* and *Pseudomonas cepacia*, *J. Bacteriol.*, 108:799–803 (1971).
61. Selander, R. K., Genetic variation in natural populations, In *Molecular Evolution*, F. J. Ayala (Ed.). Sinauer, Sunderland, Mass., pp. 21–46 (1976).

62. Holzschu, D. L., Molecular taxonomy and evolutionary relationships among cactophilic yeasts, Ph.D. thesis, University of California, Davis (1981).

63. Baptist, J. N., and C. P. Kurtzman, Comparative enzyme patterns in *Cryptococcus laurentii* and its taxonomic varieties, *Mycologia*, 68:1195–1203 (1976).

64. Yamazaki, M., and K. Komagata, Taxonomic significance of electrophoretic comparison of enzymes in the genera *Rhodotorula* and *Rhodosporidium*, *Int. J. Syst. Bacteriol.*, 31:361–381 (1981).

65. Aulakh, H. S., S. E. Straus, and K. J. Kwon-Chung, Genetic relatedness of *Filobasidiella neoformans* (*Cryptococcus neoformans*) and *Filobasidiella bacillispora*(*Cryptococcus bacillisporus*) as determined by deoxyribonucleic acid base composition and sequence homology studies, *Int. J. Syst. Bacteriol.*, 31:97–103 (1981).

66. Phaff, H. J., M. W. Miller, and M. Miranda, *Hansenula alni*, a new heterothallic species of yeast from exudates of alder trees, *Int. J. Syst. Bacteriol.*, 29:60–63 (1979).

67. Mendonça-Hagler, L. C., and H. J. Phaff, Deoxyribonucleic acid base composition and DNA/DNA hybrid formation in psychrophobic and related yeasts, *Int. J. Syst. Bacteriol.*, 25:222–229 (1975).

68. Kurtzman, C. P., M. J. Smiley, C. J. Johnson, and M. J. Hoffman, Deoxyribonucleic acid relatedness among species of *Sterigmatomyces*, Abstr. Int. Symp. Yeasts, 5th Y-5.2.5(L), p. 246 (1980).

2
Yeast Selection in Nature

Marc-André Lachance
University of Western Ontario
London, Ontario, Canada

I. INTRODUCTION

The desire to develop selection strategies for the isolation of yeast strains most suitable for one process or another implicitly recognizes the existence of selective forces that act on yeasts in their natural habitats. Selection is as much an ecological problem as it is a technical one (1). It is therefore with an ecological perspective that I shall discuss the factors affecting the fitness and the survival of yeast species in nature, with little emphasis on how we may apply that knowledge in our search for useful yeast strains. My other bias will be in favor of associations involving plants.

I cannot think of a better summary of the idea of yeast selection in nature than the following sentences written by Do Carmo-Sousa (2) in her review of yeast ecology:

> Yeast populations of several hundred species are continuously building up and dying off in terrestrial as well as in aquatic environments. They play their part in the dynamics of the biological and chemical turnover in soil, plants, animals and water, where they are active as competitors for nutrients, antagonists or symbiotic associates or as victims of the behavior of their neighbors.

Although they are never motile, yeasts are probably not limited to an appreciable degree in their physical mobility. Conceivably, like other microorganisms, they can be dispersed as aerosols and vectored by animals of all sorts. For this reason, there is a risk that yeasts may be regarded by some as being simply ubiquitous. Taken to the extreme, this view might imply that almost any yeast species can be recovered from almost any substrate, provided the right techniques are used. This is probably not the case.

Numerous surveys of substrates that were suspected to harbour yeasts, because of the availability of the necessary nutrients for their growth, have revealed that yeasts appear to be far less ubiquitous than many bacterial species. Specialization for habitat appears to be the rule rather than the exception (3).

Unlike bacteria, yeasts are relatively modest in their physiological abilities. They are strictly organotrophic, which restricts the range of habitats in which yeasts may be recovered in some abundance. Within this range, however, it is becoming clearer and clearer that a great deal of selectivity operates, from one yeast habitat to another, in determining the composition of the yeast communities associated with those habitats.

The view that yeasts are ubiquitous has some wisdom, in that it stimulates the search for methodological approaches aimed at the isolation of microorganisms endowed with very precise attributes, using media or conditions tailored to match these attributes. Early in this century, Winogradsky (4) had coined the terms "autochthonous" and "allochthonous," respectively, to distinguish between microorganisms that dominate a particular ecosystem because they always find adequate nutrients in it and organisms that are present essentially as contaminants. The methodological approach to microbial ecology places much emphasis on the allochthonous component of the microflora. Davenport (5), not unlike Winogradsky, recognizes two kinds of yeast florae in each habitat. One is the resident flora, presumably present because it possesses the ability to survive under that set of conditions, and the other is a transitory flora, presumably present because it was vectored there.

While the above considerations are indeed based on accurate observations, equally well founded is the alternate view that each yeast has a habitat and each habitat has its community. The latter concept stresses the need to exercise care in selecting the inoculum material, with considerably less emphasis on the isolation procedure itself. Conceptually, this second view makes the important distinction between the habitat of a species (that place where a species is usually found) and the niche of that species (the multidimensional array of factors that allow or favor the existence of a species in a certain habitat).

Two major approaches to our understanding of yeast ecology have thus been defined. They are not contradictory, but they reflect different visions. It may have been useful in the past, and it may still be useful to consider the biosphere as a large culture collection from which one may screen for useful organisms. As a biologist, I believe that it is equally worthwhile to seek a better understanding of selection as it takes place within yeast communities, with the possible benefit that it will enable us to devise better goal-oriented strategies of practical yeast selection.

II. THE YEAST COMMUNITY IN ITS HABITAT

A. Definitions

Community and habitat define each other reciprocally. The community is a group of organisms that, because they share certain niche attributes, are found to share a common habitat. The habitat is a place or collection of places that share sufficient similarities to result in their tendency to harbor similar yeast communities. The niche is somewhat more complex.

In reviewing the notion of niche as it applies to fungi, McNaughton (6) reminds us of the important distinction, made earlier by Hutchinson, (7) between the fundamental and the realized niche of a species. The fundamental niche is the range of conditions under which an organism can exist. For a yeast, these conditions include a certain spectrum of temperatures, pH values, salinities, osmotic pressures, and nutrient compositions as defined in the laboratory using culture media (8). The realized niche is the actual range of a species in nature, after taking into account its dispersal and its interactions with other organisms. It is a set of "physical and biological conditions within which a population occurs" (6). This aspect of yeast ecology is generally less well understood, for the obvious reason that it is not so easily studied. To define the realized niche of a yeast species, data must be acquired on the chemistry and microclimate of its habitats, the distribution of the habitats themselves, the connection between habitats by dispersal agents, and the interactions between the different yeast species sharing the same habitat (8).

Yeast habitats have less of a geographic component than plant or animal habitats. The distribution of yeast species in nature is most intimately tied to the distribution of host plants, and in some cases host animals, in their own habitats. Yeast habitats are often plant tissues that, for one reason or another, have become the interface between the plant's soluble nutrients, in particular sugars, and the septic world. The initial exposure of the plant's nutrient-rich liquids to yeasts frequently involves insects or other animals, but it may come as the result of a mechanical trauma or of a primary infection by bacteria or filamentous fungi which are better equipped enzymatically to disrupt the integrity of plant tissues. A notable exception to the above generalization is the small number of yeast species recognized as animal parasites or commensals.

B. Types of Yeast Habitats

Do Carmo-Sousa (2) recognizes the existence of four major terrestrial yeast habitat types—namely, soil, plants, animals, and the atmosphere. Aquatic habitats can probably be subdivided in a similar fashion, but in view of the nature of past sampling efforts, seawater is generally treated as a single habitat type. Phaff and Starmer (3) give a list of yeast genera along with specific habitats of their species.

Their publication should be consulted for an exhaustive compilation. The soil and the sea have unique features that deserve special mention of them at this point. It is important, when dealing with soil, to recognize the clear distinction that exists between soil as a genuine environment that supports the growth of yeasts, and soil as a repository of most life forms participating in the final steps of decay. As stated by Do Carmo-Sousa (2), "For some yeast species, the soil may be only a reservoir where they can survive protected against desiccation and drought until dispersed by animals, growing plants and wind to suitable substrates."

Good evidence exists that *Lipomyces* is a genuine soil yeast genus. In particular, Danielson and Jurgensen (9) have shown that abundant populations of *Lipomyces* species are found in soils that receive a fresh input of carbohydrates, but no nitrogenous supplementation. Thus, by virtue of their nitrogen scavenging properties, *Lipomyces* species are able to outnumber other microorganisms provided extraneous, easily metabolized nitrogen sources are not available.

Fell (10) has reviewed past work on the ecology of seawater yeasts. A direct cause-and-effect relationship between the physical-chemical parameters of seawater and yeast composition is not likely to be established because most marine yeasts are probably distributed as a function of their animal vectors. Nonetheless, patterns do exist. Fell outlines two major aspects defining the ecology of sea-water yeasts. One is their twofold ability to assimilate a diverse array of nutrients and to survive in seawater for extended periods of time. The second aspect is more interesting and much more complex, and requires a more detailed discussion.

Fell (10) shows that the distribution of yeasts in near polar seawater follows patterns that are correlated with the types of water masses from which the yeasts are collected. Bidimensional "T-S" diagrams opposing water salinity (S) to water temperature (T) are used to characterize seawater masses from different depths and latitudes. The yeasts collected from each water mass may be superimposed onto T-S diagrams in a meaningful way. Some yeasts are present in various numbers in seawater samples with temperatures from −3 to 13°C and salinities from 33.9 to 35.5%, but certain species are restricted to narrower ranges of T-S values. For example, *Candida natalensis* appears to be narrowly distributed in "lower deep circumpolar waters" with temperatures near 2°C and salinities near 34.7%, whereas other widespread species such as *Candida norvegica* are found over the entire T-S range. By contrast, the exclusively marine species *Sympodiomyces parvus* is absent from the 2°C/34.7% waters.

C. Types of Yeast Communities

Davenport (5) subdivides the yeasts according to their degree of habitat specialization, contrasting those genera likely to be found in many different habitats (general) from those whose distribution is "restricted." He further distinguishes, within the latter group, (a) genera confined to single or few habitats, (b) monospecific genera confined to single or few habitats, and (c) species found in a single habitat. The existence of such categories is linked to the idea that the patterns of presence or absence of particular yeasts in particular habitats may be accounted for by different factors.

The perception of yeasts as members of communities profoundly influences the methodological approaches used to study natural yeast populations. In view of the above typification of yeasts and their habitats, Davenport (5) proposes that an ecological study should entail a simplification of the identification procedure to fit each particular situation. This has the advantage that "only a few tests are required and the time involved is minimal" (5). Phaff and Starmer (3) express the somewhat opposing view that "an accurate picture of yeast habitats cannot be assembled if the yeast species cannot be correctly identified."

Yeast communities may also be defined in terms of their nutritional breadth, which is most certainly a fundamental parameter linking yeast community structure to habitat characteristics. It is possible to distinguish, within yeast communities, "generalistic" communities associated with nutritionally dilute and diverse habitats such as the phylloplane or aquatic habitats. The component yeasts may be diverse, but more importantly they belong to species with broad nutritional spectra. The term "polytrophic" has been used to describe such yeasts, both as individual species and as communities (11). By contrast, fewer species of nutritionally specialized yeasts are typical of more "specialized" communities such as those found in sugar-rich plant exudates, slime fluxes, nectars, or succulent necroses. The term "oligotrophic" could apply to yeasts that typically utilize only a few different nutrients, although the same word is used by aquatic biologists to refer to certain unpolluted lakes. Oligotrophic yeasts are presumed to possess the competitive advantage of accrued efficiency over other species that could potentially share the same habitats.

To obtain a quantitative and potentially more objective view of the nutritional aspects of yeast community structure, it is useful to typify a community as the average nutritional profile of its component yeast species. This approach has been used to demonstrate that the nutritional properties of yeast communities are correlated with the putative evolutionary relationships among plants known to serve as yeast habitats (12). Furthermore, the physiological expression of a yeast community has been shown to reflect intrinsic host plant properties more accurately than patterns of yeast species composition, which itself is more subject to geographical influences (13).

A rigorous method has been devised by Lachance and Starmer (14) to characterize yeast communities physiologically by comparison with the expected physiological structure of a random assemblage of yeasts. The technique has been applied heuristically to characterize yeast communities associated with a single community of oak trees (11), as well as to yeast communities recovered from prickly pear cactus communities of worldwide distribution (15) .

III. SELECTIVE FACTORS

Let us now turn to some of the factors that are either known or presumed to influence yeast community composition—in other words, the factors of *selection*. It is convenient to recognize the chemical and physical elements of habitats as part of a whole. The inanimate or "abiotic" selection factors, as they are known, are those that mirror the fundamental niche of the species. Factors involving other living organisms and ultimately defining the realized niche of the species are grouped together as "biotic" factors.

A. Abiotic Factors

1. Temperature and Heat

Davenport (5) has compiled data relevant to how yeasts respond to temperature and heat. He devised a general classification of yeasts based on their cardinal growth temperatures. Generalistic species are able to grow from 0 to 37°C. Yeasts with restricted temperature ranges can be categorized as obligate or facultative psychrophiles, generally growing from 0 to 15–25°C, while others are psychrophobic, being unable to grow at temperatures below 20–28°C, but growing well at temperatures up to 42–45°C. Yeasts growing well at 45°C are considered thermoduric. For a detailed compilation of maximum growth temperatures of yeast species, the extensive survey performed by Vidal-Leiria and co-workers (16) should be consulted.

Temperature is clearly an important physical niche attribute, although the relationship between the place of isolation of certain yeasts and their maximum temperature of growth is sometimes unexpected, as in the case of *Sporopachydermia* species (17). While *Sporopachydermia lactativora* was initially recovered in seawater near Antarctica, its maximum growth temperature exceeds 42°C. *Sporopachydermia quercuum*, by contrast, was found in trees growing in a temperate climate, and its maximum growth temperature on agar media is slightly below 37°C. This apparent contradiction was probably resolved by the subsequent isolation of *S. lactativora* almost exclusively from warm-blooded animals, which probably were the true habitats of the polar isolates in the first place.

More in line with what one would normally expect, the great majority of yeasts isolated from trees in the Sonoran desert grow at 37°C or above, whereas most isolates recovered from trees in the Pacific Northwest do not, and occasionally fail to grow even at 30°C (3). By contrast, yeasts recovered from seawater rarely grow at 30°C, and seawater samples collected near polar regions often have maximum growth temperatures near 15°C (10).

The heat resistence of yeasts is in no way comparable to that encountered in certain bacteria. According to data assembled by Davenport (5), ascospores may resist heat treatments at temperatures of about 10°C above the resistance level of vegetative cells. Several species in genera as diverse as *Debaryomyces, Nadsonia, Saccharomycodes, Saccharomyces,* or *Kluyveromyces* have been compared as to the difference in heat tolerance between vegetative and sporulating cultures (M. J. Butler and M. A. Lachance, unpublished data). A wide variation was found, such that in some cases, no significant differences were found between cultures with and without ascospores, while in other cases, differences of 12°C were noted. A vague correlation appeared to exist between the formation of a single ascospore (as opposed to four per ascus) and increased differential resistance.

2. pH

Yeasts generally grow best between pH values of 3.5 and 6.5, although their growth range extends from pH 3 to 11 (5). Some yeasts, for example, species of *Schizosaccharomyces*, have a more stringent pH requirement. Davenport (5) makes the important point that the effect of pH on growth should not be considered in isolation, because it is affected by other physical-chemical factors such as temperature and water activity. An example of this is the ability of certain psychrophobic species to grow at pH 1 provided the temperature is optimum (37°C).

3. Water Activity

The atmosphere, the phylloplane, and soil are habitats in which water content fluctuates, sometimes rapidly, from complete saturation to extensive desiccation. In such environments, a strong selective pressure is conceivably exerted upon yeasts, with the result that the species more frequently recovered from them often form abundant slimy capsules (2).

A number of yeasts are particularly well adapted to growth in environments with low water activities, or high solute concentrations. Such yeasts may be referred to as osmotophilic, osmotolerant, osmoduric, osmotrophic, xerophilic, osmophilic, or xerotolerant. The last two terms are preferred over others (18), although their profound physiological significance is not clear. It is generally agreed that xerotolerant yeasts are capable of growth on media containing sugar or salt concentrations equivalent, in water activity, to 50% (w/w) glucose. In a preliminary study of the yeast flora associated with milkweed nectaries, I have observed that although several yeast species may be isolated from unopened flower buds or various parts of the surface of any flowers, the nectar itself contains almost exclusively xerotolerant species (e.g., *Metschnikowia* spp.) regardless of the nectar sugar concentration, which varies to a considerable extent.

4. Physical Dispersal Agents

Yeasts of the phylloplane often discharge their buds forcibly. The interplay between ballistospore dispersal and atmospheric parameters is probably much more complex than the simple effects of wind on showers of ballistospores. Less obvious but no less significant is the influence of climactic parameters that undergo daily fluctuations on the intensity of ballistospore release. These aspects have been discussed at length by Last and Price (19) .

An unusual clue as to the importance of various modes of yeast dispersal comes from comparing the physiological profiles of yeast communities from exudates of various tree families (12). Although the similarity patterns among yeasts from angiosperms are somewhat consistent with the presumed relationships among families of those host plants, unexpected patterns arise when yeasts from angiosperms and gymnosperms are considered together. Conifer-associated yeast communities do not stand out as a separate group, but instead, they appear most similar to the yeast communities associated with those angiosperm taxa that, like conifers, are wind-pollinated. It might be presumptuous to regard this as evidence that wind acts as a very selective agent of yeast dispersal. Considered from the standpoint of yeasts, however, wind pollination is equivalent to absence of biotic dispersal through insect pollinators.

5. Light

Yeast of the phylloplane are presumed to respond to exposure to visible and ultraviolet light. I am not aware of any study providing evidence of differential survival of naturally occurring yeast species as a result of their irradiation by the sun, but it is quite clear that light affects phylloplane yeasts under artificial conditions (20). It is often presumed that pigmentation may act to protect yeasts from the lethal effects of sunlight. In the laboratory, germicidal ultraviolet light does not seem to have the expected differential effect when hyaline yeast cells (both natural strains and mutants of pigmented strains) are compared to carotenoid or melanin pigmented yeast cells (M. J. Butler and M. A. Lachance, unpublished data).

Light is unique in that it is probably one of the rare agents of natural selection that has the potential to affect an organism directly at the level of its genome. It is not unreasonable to speculate that yeasts that spend most of their cell cycle in the haploid state (haplonts) are more vulnerable to irradiation than yeasts with higher ploidies, in view of the "target" nature of lethal radiation effects (20). Similarly, it is conceivable that yeasts whose genomes contain a high molar proportion of guanine and cytosine might be less affected by ultraviolet light, in view of the involvement of thymine in ultraviolet-induced mutagenesis. Indeed, phylloplane yeasts are most frequently anamorphs of basidiomycetous species whose guanine plus cytosine content is generally high. This sort of reasoning is implicit in studies reported by Miller and Baharaeen (21) on species of *Hansenula*.

A significant negative correlation (−0.56) exists between the intensity of photoreactivation (repair of thymine dimers) and the DNA base composition of 33 species studied.

B. Biotic Factors

1. Competition for Nutrients

According to Do Carmo-Sousa (2), "Competition for nutrients is probably the most important factor in yeast ecology." Although one would intuitively tend to agree with this notion, direct experimental evidence in its support is generally lacking. Almost as a matter of routine, upon completing a yeast collection, I subject the data to an analytical procedure aimed at identifying possible associations or exclusions within communities from similar habitats. In no case have I ever identified clear evidence that the presence of one yeast species is conditional on the absence of another yeast species which otherwise would be equally well adapted for growth in the same substrate.

Oligotrophic species (e.g., *Kloeckera* or *Pichia* spp.) are conspicuously absent from habitats such as the phylloplane, in which nutrients are complex and dilute, whereas polytrophic species (e.g., *Rhodotorula* or *Cryptococcus* spp.) are rare in rich but less diverse habitats such as fruit, nectar, or cactus flesh. Can such separation be attributed to competitive exclusion? More likely, this kind of pattern is a consequence of the fundamental niche of each kind of yeast, rather than the result of interspecific competition.

The dominance of other yeasts over *Lipomyces* (9) in soils that have been supplemented with simple nitrogen sources has been discussed above. It should probably be regarded as an example of competitive exclusion by yeasts.

2. Insect Vectors

Several examples of selection of particular yeast species by insects are known. Generally the yeasts serve as food for one or more developmental stages of the insects. Do Carmo-Sousa (2) listed a number of such interactions, stating that considerable variation existed in the amount of specificity associated with each interaction. She stated that *Drosophila* species choose yeasts in a more or less random way, whereas very strict types of associations were observed in other cases—for example, the intracellular symbiosis in Cerambycidae bark bettles. The latter contain certain well-defined *Candida* species.

The bark beetle *Ips typographicus* is known to carry a number of yeasts prior to attacking a new tree. Some of the yeasts are able to convert the pheromone *cis*-verbenol into verbenone during the early phase of attack (22). The latter compound acts as a congregating signal for beetles, thus enhancing their dispersal to newly infected trees. Whether or not the insect actively chooses one yeast species over another is not clear, but as we shall see later, yeast selection does take place.

Phaff and Miller (23) identified the highly specific vectoring of *Candida guilliermondii* into figs by the fig wasp, *Blastophaga psenes*. It is not known whether selection of the yeast species is a function of the insect's behavior or controlled by the yeast's resistance to some compounds present in the fig. It should be noted, however, that the yeast is atypical of its assigned species, indicating that it may be completely unique to the wasp-fig interaction.

Interactions involving *Drosophila* are far from being completely haphazard. Cooper (24) reviewed the several instances where *Drosophila* species had been shown to exhibit preferences for one yeast over another, and determined that preferences differ when larvae and adult flies are compared. A correlation was also established between preference for a yeast and the nutritional value of that yeast. More recently, Fogleman and co-worker (25) have demonstrated clearly that larvae of *Drosophila mojavensis* are able to distinguish between different yeast species frequently found in the same host plant. The larvae, when given the choice, seek and feed on *Pichia cactophila* rather than *Clavispora lusitaniae* or *Candida sonorensis*. This preference takes place in the natural habitat as well as in the laboratory.

Vector selectivity was evident when all the insects observed to land in the vicinity of a necrotic pocket on a giant saguaro cactus were collected and analyzed for the yeasts they carried (26). The flora carried by nereids and muscoids was not exceedingly different from that of *Drosophila* species captured on the same plant, but other insects including dolichopodids, syrphids, wasps, and others were essentially devoid of yeasts.

3. Host Plant Toxicity

Many plants contain a number of compounds whose adaptive significance may be the defense of the plant against invertebrates and microorganisms (27), or even mammalian herbivores (28). These compounds also act, in some cases, as selective agents which shape yeast community composition. As stated by Phaff and Starmer (3), "Several species of *Hansenula* and *Pichia* that are associated with the bark beetles that attack pine trees grow vigorously in the sap of these trees in the presence of high concentrations of oligoterpenes that are inhibitory to the great majority of other yeasts."

The yeasts carried by the bark beetle *Ips typographicus* (see Section B.2) are present as a mixed flora in which *Candida diddensii* and *Hansenula holstii* are more or less dominant. *Hansenula capsulata* and *Pichia pinus* are present in significant numbers as well, in addition to a number of other incidental species. Leufven and Nehls (29) have demonstrated quite clearly that the monoterpene hydrocarbons of spruce trees, by their microbial toxicity, eliminate much of the yeast flora initially vectored by the beetle, and that *Candida diddensii* acquires dominance in this community by virtue of its ability to adapt to relatively high concentrations of the toxic compounds.

The ability to metabolize a given class of plant compounds is rarely found to influence yeast species composition directly. An exception is the observation (13) that, considered as a whole, communities associated with exudates of poplars tend to contain yeasts able to metabolize ß-glucosides more efficiently than similar communities recovered from other trees such as elms, oaks, and Douglas firs. Poplars and other members of the family Salicaceae are known to contain several ß-glucosides (e.g., salicin). The exact mechanism by which poplars favor a yeast flora endowed with ß-glucosidase activity has not been investigated, but it is unlikely that the simple ability to assimilate specific glucosides is at stake. The diversity of phenolic glycosides in trees such as birch, willow, or poplar (28) is such that it is likely that ß-glucosidases might release, by hydrolysis of components in those trees, toxic moieties to which the poplar yeasts are resistant, thus eliminating other yeast species. This kind of selection may also be operating in plants rich in nontoxic organic cyanide derivatives (e.g., amygdalin in *Prunus* species), although this hypothesis awaits substantiation.

I have observed that yeasts recovered from oak and cherry trees tend to assimilate crude tannic acid in the laboratory, more than expected from equivalent random assemblages of yeasts, but I have not systematically investigated the toxic properties of plant tannins and phenolics as they affect yeast composition. In one instance (M. J. Butler and M. A. Lachance, unpublished data), resistance to various phenolics was assessed in several yeast strains, including the constitutively melanic yeast *Phaeococcomyces* sp. and an albino mutant of that black yeast. We were unable to validate the hypothesis that melanin biosynthesis may act as a means of detoxifying phenolics. Wild-type and mutant strains responded alike, and unpigmented yeasts such as *Saccharomyces* sp. exhibited more tolerance to the toxic compounds than pigmented ones.

Starmer (30) has studied the significance of the lipids found in organ-pipe cactus as selective agents affecting yeast composition. The lipids include medium-chain fatty acid esters of sterol diols and triterpene diols, and altogether they account for 10–15% of the cactus tissue. The medium-chain fatty acids include caproic (C 6), caprylic (C 8), capric (C 10), and lauric (C 12) acids. Free medium-chain fatty acids are more toxic and less utilized by most yeasts than the unhydrolyzed lipid fraction present in cactus. *C. ingens*, a highly lipolytic yeast, is very resistant to the toxic effects of the lipids, and it also assimilates both hydrolyzed and unhydrolyzed fats efficiently.

Essential oils produced by some plants may, in some cases, act as selective agents. In a study of the effect of essential oils on the recovery of heat-stressed yeasts, Conner and Beuchat (31) have shown that different yeast species respond differently to different oils after having been subjected to a heat shock. The relevance of their work in food preservation is evident, but some of the results may be significant in an ecological context as well. Indeed, yeasts not subjected to stress tend to be inhibited significantly by garlic oil, but not so much by allspice, cinnamon, clove, oregano, and thyme oils. A strain of *Saccharomyces cerevisiae* has been found to be most susceptible compared with strains of a number of other species. Unlike other yeasts, *Sacch. cerevisiae* is also inhibited by onion oil.

By far the most interesting case where host plant compounds unequivocally act as selective agents against certain yeasts has been uncovered by Starmer and co-workers (32). The two described varieties of *Pichia amethionina* and the two hybridizable species formerly treated as varieties of *Pichia opuntiae* are generally recovered in a number of different cactus species. In either case, the aforementioned yeast taxa can be subdivided into one subtaxon that is resistant to triterpene glycosides made by cacti in the substribe stenocereinae, and one subtaxon that is not. The yeast subtaxa also occupy separate habitats. Althought their geography may be coincident, the varieties of *P. amethionina* are separated, in the expected way, between cactus species that form triterpene glycosides and other cactus species that do not. *P. opuntiae* and its former variety *P. thermotolerans* are separated not only according to host but also along geographic lines, although in this case the effects of triterpene glycosides appear to be superseded by other effects. The active agent is apparently similar to digitonin, itself known to be related, at least in origin, to so-called cardiac glycosides.

Phytoalexins are specifically synthesized by plants in response to invasion by microorganisms, and are capable of interfering with fungal pathogens. West (33) has reviewed the various fungal molecules capable of inducing the phytoalexin response of plants. Some of the elicitors are polysaccharides similar to those found in most yeast cell walls. Indeed, the purified ß-glucan of *Saccharomyces cerevisiae* elicits the production of glyceollin by soybeans (34). Whether or not phytoalexins are agents capable of affecting the composition of yeast communities of plants remains to be investigated.

4.   Predation

Numerous organisms from the animal, fungal, prokaryotic kingdoms produce hydrolytic enzymes capable of disrupting the integrity of yeast cell walls. Indeed, a number of commercial lytic enzyme preparations originate from various such sources (e.g., snail digestive juices). Tree exudates, well known to be a major yeast habitat, are presumed to be initiated and maintained by the action of an anaerobic bacterial flora that excretes enzymes capable of causing impairment, but not death, of the host trees (35). Yeasts are suspected to live off the nutrients present in the resulting sap flow, which probably contains at least some substances, including bacterial enzymes, endowed with selective action against the yeasts.

Martin and co-workers (36) have examined enzyme activities of the intestinal tracts of various beetles, comparing insectivorous and fungivorous species. Most species of fungivores examined are equipped with substantial amounts of α-1,

4-amylase, laminarinase (1,3- or 1,6-ß-glucanase), and protease activities, and a few possess chitinase as well, but they are devoid of cellulases (ß-1-4-glucanases), xylanases, and pectinases, and the gut pH is nearly neutral. This is in sharp contrast to the digestive juices of herbivorous beetles which often contain enzymes able to digest plant polysaccharides (cellulose, xylan, pectin) but little, if any enzyme activity directed toward fungal polysaccharides (1,3- or 1,6-ß-glucanases, chitinase). In addition, beetles that feed on plants have an alkaline gastric juice, which helps to process proteins complexed with plant phenolics.

Melanin synthesis by certain yeasts may confer accrued fitness by endowing those yeasts with the ability to resist enzymatic attack. *Phaeococcomyces* sp. cultures grown under certain conditions fail to synthesize melanin, and as a result lose their imperviousness to hydrolysis by Zymolyase, a commercial lytic preparation of bacterial origin (37). Whether or not this reflects a naturally significant mechanism remains to be explored.

5. Toxicity of Animal Secretions

Certain insects secrete antimicrobial substances. For example, the mandibular secretions of an ant *(Calomyrmex* sp.) have strong bacteriocidal and moderate fungistatic properties (38), and two compounds present in secretions of the honeybee, geraniol and citral, have strong fungistatic activities (39). In either case, it is not known whether yeasts are potentially affected. Some preliminary observations I have made on the yeasts associated with ants or with honeybees suggest that strong selection is taking place. Although ants are important yeast vectors (40), their digestive system appears to be totally devoid of yeasts, even when recovered from habitats known to contain yeasts in large numbers. Bees vector a very restricted number of yeast species through their proboscis.

6. Antibiosis

Taken loosely, the term antibiosis may be considered to apply to some of the toxic interactions discussed above. I shall restrict its use here to the production, by microorganisms other than yeasts, of substances that are selectively lethal to some yeasts, presumably by interfering specifically with certain metabolic processes. A few antibiotics active against yeast are available commercially, the better known being cycloheximide (Actidione), nystatin (Mycostatin), and amphothericin B (Fungizone). There is an ongoing debate as to whether antibiosis is indeed a naturally significant process or just a phenomenon restricted to laboratory conditions.

The literature on yeast antibiosis (antizymosis?) is surprisingly scant. Di Menna (41) has compared the yeast composition of leaves and soil, in parallel with a comparison of their responses to antibiotics produced by leaf and soil bacteria, mainly species of *Streptomyces*. Her data cannot be interpreted unequivocally as supporting the contention that antibiosis is a meaningful selection factor.

Lachance and Boivin (42) have claimed that the presence of one particular yeast species to the exclusion of many other species in the beetle frass of an oak tree is accountable to the presence, in that tree, of an antibiotic-producing strain of *Streptomyces*. Subsequent to that report, I have examined several other frass samples and found that the variation among their yeast florae is to some extent explainable by the presence of various combinations of *Streptomyces* strains endowed with diverse antibiotic spectra. The fundamental problem is to determine which bacterial isolates were actively releasing what antibiotics at the time that the yeasts were actively growing on nutrients present in the frass. The mere coisolation of a yeast and an antibiotic-producing bacterium does not necessarily mean that they were mutually compatible in that same habitat under conditions that no longer exist in that habitat.

7. Killer Toxins

Some yeasts excrete substances that are specifically lethal to other yeasts. The killer phenomenon, as it is known, has been well studied at the molecular and genetic levels, but the role of killer toxins as agents of natural selection has received little attention. The frequency of killer-sensitive and killer-producing yeasts in their natural habitats is believed to be in the order of 10 and 20%, respectively (43), but the importance of killers in nature has remained strictly the object of speculation until recently (44). In one instance, significant exclusion has been identified between *Pichia kluyveri*, a yeast species known to produce a strong killer toxin, and *Cryptococcus cereanus*, another species known to be sensitive to it (55). Those two yeasts share the same cactus as habitat, but in localities where *P. kluyveri* is abundant, *Cr. cereanus* tends to be less frequent.

IV.   PRACTICAL SELECTION FROM NATURAL HABITATS

As all media, by their very nature, are selective, it is highly likely that a considerable portion of the yeast community is ignored with the present techniques (10).

A.   Elective Media

Media are considered elective if they favor the growth of certain microbial types without directly preventing the growth of other types. The distinction between elective and selective media is not always clear.

1.   General Yeast Enrichments

Wickerham (45) used various liquid enrichment media to recover yeasts from natural substrates. One of these is D20 (dextrose 20%), a mixture of yeast nitrogen base, malt extract, yeast extract, and glucose. I have used a modification referred to as D20T (46). Tween 80 (0.01%) is added to improve the medium's wetting properties, which is useful when screening samples containing molds. Also, the addition of a suitable indicator (e.g., Bromo-phenol blue) allows one to monitor changes in pH and to make periodic adjustments if desired. Yeasts from dilute or contaminated samples are isolated by placing small amounts (less than 0.1 g) of sample in 15 ml of D20T in a large culture tube, and subsequently streaking a loopful of medium (after signs of yeast growth appear) onto a plate of acidified YM agar. The use of D20T in shake flasks is not equally successful. A special version of D20T containing D-xylose instead of D-glucose has been used (47) to recover yeasts able to ferment D-xylose from woody substrates. In that case, it was specially critical to maintain a very small inoculum size in order to minimize the introduction of sugars other than D-xylose present in the natural substrate.

2. Specific Enrichments

Other liquid media used by Wickerham (45) favor yeasts with certain nutritional profiles. The media are based on yeast nitrogen base and yeast carbon base supplemented, respectively, with specific carbon or nitrogen sources. Van Dijken and Harder (48) used a liquid medium containing methanol as the sole carbon source and cycloserine and penicillin as antibacterial agents to enrich samples for methanol utilizing yeasts.

B. Selective Media

1. Selection Against Other Microorganisms

Virtually all agar media used in the isolation of yeasts from nature must be selective against bacterial growth. The addition of antibiotics or other inhibitors to multipurpose agar media is effective in the isolation of yeasts from samples known to contain large populations of bacteria or molds, but suspected or confirmed inhibition of yeasts by antibacterial or antifungal substances makes this a dangerous practice (49). Nonetheless, the interested reader should consult Davenport's (50) list of formulations of antibacterial, antifungal, and antiyeast agents, along with media in which they have been found effective.

The use of acidified agar is adequate for isolation of yeasts from terrestrial substrates that contain bacteria. Hydrochloric or sulfuric acid at a final concentration of about 7 mEq/liter is generally sufficient to lower the pH of YM agar to about 3.7. Most, but not all, bacteria fail to grow at such high acidity. Organic acids have been used by some to inhibit bacteria, molds, or both (51). This is not a recommendable practice, in view of the fact that certain yeasts are inhibited by several carboxylic acids (3,49). Fell (10) has found acidified media unsatisfactory for eliminating bacterial growth in certain samples of seawater, and favors the use of media supplemented with chloramphenicol instead.

Van der Walt and Yarrow (49) suggested that in some instances, it may be useful to inoculate samples into broth medium and to cover the medium with a 1-cm plug of paraffin wax. If selection against strict aerobes is acceptable, then this method is very efficient in obtaining yeasts from very moldy samples. Another suggestion (49) meriting exploration is the use of agar media supplemented with 30–50% glucose and acidified to pH 4.5–5.0.

2. Selection of Specific Yeasts

Cycloheximide is the only antibiotic whose spectrum of activity includes certain yeast species and whose use as a selective agent is feasible. This is facilitated by the fact that cycloheximide resists autoclaving to an appreciable degree. Media containing cycloheximide are used by brewers to assess contamination of pitching yeast (52). Van der Walt and Van Kerken (53) used a medium supplemented with cycloheximide and sorbic acid, and acidified to pH 4.8, to enumerate *Dekkera* (or *Brettanomyces*) in wine samples containing mixtures of yeasts.

Selective media containing defined nutrients known to support only the growth of certain yeasts may have their usefulness in studies of habitats that have already been characterized to some degree. For instance, selective counts of *Cryptococcus cereanus, Candida sonorensis, Pichia cactophila,* and *Pichia heedii* could be obtained by plating dilutions of cactus tissue onto media made up of yeast nitrogen base and, respectively, inositol, methanol, glucosamine, or xylose as sole carbon source (3). The usefulness of analogous media to screen for yeasts from poorly defined habitats is doubtful at best.

Phaff and Starmer (3) have listed a number of yeasts that are unique by virtue of special physiological idiosyncrasies. These include, among others, *Schizosaccharomyces octosporus*, which naturally requires adenine; *Pichia amethionina* (now known to comprise two or more different species), which is a natural auxotroph for sulfur aminoacids; species of *Dekkera* or *Brettanomyces* requiring thiamine; *Candida pintolopesii* requiring choline; *Metschnikowia bicuspidata*, some strains of which require salt; and *Zygosaccharomyces rouxii*, which, upon isolation, requires media of high osmotic pressure. In each case, the unique property is a requirement that cannot be used directly to formulate a medium that would allow recovery of such yeast from a mixture. The requirements would only allow recovery of *other yeasts* in samples that for one reason or another are heavily contaminated with one of the specified yeasts. Cross-feeding would also take place.

Florenzano and co-workers (54) successfully formulated a selective procedure to isolate species of *Schizosaccharomyces* from grapes. Vitamin-free yeast base (1.67%) is supplemented with glucose (9%), malic acid (1%), cycloheximide (10 mg/liter), $K_2S_2O_3$ (50 mg/liter), thiamine hydrochloride (1 mg/liter), nicotinamide (0.5 mg/liter), calcium pantothenate (0.5 mg/liter), biotin (0.05

mg/liter), inositol (7 mg/liter), and agar (2%); the pH is adjusted to 4. After inoculation, agar plates are inculbated under anaerobic conditions for 8–12 days. Colonies of *Schizosaccharomyces* are recognizable by their larger size and ocher coloration in the central part of colonies. In this case, the utilization of a selective medium is warranted by the fact that other methods of estimating the distribution of *Schizosaccharomyces* in nature would severely underestimate its abundance.

V. CONCLUDING REMARKS

I was once faced with the challenging task of designing a culture medium that would be selective for a yeast that had been found to possess desirable characteristics that could not be correlated directly with a specific trait of yeast nutrition. I approached the problem through the prejudice of a yeast community ecologist. A relatively simple computer algorithm allowed me to compare the mean nutritional profile of the desired yeast with the profiles of all described yeasts in such a way as to produce a list of the physiological tests normally performed in the course of yeast identification, along with the percentage of all known yeasts not sharing the same response, for each trait, as the desired yeast. That percentage is also a measure of the potential selectivity of a medium formulated to be selective for that trait. The algorithm further allowed the selection of pairs or even triads of traits, to see what the selectivity of those combinations would be. The result was the formulation of an agar medium on which a vigorously growing yeast colony had 96% chance of belonging to the right species. The medium was evaluated and found specially useful in that the desired yeast acquired a very distinct colony morphology when growing on it. The isolation of that yeast was therefore simplified. In subsequent screenings using the newly formulated medium, the yeast in question was recovered successfully, but almost exclusively from sources already known to constitute habitats of that yeast. When substrates from which that yeast had not been recovered before were examined, they still failed to yield even small numbers of the yeast. This, I believe, illustrates quite well the need to take habitats into account as preponderant elements in planning any yeast screening program.

In summary, I have attempted to review some of the information available relative to selection as it affects yeast in nature, hoping that such an approach can help generate useful ideas for those interested in recovering new strains to perform specific tasks. Yeasts have been used as domestic microorganisms for almost as long as the history of mankind is known, and there is no reason to believe that they will cease to generate new benefits as we further our understanding of the interactions of yeasts with their natural habitats.

ACKNOWLEDGMENTS

I gratefully acknowledge the help of William T. Starmer, Syracuse University, for providing me with the hospitality of his laboratory during my sabbatical leave, and for his invaluable assistance in reviewing the literature pertinent to this paper. I am indebted to the Natural Science and Engineering Research Council of Canada for research funding.

REFERENCES

1. Collins, J. P., *Evolutionary ecology* and the use of natural selection in ecological theory, *J. Hist. Biol.*, 19:257–288 (1986).
2. Do Carmo-Sousa, L., Distribution of yeasts in nature, in *The yeasts*, A. H. Rose and J. S. Harrison (Eds.), Volume I, Academic Press, London, pp. 79–105 (1969).
3. Phaff, H. J., and W. T. Starmer, Specificity of natural habitats for yeasts and yeast-like organisms, in *Biology and Activities of Yeasts*, F. A. Skinner, S. M. Passmore, and R. R. Davenport (Eds.), Academic Press, London, pp. 79–102 (1980).
4. Schlegel, H. G., *General Microbiology*, Cambridge University Press, Cambridge, U.K. (1986).
5. Davenport, R. R., An introduction to yeasts and yeast-like organisms, in *Biology and Activities of Yeasts*, F. A. Skinner, S. M. Passmore, and R. R. Davenport (Eds.), Academic Press, London, pp. 1–27 (1980).
6. McNaughton, S. J., Niche: Definition and generalization, in *The Fungal Community*, D. T. Wicklow and G. C. Carroll (Eds.), Marcel Dekker, New York, pp. 79–88 (1981).
7. Hutchinson, G. E., Concluding remarks. *Cold Spring Harbor Symp. Quant. Biol.*, 22:415–427 (1958).
8. Starmer, W. T., An analysis of the fundamental and realized niche of cactophilic yeasts, in *The Fungal Community*, D. T. Wicklow and G. C. Carroll (Eds.), Marcel Dekker, New York, pp. 129–156 (1981).
9. Danielson, R. M., and M. F. Jurgensen, The propagule density of *Lipomyces* and other yeasts in forest soils, *Mycopathol. Mycol. Appl.*, 51:191–198 (1973).
10. Fell, J. W., Yeast in oceanic regions, in *Recent Advances in Aquatic Microbiology*, E. B. Gareth-Jones (Ed.), Halsted Press, New York, pp. 93–124 (1976).
11. Bowles, J. M., and M. A. Lachance, Patterns of variation in the yeast florae of exudates in an oak community, *Can. J. Bot.*, 61:2984–2995 (1983).
12. Lachance, M. A., and W. T. Starmer, Evolutionary significance of physiological relatedness among yeast communities associated with tree exudates, *Can. J. Bot.*, 60: 285–293 (1982).
13. Lachance, M. A., B. J. Metcalf, and W. T. Starmer, Yeasts from exudates of *Quercus*, *Ulmus*, *Populus*, and *Pseudotsuga*: New isolations and elucidation of some factors affecting ecological specificity, *Microb. Ecol.*, 8:191–198 (1982).
14. Lachance, M. A., and W. T. Starmer, The community concept and the problem of non-trivial characterization of yeast communities, *Coenoses* 1:21–28 (1986).
15. Starmer, W. T., H. J. Phaff, and M. A. Lachance, A comparison of yeast communities found in necrotic tissue of cladodes and fruits of *Opuntia stricta* on islands in the

Caribbean Sea and where introduced into Australia, *Microb. Ecol.*, 14:179–192 (1987).

16. Vidal-Leiria, M., H. Buckley, and N. van Uden, Distribution of the maximum temperature for growth among yeasts, *Mycologia*, 71:493–501 (1979).

17. Lachance, M. A., *Sporopachydermia quercuum*, a new yeast species from exudates of *Quercus rubra*, *Can. J. Microbiol.* 28:567–571 (1982).

18. Tilbury, R. H., Xerotolerant (osmophilic) yeasts, in *Biology and Activities of Yeasts*, F. A. Skinner, S. M. Passmore, and R. R. Davenport (Eds.), Academic Press, London, pp. 153–179 (1980).

19. Last, F. T., and D. Price, Yeasts associated with living plants and their environs, in *The Yeasts*, A. H. Rose and J. S. Harrison (Eds.), Volume I, Academic Press, London, pp. 183–218 (1969).

20. Pennycook, S. R., and F. J. Newhook, Ultraviolet sterilization in phylloplane studies, *Trans. Br. Mycol. Soc.*, 78:360–361 (1982).

21. Miller, G. R., and S. Baharaeen, DNA base compositions and photoreactivation capabilities of six *Hansenula* species, *Antonie van Leeuwenboek*, 45:365–368 (1979).

22. Leufven, A., G. Bergstrom, and E. Falsen, Interconversion of verbenols and verbenones by identified yeasts isolated from the spruce bark beetle *Ips typographicus*, *J. Chem. Ecol.*, 10:1349–1361 (1984).

23. Phaff, H. J., and M. W. Miller, A specific microflora associated with the fig wasp, *Blastophaga psenes* Linnaeus, *J. Ins. Pathol.*, 3:233–243 (1961).

24. Cooper,D.M., Food preferences of larval and adult *Drosophila*, *Evolution*, 14:41–55 (1960).

25. Fogleman, J. C., W. T. Starmer, and W. B. Heed, Larval selectivity for yeast species by *Drosophila mojavensis* in natural substrates, *Proc. Natl. Acad. Sci. U.S.A.*, 71:4435–4439 (1981).

26. Starmer, W. T., H. J. Phaff, J. M. Bowles, and M. A. Lachance, Yeasts vectored by insects feeding on decaying saguaro cactus. *Southw. Nat.*, 33:362–363 (1987).

27. Robinson, T., Metabolism and function of alkaloids in plants, *Science*, 184:430–435 (1974).

28. Palo, R. T., Distribution of birch (*Betula* spp.), willow (*Salix* spp.), and poplar (*Populus* spp.) secondary metabolites and their potential role as chemical defense against herbivores, *J. Chem. Ecol.*, 10:499–520 (1984).

29. Leufven, A., and L. Nehls, Quantification of different yeasts associated with the bark beetle, *Ips typographicus*, during its attack on spruce tree, *Microb. Ecol.*, 12:237–243 (1986).

30. Starmer, W. T., Associations and interactions among yeasts, *Drosophila* and their habitats, in *Ecological Genetics and Evolution*, J. S. F. Barker and W. T. Starmer (Eds.), Academic Press, Sydney, pp. 159–174 (1982).

31. Conner, D. E., and L. R. Beuchat, Sensitivity of heat-stressed yeasts to essential oils of plants, *Appl. Environ. Microbiol.*, 47:229–233 (1984).

32. Starmer, W. T., H. W. Kircher, and H. J. Phaff, Evolution and speciation of host plant specific yeasts, *Evolution* 34:137–146 (1979).

33. West, C. A., Fungal elicitors of the phytoalexin response in higher plants, *Naturwissenschaften*, 68:447–457 (1981).

34. Hahn, M. G., and P. Albersheim, Host-pathogen interactions. XIV. Isolation and partial characterization of an elicitor from yeast extract, *Plant Physiol.*, 62:107–111 (1978).

35. Schink, B. J., J. C. Ward, and J. G. Zeikus, Microbiology of wetwood: Role of anaerobic bacterial populations in living tress, *J. Gen. Microbiol.*, 123:313–322 (1981).
36. Martin, M. M., J. J. Kukor, J. S. Martin, T. E. O'Toole, and M. W. Johnson, Digestive enzymes of fungus-feeding beetles, *Physiol. Zool.*, 54:137–145 (1981).
37. Butler, M. J., and M. A. Lachance, Inhibition of melanin synthesis in the black yeast *Phaeococcomyces* sp. by growth on low pH ascorbate medium: Production of spheroplasts from "albinized" cells. *Can. J. Microbiol.*, 33:184–187 (1987).
38. Brough, E. J., The antimicrobial activity of the mandibular gland secretion of a formicine ant, *Calomyrmex* sp. (Hymenoptera: Formicidae), *J. Invertebr. Pathol.*, 42:306–311 (1983).
39. Gochnauer, T. A., R. Boch, and V. J. Margetts, Inhibition of *Ascosphaera apis* by citral and geraniol, *J. Invertebr. Pathol.*, 34:57–61 (1979).
40. Slavikova, E., and A. Kockova-Kratochvilova, The yeasts of the genus *Debaryomyces* transferred by insects on the lowlands of Zahorie, *Ces. Mykol.*, 34:21–28 (1980).
41. Di Menna, M. E., The antibiotic relationships of some yeasts from soil and leaves, *J. Gen. Microbiol.*, 27:249–257 (1962).
42. Lachance, M. A., and M. F. Boivin, Antibiosis in a yeast community. *Sixth International Symposium on Yeasts*, Montpellier, France (1984).
43. Stumm, C., J. M. H. Hermans, E. J. Middelbeek, A. F. Croes, and G. J. M. L. de Vries, Killer-sensitive relationships in yeasts from natural habitats, *Antonie van Leeuwenhoek*, 43:125–128.
44. Radler, F., P. Pfeiffer, and M. Dennert, Killer toxins in new isolates of the yeasts *Hanseniaspora uvarum* and *Pichia kluyveri*, *FEMS Microbiol. Lett.*, 29:269–272.
45. Wickerham, L. J., New homothallic taxa of *Hansenula*, *Mycopathol. Mycol. Appl.*, 37:15–32 (1969).
46. Lachance, M. A., Yeasts associated with the black knot disease of trees, in *Current Developments in Yeast Research*, G. G. Stewart and I. Russel (Eds.), Pergamon, Toronto, pp. 607–613 (1980).
47. Nigam, J. N., A. Margaritis, R. S. Ireland, and M. A. Lachance, Isolation and screening of yeasts which ferment of D-xylose directly to ethanol. *Appl. Environ. Microbiol.*, 50:1486–1489 (1985).
48. Van Dijken, J. P., and W. Harder, Optimal conditions for the enrichment and isolation of methanol-assimilating yeasts, *J. Gen. Microbiol.*, 84:409–411 (1974).
49. Van der Walt, J. P., and D. Yarrow, Methods for the isolation, maintenance, classification and identification of yeasts, in *The Yeasts, a Taxonomic Study*, N. J. W. Kreger–Van Rij (Ed.), Amsterdam, Elsevier, pp. 45–104 (1984).
50. Davenport, R. R., An outline guide to media and methods for studying yeasts and yeast-like organisms, in *Biology and Activities of Yeasts*, F. A. Skinner, S. M. Passmore, and R. R. Davenport (Eds.), Academic Press, London, pp. 261–278 (1980).
51. Hertz, M. R., and M. Levine, A fungistatic medium for enumeration of yeasts, *Food Res.*, 7:430–441 (1942).
52. Ingledew, W. M., and G. P. Casey, The use and understanding of media used in brewing mycology. I. Media for wild yeasts, *Brew. Dig.*, 57(3):18–22 (1982).

53. Van der Walt, J. P., and A. E. van Kerken, The wine yeasts of the Cape. V. Studies on the occurrence of *Brettanomyces intermedium* and *Brettanomyces schanderlii*, *Antonie van Leeuwenhoek*, 27:81–90 (1961).

54. Florenzano, G., W. Balloni, and R. Materassi, Contributo all ecologia dei lieviti *Schizosaccharomyces* sulle uve, *Vitis*, 16:38–44 (1977).

55. Starmer, W. T., P. F. Ganter, V. Aberdeen, M.-A. Lachance, and H. J. Phaff, The ecological role of killer yeasts in natural communities of yeasts, *Can. J. Microbiol.*, 33:783–796 (1987).

3
Classical Approaches to Yeast Strain Selection

Carl A. Bilinski
Labatt Brewing Co. Ltd.
London, Ontario, Canada

Nelson Marmiroli*
University of Parma
Parma, Italy

I. INTRODUCTION

The yeast *Saccharomyces cerevisiae* has clearly proved to be a model eukaryotic organism for investigation of fundamental problems in cell biology. Dissection of any cellular process demands use of an organism from which mutants can be readily secured and a genetic system to allow assessment of not only the number of gene functions involved but also the sequence of their expression. Yeast as a unicellular system provides the framework to address these and other complex issues, and it is the intent of this review to provide the reader with general information pertaining to yeast life cycle, mutant isolation, and approaches to strain construction. Aspects of strain selection surveyed are restricted primarily to haploids, with some attention given to the more complex genetic systems of polyploids when necessary. Protocols for mutagenesis have been recently reviewed in detail elsewhere and will not be reviewed here (1).

**Present affiliation:* University of Lecce, Lecce, Italy.

II. YEAST LIFE CYCLE

A. Alternation of Generations

In the life cycle of *Saccharomyces cerevisiae* vegetative cells multiply by mitotic nuclear division and budding in a suitable environment containing assimilable carbon and nitrogen sources, vitamins and salts supplying trace elements (2). Sporulation is a developmental process initiated on nutrient deprivation in cells heterozygous (MATa/MATα) for the mating type locus (3,4). When MATa/MATα diploids are induced to sporulate, the two meiotic nuclear divisions occur and cells differentiate into tetranucleate asci containing one to four uninucleate haploid ascospores. Thus, an ascus containing a tetrad of spores harbors two spores that are MATa and two that are MATα. When individual haploid ascospores are isolated by any one of several techniques (1) and returned to conditions favoring vegetative cell multiplication, the spores swell and germinate, giving rise to clones of haploid cells that propagate by mitotic nuclear division and budding.

Much attention has been given to the mating behavior of the vegetative cells originating from germinated ascospores in *Saccharomyces cerevisiae*, and strains have been classified as heterothallic or homothallic with respect to mating capacity (5,6). In heterothallic strains individual spores give rise to haploids displaying either MATa or MATα phenotypes, and diploidization does not occur unless cells of opposite mating type are permitted to fuse. In contrast, diploidization in homothallic strains occurs through fusions between daughter cells of opposite mating type, both descendants of the same haploid ascospore; consequently, sporogenic MATa/MATα diploids result which are isogenic for all loci except mating type.

With genetical and cytological evidence, Winge (7) and Winge and Laustsen (8) demonstrated the alternation in haploid and diploid generations in the yeast life cycle. Lindegren (9) established existence of a mating system, and Roman and Sands (10) showed that heterozygosity for the mating type alleles is a necessary prerequisite for occurrence of meiosis and hence of sporulation.

B. The Mating System

1. Physiological Requirements

Mutual interaction between MATa and MATα cells is necessary for conjugation in *Saccharomyces cerevisiae* (11). Each mating type produces a specific cell surface glycoprotein which facilitates this interaction and mating pheromones enhance agglutinability (11,12). A medium favoring conjugation designed by Brock (13) lacks a nitrogen source and consists of 1% glucose and 0.1% $MgSO_4$ in 0.05 M potassium phosphate buffer (pH 5.7). Conjugation is usually completed in cultures maintained for 15 hr at 20°C, the optimum temperature for the process. In experiments in which individual medium constituents were omitted, it has been shown that glucose and inorganic salts are essential for conjugation and that karyogamy following cell fusion can be influenced by the further inclusion of an assimilable nitrogen source such as yeast extract (14). Aeration is required for conjugation (15,16), and high concentrations of glucose can enhance mating efficiency (17).

In Jakob's mating procedure (15), MATa and MATα haploids are mixed and aerated vigorously for 2 hr. Cells are pelleted and maintained for 30 min in this form to facilitate contact and hence to initiate conjugation. Cells are then resuspended in fresh culture medium and aerated for 3 hr to induce zygote formation. Several rounds of subculturing and of hybrid cell micromanipulation conclude the procedure.

A simpler method devised by Haefner (18) is even more effective. Equivolumes of haploid cultures possessing complementing auxotrophies are combined on the surface of a complete agar medium. Following 24–36 hr of incubation, growth obtained is streak-plated onto a minimal medium, and after several days incubation plates are examined for prototrophic colonies. A temperature decrease from 30 to 4°C for several hours increases the percentages of zygotes formed with this procedure.

2. Mutations Affecting Mating

The mating system can be subjected to mutational changes that confer sterility (19–21). In homothallic strains, sterility can also result from mutations in modifying genes other than MATa or MATα. The third chromosome of *Saccharomyces cerevisiae* contains three genetic loci, designated HMR, HML, and MAT, capable of determining cell type by regulation of various cell type-specific genes but only information resident at MAT is expressed and HMR and HML which encode silent MATa and MATα information are not expressed (23). In contrast to heterothallic strains, homothallic strains harbor a dominant homothallism gene (HO) which permits interconversion of MAT alleles with extreme fidelity during mitotic reproduction of vegetative progeny derived from single germinated asocospores (23). This interconversion is achieved through transposition of a copy of sequences from the HMR or HML locus to the MAT locus (24–27). MAR (mating type regulator), SIR (silent information regulator), and CMT (change of mating type) loci have been identified that regulate expression of genes controlling mating type (28–32). Of particular interest are the four *trans*-acting SIR loci which encode products that lead to expression of silent MAT information and hence in generation of a nonmating phenotype in a haploid (22). HMR and HML are flanked by *cis*-acting E and I sequences, elements in which mutation can also cause expression of HMR and HML loci (33,34). It appears that some of the SIR gene products are DNA-binding proteins that recognize E and I sites whereas other SIR gene products transmit regulatory signals to promoters downstream from these sites. Genes controlling homothallism have also been found in other yeasts: *Schizosaccharomyces pombe* (12,35), *Saccharomyces chevalieri* (36), *Saccharomyces oviformis* (37,38), and *Saccharomyces lactis* (39).

In addition to sterile strains, hyperfertile strains can spontaneously arise that are capable of conjugation with both MAT*a* and MATα cells (40). The mating between opposite cell types from different strains can sometimes be restricted in that MAT*a* of one strain can be mated with MATα of another strain but not vice versa. Difficulties in hybridization resulting from weak mating reactions can be overcome by inbreeding (41,42). Diploids of either *aa* or αα type can sometimes arise by illegitimate mating, especially in old cultures (40,43–45).

C. Cell Division Cycle

A number of excellent reviews are available that describe in detail the yeast cell division cycle (46–51). The mitotic cell cycle is conventionally divided into a G1 phase, which precedes initiation of nuclear DNA synthesis; an S phase, during which chromosomal DNA is replicated; a subsequent G2 phase; and an M phase during which mitosis and nuclear division occur. Hartwell (46) described in detail the isolation of conditionally temperature-sensitive cell division cycle (cdc) mutants blocked at specific stages in this cycle, and Mitchison (52) has reviewed various methods available for culture synchronization and age fractionation in cell cycle studies.

D. Sporulation

1. Physiological Requirements

Culture conditions for induction of sporulation in *Saccharomyces cerevisiae* have been the subject of a recent review (3). Physiological requirements for maximum ascosporogenesis can vary dependent on the genetic background of the yeast strain under study (53). In diploid strains sporulation has been reported to be optimal at 27–30°C using stationary phase glucose-grown cells or exponential phase acetate-grown cells (54,55). In a polyploid *Saccharomyces uvarum (carlsbergensis)*, ascus formation has been optimized at 21°C when acetate-grown cells harvested from stationary phase are inoculated into sporulation medium and no ascus formation occurs following incubation of presporulation and/or sporulation cultures at 27°C (56). In a polyploid *Saccharomyces cerevisiae* exponential phase acetate-grown cells have been found to also give markedly higher sporulation percentages on incubation of sporulation cultures at 21°C instead of 17°C (57,58). Several approaches to isolation of conditional sporulation-deficient (*spo*) mutants are given below.

2. Isolation of Mutants

Use of Homothallic Strains. Esposito and Esposito (59) exploited the self-diploidization process characteristic of homothallic yeast to isolate from ultraviolet or x-irradiated haploid spores diploid mutants that harbor temperature-sensitive (ts) recessive mutations conferring a *spo* phenotype for use in genetic and physiological studies. In efforts to isolate additional *spo* mutants, Dawes and Hardie (60) performed random spore analysis on sporulated cultures of homothallic diploids subjected to chemical mutagenesis during presporulation growth. Thus homothallic strains have been particularly useful in isolation of mutants blockied at various landmark stages of meiosis and ascospore formation (61–63).

Use of Strains Disomic for Chromosome III. Roth and Fogel (64) demonstrated that MAT*a*/MATα haploids disomic for chromosome III fail to sporulate but do undergo premeiotic DNA syntheisis; moreover, on introduction of markers in heterozygous or heterallelic condition on the homologous chromosome III pair, high recombination frequencies were detected. This together with the fact that all other chromosomes were present in single copies made possible the application of EMS mutagenesis to facilitate subsequent recovery of dominant and recessive mutations affecting meiotic recombination.

Identification of Naturally Occurring Mutants. In an extensive survey of various *Saccharomyces cerevisiae*, Grewal and Miller (65) identified 17 naturally occurring strains that formed few or no asci with more than two spores. Fourteen of the strains underwent normal meiosis during sporulation, since the asci always contained four nuclei. Three of the strains (ATCC 4117, ATCC 4098, 19el), however, were, unusual in that the asci never contained more than two nuclei. On further analysis they showed that the nuclei of the spores were diploid and that clones derived from germinated ascospores were competent to sporulate in the absence of prior matings; moreover, the single nuclear division preceding ascospore formation was not glucose-inhibited as in meiosis I of normally four-spored strains (66). Such yeasts are apomictic, since, as in apomictic parthenogenesis in higher eukaryotes, the sexual cycle bypasses gamete formation and the requirement for fertilization has been lost (67–69).

In a genetic analysis of strain 4117, Klapholz and Esposito (70) identified two recessive mutations, designated spo12-1 and spo13-1, either of which alone could confer an apomictic phenotype; in addition, an analysis of marker segregation (71) substantiated ultrastructural observations indicating that meiosis I chromosome segregation was bypassed and that a meiosis II-like division occurred with consequent formation of two diploid nuclei per ascus (72–74). Physiological conditions have been identified for rescue of a normal meiotic phenotype in apomictic mutants (75–77), thus providing a model system for investigation of events controlling meiotic and apomictic development (77–79). Other isolates from nature may provide an additional source of mutations leading to altered sporulation phenotypes in *Saccharomyces cerevisiae.*

E. Map Characteristics

Given the alternation of generations in the life cycle of *Saccharomyces cerevisiae*, an ideal experimental system has been made available for use in classical genetic studies. In addition to a well-defined genetic system composed of genes on some 17 chromosomes, *Saccharomyces cerevisiae* has been the focus of attention in application of recombinant DNA methodology and chromosome replacement. About 300 loci have been mapped including centromere-linked markers for all but one chromosome (80). The haploid yeast genome has a size of about 14,000 kilobase (kb) pairs (81) and contains approximately 4600 centimorgans (cM) of genetic information with a map correspondence that ranges from 2.7 kb/cM for chromosome III (82) to 10 kb/cM for other parts of the genome (83). Thus high recombination frequencies per number of base pairs are evident in comparison to other organisms.

Complementation analyses provide the means to distinguish whether mutations responsible for a given mutant phenotype are in the same or different genes. Mutations within a gene do not complement, and thus when two haploids that harbor a mutation within the same locus are crossed, the resultant diploid will still display the mutant phenotype. Mutations in different genes do complement in a diploid, and, in contrast to the parents, a wild-type phenotype is observed. Once mutations affecting a given process have been assigned to different complementation groups and the terminal phenotypes of each complementation group have been distinguished, an analysis of dependency pathways in gene function can be performed. Comparison of the terminal phenotypes of single and double conditional mutants under nonpermissive conditions has allowed ordering of gene functions in the mitotic cell division cycle (46–51), sporulation (61–63), and metabolism (84,85).

In gene mapping the important task is to initially assign a given marker to a particular chromosome and then to precisely define its location by fine-structure mapping. Classical approaches to mapping by tetrad analysis in *Saccharomyces cerevisiae* have been thoroughly described (1,86–88). Other methods employ genetically and chemically induced mitotic chromosome loss (1,88,89), supertriploids (90), and mutations such as spo11-1 that affect chromosome segregation during meiosis (91,92).

III. APPROACHES TO MUTANT ISOLATION

A. Enrichment Procedures

Construction of a genetic map requires genetic markers and a good system for studying recombination. Although classical mutagenesis or ionizing radiations can be employed to isolate nutritional and temperature-sensitive (ts) mutants directly from haploid strains, mutations at several loci within a given cell can result, and consequently these methods are not reliable in the construction of a series of strains that differ at single loci but otherwise remain isogenic. Such difficulties are not encountered with spontaneously arising mutants, but these occur at very low frequencies in yeast (less than 10). Regardless of experimental requirements, retrieval of mutants of interest is often hindered by the persistent presence of prototrophic clones. Enrichment procedures, however, are available to discriminate against survival of prototrophs with little, if any, effect on survival of spontaneously arising or mutagen-induced auxotrophs and ts mutants. Pringle (93) has reviewed in detail conditions for induction and selection of its and other conditional mutants from yeast. Spencer and Spencer (1) have reviewed various classes of mutagens and their application. This section will focus primarily on enrichment procedures available for isolation of auxotrophic mutants.

1. Nystatin

Moat et al. (94) demonstrated that the drugs amphotericin B, edomycin, and nystatin can be added to an adenine-free medium in order to eliminate prototrophs from artificial mixtures of adenine-requiring and wild-type cells, thereby increasing the proportions of adenine-requiring survivors. Snow (95) extended their approach employing nystatin under conditions of amino acid deprivation to enrich for spontaneous and EMS-induced amino acid auxotrophs from haploids. Amino acid-requiring mutants have been isolated from polyploid strains by application of nystatin enrichment following EMS mutagenesis and exposure to the fungicide, benomyl, the active component of which induces chromosome loss (96).

In addition to known limitations of nystatin enrichment summarized by Littlewood (97), which incidentally may also apply to other enrichment techniques described below, elimination of mutants by cross-feeding is another possible limitation given that nystatin breaks open the cell walls of actively dividing yeast (95). Similar difficulties are encountered when other inhibitors of cell wall biosynthesis such as echinicidin (98) and amphotericin B (N. Marmiroli, unpublished) are applied in enrichment procedures. In this regard, it is of interest that cross-feeding is not encountered when the antibiotic neotropsin is employed as a counterselecting agent (99).

2. Inositol-less and Fatty Acid-less

A protocol for retrieval of spontaneously arising and EMS-induced auxotrophs has been developed based on the sensitivity of inositol-requiring (ino⁻) mutants to inositol-less death (100). Following introduction of inositol markers (ino-1 and ino-4) into the desired genetic background, the resultant inositol auxotrophs are cultivated in complete medium, harvested from exponential growth and incubated in a "prestarvation medium" containing inositol but lacking the amino acid for which auxotrophy is desired. Subsequent transfer of cells to a "starvation medium" lacking in addition, inositol, enriches for the amino acid auxotroph while eliminating inositol-requiring cells that are not auxotrophic for the particular amino acid of interest. Putative auxotrophs are then analyzed further by replica plating from complete medium to minimal medium supplemented with inositol.

Using a similar rationale to inositol-less enrichment, Henry and Horowitz (101) devised a procedure for selection of auxotrophs employing *olel fasl* haploids that require both saturated and unsaturated fatty acids. Since inclusion of cycloheximide under fatty acid starvation conditions was found to protect these mutants against fatty acid-less death, it was found possible to adapt the procedure for enrichment of ts mutants defective in macromolecular synthesis.

3. Thymineless

Barclay and Little (102) devised an enrichment based on the resistance of auxotrophs to thymineless-induced death. In contrast to inositol-less and fatty acid-less enrichments, which are strictly dependent on prior introduction of appropriate markers (100,101), thymineless enrichment can be applied directly to wild-type strains in selection of spontaneous auxotrophs. The biosynthetic pathways for adenine, histidine, methionine, and deoxythymidine 5'-monophosphate (dTMP) are tetrahydrofolate-dependent. Consequently cultures depleted of tetrahydrofolates by exposure to a combination of sulfanilimide and aminopterin or methotrexate (B. Barclay, personal communication) are rendered sensitive to thymineless death when starved for dTMP in the presence of adenine, histidine, and methionine. Since auxotrophs are much less sensitive to effects of dTMP starvation than prototrophs, it is possible to increase their proportions by cycling cells between dTMP starvation and complete media. Putative auxotrophs are then characterized in the usual manner by replica plating of survivors from complete to supplemented and unsupplemented minimal media. Since methionine offers some protection against thymineless-induced death, the protocol can be modified for enrichment of ts mutants blocked in macromolecular synthesis.

4. Tritium Suicide

Tritium suicide death is caused by decay of tritium incorporated into macromolecular components (103) and is an enrichment scheme originally used to select auxotrophs in *Escherichia coli* (104). Mutagenized cells are labeled in minimal media under conditions in which surviving prototrophs incorporate lethal doses of tritiated precursors. Littlewood and Davies (105) applied this enrichment in an effort to isolate auxotrophic and ts mutants from *Saccharomyces cerevisiae*. By appropriate choice of tritiated precursors, ts mutants harboring specific defects in macromolecular synthesis were isolated. Thus, storage of cultures at 4°C following application of [^{3}H]uridine at a high temperatures (38°C) facilitated subsequent retrieval of ts mutants blocked in RNA synthesis, whereas, exposure to [^{3}H]amino acids instead of [^{3}H]uridine enriched for ts mutants blocked in protein synthesis.

In selection of auxotrophs, mutagenized cultures were administered [^{3}H]-labeled precursors at 28°C and selected following storage at 4°C.

5. Other Uses of Mutants

The inositol-less and fatty acid-less enrichment procedures are only two examples where known mutants can be exploited to isolate other mutants. Several other procedures have been described to enrich for auxotrophs by use of ts mutants that undergo extensive cell death at their restrictive temperatures (106–108). These methods provide for selection of large numbers of independent mutants with specific requirements and can be used in direct selection on plates or in liquid culture. Littlewood (106) employed ts suicide mutants designated "kamikaze" in order to isolate antibiotic-sensitive mutants, and Hardie and Dawes (107) found ts suicide mutants suitable in isolation of spontaneous auxotrophs.

Selection of nutritional mutants based on a visible phenotypic alteration is the exception and not the rule in *Saccharomyces cerevisiae*. The only examples are the *ade 1* and *ade 2* mutants blocked in adenine biosynthesis, which can be distinguished as red colonies in media containing low adenine concentrations. Nonetheless, a procedure that employs these mutants has been recently devised to isolate amino acid auxotrophs (109). Mutagenized *ade 1* or *ade 2* haploids are cultivated on minimal agar medium containing a suboptimal adenine concentration for pigment production and supplemented with a low concentration of amino acid for which auxotrophy is sought. Putative auxotrophs appear as white colonies against a background of red ones because their growth ceases on depletion of the particular amino acid prior to the onset of adenine limitation. These white colonies are then subcultured on complete medium and replica plated onto synthetic media supplemented with adenine, with and without the appropriate amino acid.

6. Selective Cell Wall Lysis

It is well known that spheroplasts can be readily obtained by snail enzyme digestion from exponential but not from stationary phase cultures (110,111). Applying this rationale, Ferenczy (112) found it possible to enrich for auxotrophs and ts mutants. Under selective conditions, the mutants, like stationary phase cells, are much more resistant to snail enzyme digestion because of their incapacity to divide. In contrast, actively dividing prototrophs are sensitive to digestion and hence to cell lysis in an osmotically stabilized medium. Piedra and Herrera (113) applied this procedure successfully in isolation of auxotrophs from mutagenized cultures of *Saccharomyces cerevisiae* and obtained auxotroph yields of approximately 66% in a population that had gone through only five cell divisions prior to application of snail enzyme.

Applying a rationale similar to selective cell wall lysis, Walton et al. (114) exploited the sensitivity of exponential phase cells to killing by heat as a very efficient enrichment procedure for isolation of various yeast mutants.

B. Sensitivity and Resistance Mutants

Mutagenesis and selection have been applied to isolate stable mutants sensitive to cations (115–117), ionizing radiation (118,119), chemical mutagens (120,121), drugs (122,123), *a*- and α-pheromones (124), and phenethyl alcohol (125). Mercury-sensitive mutants have been useful in evaluating factors affecting metal toxicity (116) and in the study of catabolite regulation (126). Calcium-sensitive mutants have been employed to analyze regulatory roles of the divalent cation in the yeast cell division cycle (127).

An array of mutants have been isolated and characterized for resistance to antibiotics (128–134), fungicides and herbicides (135,136), cations (137-141), and mutagens (118,142). Dominant resistance markers have served as valuable tools in transformation of novel metabolic capabilities into polyploid industrial yeast strains.

IV. APPROACHES TO STRAIN CONSTRUCTION

A. Classical Hybridization

Hybridization of MAT*a* and MATα haploid cells by micromanipluation was first described by Chen (143). Zygote formation can be followed by direct light microscopic observation. Zygotes can also be selected through complementation of auxotrophies present in the two parents (144). Clearly the effectiveness of these procedures in giving hybrids depends on the mating capacity of the two haploid parents. On occasion this can be so weak that no fusions occur and consequently zygotes are not formed; on the other hand, even when cell fusion occurs, nuclear fusion does not necessarily follow (145). Diploids that arise from matings are usually larger in size than haploids. They should have the capacity to sporulate and should be confirmed to be true hybrids by tetrad analysis.

Hybridization can also be achieved by mass mating, and this procedure has been recommended for use with commercial strains (43,44). A method was designed by Fowell (145), as a modification of Lindegren's techniques. A mass mating incubated overnight was subcultured several times in order to enrich for hybrids. Large oval cells were then isolated by micromanipulation and transferred to complete medium. The culture was then checked for purity by direct observation and/or genetic analysis. Commercial strain selection based on genetic complementation is often impractical, because mutagenic treatments required for induction of auxotrophies can impair strain performance (96).

B. Cytoduction and Single Chromosome Transfer

Karyogamy-defective (*kar*) mutants have been described in *Saccharomyces cerevisiae* that affect nuclear fusion following conjugation (146). In crosses in which one of the parents harbors the *dar 1* mutation, mating proceeds in the usual fashion except subsequent nuclear fusion does not occur. This yields dikaryotic cells which can be resolved into either heteroplasmons, in which the two nuclei are segregated from each other on budding, or heterokaryons, which contain a mixture of both nuclei in individual cells. In both cases the mixing of cytoplasms occurs, and in the case of heteroplasmons, an opportunity is given to investigate nucleomitochondrial interaction, for with a suitable selection scheme cell lines can be obtained that harbor the nucleus of one parent but the mitochondria of the other parent. The transfer of single chromosomes (147,148) and of elements such as 2 μ DNA (149), killer particles (150–152), and composite plasmids (153) can also be achieved through use of *kar* mutants. The *kar 1* mutation has also been useful in assessing whether mating functions act via the cytoplasm in sporulation (154).

C. Spheroplast Fusion

Methods employed for fusion of plant protoplasts (155) can also be applied in fusion of yeast spheroplasts where mating is not feasible (1,156). In selection procedures for fusion products it is important to determine whether the two parents can be readily distinguished. In some instances, partners can be discriminated on the basis of their substrate utilization abilities (carbon and nitrogen sources) while others can differ in giant colony morphology and/or growth ability at higher temperatures (157). Dominant mutations conferring cation resistance or drug resistance are particularly useful especially in strains of higher ploidy where recessive markers are not readily secured. Care should be taken to confirm that desirable traits are not lost in isolation of derivatives harboring markers. Ideally partners should be multiply marked since chromosome loss, and thus loss of a given marker, is not uncommon following fusion (158). Typical protocols for spheroplast fusion have been reviewed (1).

Mitochondrial function is essential for survival in petite negative but not in petite positive yeasts (159,160). Fusion products between such yeasts can serve as model systems for investigation of nucleomitochondrial interactions. Respiratory-deficient petites have been useful in selection of fusion products involving *Saccharomyces cerevisiae* and *Saccharomyces diastaticus*, which are petite positive yeasts (161,162). A study of mitochondrial recombination has demonstrated that fusion products will accept the mitochondrial genome from a *rho* parent but not from a neutral petite (163,164).

Spheroplast fusion provides the means to make zygote formation independent of mating type control since strains of identical mating type can be fused with production of true fusion products in which karyogamy has occurred (163–166). Interspecific fusion has been performed successfully between fission yeasts (167) and between budding yeasts (161). Intergeneric fusion products have been obtained between *Saccharomyces cerevisiae* and *Pichia membranefaciens, Hansenula capsulata (168),* and between *Saccharomyces cerevisiae* and *Schizosaccharomyces pombe* (169).

Despite the degree of taxonomic relatedness between fusion partners, genomic differences can lead to undesirable growth and morphological and/or physiological alterations. Loss of genetic markers contributed from one of the partners is frequently observed in newly formed fusants following several cell divisions, and often there is a requirement for maintenance on a selective medium for stability to be retained (167). As with karyogamy-defective mutants described earlier, heterokaryons and heteroplasmons can result from spheroplast fusion but fusions with certain yeasts tend to yield products in which nuclear fusion does occur (170).

The classical techniques employed in strain selection and strain construction will continue to play a key role in advancing understanding of cell proliferation, metabolism, and differentiation. Without doubt, it is the amenability of yeast to genetic manipulation that has made it the ideal organism of choice to address fundamental problems in eukaryotic cell biology.

REFERENCES

1. Spencer, J. F. T., and D. M. Spencer, Yeast genetics, in *Yeast, A Practical Approach*, I. Campbell and J. H. Duffus (Eds.). IRL Press, Oxford, U.K., pp. 65–106 (1988).
2. Silverman, S. J., Current methods for *Saccharomyces cerevisiae*, I. Growth, *Anal. Biochem.*, 164:271–277 (1987).
3. Olempska-Beer, Z., Current methods for *Saccharomyces cerevisiae*, II. Sporulation, *Anal. Biochem.*, 164:278–286 (1987).
4. Esposito, R. E., and S. Klapholz, Meiosis and ascospore formation, in *The Molecular Biology of the Yeast Saccharomyces I. Life Cycle and Inheritance*, J. N. Strathern, E. W. Jones, and J. R. Broach (Eds.). Cold Spring Harbor Laboraty, Cold Spring Harbor, NY, pp. 211–287 (1981).
5. Herskowitz, I., J. N. Strathern, J. B. Hicks, and J. Rine, Mating type interconversion in yeast and its relationship to development in higher eukaryotes, in *Molecular Approaches to Eukaryotic Genetic Systems*, F. Wilcox, J. Abelson, and C. F. Fox (Eds.). Academic Press, New York, pp. 193–202 (1977).
6. Herskowitz, I., and Y. Oshima, Control of cell type in *Saccharomyces cerevisiae*: Mating type and mating type interconversion, in *Molecular Biology of the Yeast Saccharomyces cerevisiae. I. Life Cycle and Inheritence*, J. N. Strathern, E. W. Jones, and J. R. Broach (Eds.). Cold Spring Harbor Laboratory, Cold Spring Harbor, NY, pp. 181–209 (1981).
7. Winge, O., On haplophase and diplophase in some *Saccharomycetes, C. R. Trav. Lab. Carlsberg Ser. Physiol.*, 21:77–111 (1935).
8. Winge, O., and O. Laustsen, On two types of spore germination, and on genetic segregation in *Saccharomyces*, demonstrated through single spore cultures, *C. R. Trav. Lab. Carlsberg Ser. Physiol.*, 22:99–116 (1937).

9. Lindegren, C. C. *The Yeast Cell, Its Genetics and Cytology*, St. Louis Educational Publishers, St. Louis, pp. 365 (1949).

10. Roman, H., and S. M. Sands, Heterogeneity of clones derived from haploid ascospores, *Proc. Natl. Acad. Sci. U.S.A.*, 39:171–179 (1953).

11. Levi, J. D., Mating reaction in yeast, *Nature*, 177:753-754 (1956).

12. Crandall, M., R. Egel, and V. L. MacKay, Physiology of mating in three yeasts, *Rec. Adv. Microbiol. Physiol.*, 15: 07–399 (1977).

13. Brock, T. D., Physiology of the conjugation process in the yeast *Hansenula wingei*, *J. Gen. Microbiol*, 26:489–497 (1961).

14. Iguchi, S., and T. Onobu, Conjugation in *Saccharomyces cerevisiae, Bot. Mag. (Tokyo)*, 77:181–190 (1964).

15. Jakob, H., Technique de synchronisation de la formation des zygotes chez laz levure *Saccharomyces cerevisiae, C. R. Hebd, Seanc. Acad., Sci., Paris*, 254:3909–3911 (1962).

16. Sena, E. P., D. N. Radin, J. Welch, and S. Fogel, *Methods Cell Biol.*, 11:71–88 (1975).

17. Lee, E. H., C. V. Lusena, and B. F. Johnson, A new method of obtaining zygotes in *Saccharomyces cerevisiae, Can. J. Microbiol.*, 21:802–806 (1975).

18. Haefner, K., cited in reference 42.

19. Lindegren, C. C., Yeast genetics, *Bacteriol. Rev.*, 9:111–169 (1945).

20. MacKay, V. L., and T. R. Manney, Mutations affecting sexual conjugation and related processes in *Saccharomyces cerevisiae*. I. Isolation and characterization of non-mating mutants, *Genetics*, 76:255–271 (1974).

21. MacKay, V. L., and T. R. Manney, Mutations affecting sexual conjugation and related processes in *Saccharomyces cerevisiae, Genetics*, 76:273–288 (1974).

22. Klar, A. J. S., J. N. Strathern, and J. B. Hicks, Developmental pathways in yeast, in *Microbial Development*, R. Losick and L. Shapiro (Eds.), Cold Spring Harbor Laboratory, Cold Spring Harbor, NY, pp. 151–195 (1984).

23. Harshima, S., Y. Nogi, and Y. Oshima, The genetic system controlling homothallism in *Saccharomyces* yeasts, *Genetics*, 77:639–650 (1974).

24. Hicks, J. B., J. Strathern, and I. Herskowitz, The cassette model of mating-type interconversion, in *DNA Insertion Elements, Plasmids and Episomes*, A. I. Bukhari, A. Shapiro, and S. L. Adhya (Eds.), Cold Spring Harbor Laboratory, Cold Spring Harbor, NY, pp. 457–462 (1977).

25. Klar, A. J. S., Interconversion of yeast cell types by transposable genes, *Genetics*, 95:631–648 (1980).

26. Klar, A. J. S., and S. Fogel, Activation of mating type genes by transposition in *Saccharomyces cerevisiae, Proc. Natl. Acad. Sci. U.S.A.*, 76:4539–4543 (1979).

27. Klar, A. J. S., J. N. Strathern, and J. B. Hicks, A position-effect control for gene transposition: State of expression of yeast mating-type genes affects their ability to switch, *Cell*, 25:517–524 (1981).

28. Klar, A. J. S., S. Fogel, and K. McLeod, MAR1 a regulator of *HMa* and *HMα* loci in *Saccharomyces cerevisiae, Genetics*, 92:759–776.

29. Rine, J. D., J. N. Strathern, J. B. Hicks, and I. Herskowitz, A suppression of mating type locus mutations in *Saccharomyces cerevisiae*: Evidence for and identification of cryptic mating type loci, *Genetics*, 93:877–901.

30. Hopper, A. K., and B. D. Hall, Mutation of a heterothallic strain to homothallism,

Genetics, 80:77–85 (1975).

31. Haber, J. E., and J. P. George, A mutation that permits the expression of normally silent copies of mating-type information in *Saccharomyces cerevisiae, Genetics*, 93:13–35 (1978).

32. Kassir, Y., and G. Simchen, Regulation of mating and meiosis in yeast by the mating type region, *Genetics*, 82:187–206 (1976).

33. Abraham, J. K., K. A. Nasmyth, J. N. Strathern, A. J. S. Klar, and J. B. Hicks, Regulation of mating type information in yeast: Negative control requiring sequences both 5' and 3' to the regulation region, *J. Mol. Biol.*, 176:303–331 (1984).

34. Feldman, J. B., J. B. Hicks, and J. R. Broach, Identification of sites required for expression of a silent mating type locus in yeast, *J. Mol. Biol.*, 178:815–834 (1984).

35. Leupold, U., Die verenbung von homothallie und heterothallie bei *Schizosaccharomyces pomble, C. R. Trav. Lab. Carlsberg Ser. Physiol.*, 24:381–450 (1950).

36. Winge, O., and C. Roberts, A gene for diploidization in yeast, *C. R. Trav. Lab. Carlsberg Ser. Physiol.*, 24:341–346 (1949).

37. Takahashi, T., H. Saito, and Y. Ikeda, Heterothallic behaviour of a homothallic strain in *Saccharomyces* yeast, *Genetics*, 44:249–260 (1958).

38. Takano, I., and Y. Oshima, An allele-specific and a complementary determinant controlling homothallism in *Saccharomyces oviformis, Genetics*, 57:875–885 (1967).

39. Herman, A., and Roman, H., Allele specific determinants of homothallism in *Saccharomyces, Genetics*, 53:727–740 (1966).

40. Lindegren, C. C., and G. Lindegren, Segregation, mutation and copulation in *Saccharomyces cerevisiae, Ann. Mo. Bot. Gdn.* 30:453–469 (1943).

41. Fowell, R. R., The hybridization of yeasts, *J. Appl. Bacteriol.*, 18:149–160 (1955).

42. Fowell, R. R., Sporulation and hybridization of yeast, in *The Yeasts, Vol. 1; Biology of Yeasts*, A. H. Rose and J. S. Harrison (Eds.), Academic Press, New York, pp. 303–383 (1969).

43. Lindegren, C. C., and G. Lindegren, A new method for hybridizing yeast, *Proc. Natl. Acad, Sci. U.S.A.*, 29:306–308 (1943).

44. Lindegren, C. C., and G. Lindegren, The improvement of industrial yeasts by selection and hybridization, *Wallerstein Labi Commun.*, 7:153–168 (1944).

45. Lindegren, C. C., and G. Lindegren, Selecting, inbreeding, recombining and hybridizing commercial yeasts, *J. Bacteriol.*, 46:405–419 (1943).

46. Hartwell, L. H., *Saccharomyces cerevisiae* cell cycle, *Bacteriol. Rev.*, 38:164–198 (1970).

47. Hartwell, L. H., J. Culotti, J. R. Pringle, and B. J. Reid, Genetic control of the cell division cycle in yeast, *Science* 183:46–50 (1974).

48. Simchen, G., Cell cycle mutants, *Annu. Rev. Genet.* 12:161–191 (1978).

49. Carter, B. L. A., J. R. Piggott, and E. F. Walton, Genetic control of cell proliferation, in *Yeast Genetics, Fundamental and Applied Aspects*, J. F. T. Spencer, D. M. Spencer, and A. R. W. Smith (Eds.), Springer-Verlag, New York, pp. 1–28 (1983).

50. Elliott, S. G., and C. S. Mclaughlin, The yeast cell cycle: Coordination of growth and division rates, *Prog. Nucl. Acid Res. Mol. Biol.*, 28:143–176 (1983).

51. Pringle, J. R., and L. H. Hartwell, *Saccharomyces cerevisiae* cell cycle, in *Molecular Biology of the Yeast* Saccharomyces cerevisiae. I. Life Cycle and Inheritence, J. N. Strathern, E. W. Jones, and J. Broach (Eds.), Cold Spring Harbor Laboratory, Cold

Spring Harbor, NY, pp. 97–142 (1981).

52. Mitchison, J. M., Synchronous cultures and age fractionation, in *Yeast, A Practical Approach*, I. Campbell and H. Duffus (Eds.), IRL Press, Oxford, U.K., pp. 51–63 (1988).

53. Miller, J. J., Sporulation in *Saccharomyces*, in *The Yeasts*, Vol. 3, A. H. Rose and J. S. Harrison (Eds.), Academic Press, New York, pp. 489–550 (1989).

54. Roth, R., and H. O. Halvorson, Sporulation of yeast harvested during logarithmic growth, *J. Bacteriol*, 98:831–832 (1969).

55. Fast, D., Sporulation synchrony of *Saccharomyces cerevisiae* grown in various carbon sources, *J. Bacteriol*, 116:925–930 (1975).

56. Bilinski, C. A., I. Russell, and G. G. Stewart, Physiological requirements for induction of sporulation in lager yeast, *J. Inst. Brew.* 93:216–219 (1987).

57. Bilinski, C. A., D. E. Hatfield, J. A. Sobczak, I. Russell, and G. G. Stewart, Analysis of sporulation and segregation in a polyploid brewing strain of *Saccharomyces cerevisiae*, in *Biological Research on Industrial Yeasts*, G. G. Stewart, I. Russell, R. D. Klein, and R. R. Hiebsch (Eds.), CRC Press, Boca Raton, FL, pp. 37–47 (1987).

58. Bilinski, C. A., I. Russell, and G. G. Stewart, Analysis of sporulation in brewer's yeast: Induction of tetrad formation, *J. Inst. Brew.*, 92:594–598 (1986).

59. Esposito, M. S., and R. E. Esposito, The genetic control of sporulation in *Saccharomyces*. I. The isolation of temperature-sensitive sporulation-deficient mutants, *Genetics*, 61:79–89 (1969).

60. Dawes, I. W., and I. D. Hardie, Selective killing of vegetative cells in sporulated yeast cultures by exposure to diethyl ether, *Mol. Gen. Genet.*, 131:281–189 (1974).

61. Esposito, M. S., and R. E. Esposito, Mutants of meiosis and ascospore formation, in *Methods in Cell Biology XI, Yeast Cells*, D. M. Prescott (Ed.), Academic Press, New York, pp. 303–326 (1975).

62. Esposito, M. S., and R. E. Esposito, Aspects of the genetic control of meiosis and ascospore development inferred from the study of *spo* (sporulation-deficient) mutants of *Saccharomyces cerevisiae, Biol. Cell.* 33:93–102 (1978).

63. Dawes, I. W., Genetic control and gene expression during meiosis and sporulation in *Saccharomyces cerevisiae*, in *Yeast Genetics, Fundamental and Applied Aspects*, J. F. T. Spencer, D. M. Spencer, and A. R. W. Smith (Eds.), Springer-Verlag, New York, pp. 29–64 (1983).

64. Roth, R., and S. Fogel, A system selective for yeast mutants deficient in meiotic recombination. *Mol. Gen. Genet.*, 112:295–305 (1971).

65. Grewal, N., and J. J. Miller, Formation of asci with two diploid spores by diploid cells of *Saccharomyces cerevisiae, Can. J. Microbiol.*, 18:1897–1905 (1972).

66. Miller, J. J., A comparison of the effects of several nutrients and inhibitors on yeast meiosis and mitosis, *Exp. Cell Res.*, 33:46–49 (1964).

67. Gustafsson, A., Apomixis in higher plants. I. The mechanism of apomixis, *Lunds Univ. Arsskr.*, 42:1–68 (1946).

68. Suomalainen, E., A. Saura, and J. Lokki, Evolution of parthenogentic insects, *Evol. Biol.*, 9:209–257 (1976).

69. Bilinski, C. A., N. Marmiroli, and J. J. Miller, Apomixis in *Saccharomyces cerevisiae* and other eukaryotic microorganisms, *Adv. Microbial Physiol.*, 30:23–52 (1989).

70. Klapholz, S., and R. E. Esposito, Isolation of *spo12-1* and *spo13-1* from a natural variant of yeast that undergoes a single meiotic division, *Genetics* 96:567–588 (1980).
71. Klapholz, S., and R. E. Esposito, Recombination and chromosome segregation during the single division meiosis in *spo12-1* and *spo13-1* diploids, *Genetics*, 96:589–611 (1980).
72. Moens, P. B., Modification of sporulation in yeast strains with two-spored asci *(Saccharomyces*, Ascomycetes), *J. Cell Sci.*, 16:519–527 (1974).
73. Moens, P. B., M. Mowat, M. S. Esposito, and R. E. Esposito, Meiosis in a temperature-sensitive DNA-synthesis mutant and in an apomictic yeast strain *(Saccharomyces cerevisiae)*, *Phil. Trans. R. Soc. Lond. (B) Biol. Sci.*, 277:351–358 (1977).
74. Marmiroli, N., C. Ferrari, F. Tedeschi, P. P. Puglisi, and C. Bruschi, Ultrastructural analysis of the life cycle of an apomictic (Apo) strain of *Saccharomyces cerevisiae*. I. Meiosis and ascospores development, *Biol. Cell.*, 41:79–84 (1981).
75. Bilinski, C. A., and J. J. Miller, Induction of normal ascosporogenesis in two-spored *Saccharomyces cerevisiae* by glucose, acetate and zinc, *J. Bacteriol.*, 143:343–348 (1980).
76. Bilinski, C. A., and J. J. Miller, Temperature regulation of nuclear division in apomictic yeast, *Can. J. Microbiol.*, 30:793–797 (1984).
77. Marmiroli, N., and C. A. Bilinski, Partial restoration of meiosis in an apomictic strain of *Saccharomyces cerevisiae*: A model system for investigation of nucleomito-chondrial interactions during sporulation, *Yeast*, 1:39–47 (1985).
78. Bilinski, C. A., J. J. Miller, and S. C. Girvitz, Events associated with restoration by zinc of meiosis in apomictic *Saccharomyces cerevisiae*, *J. Bacteriol.*, 155:1178–1184 (1983).
79. Bilinski, C. A., N. Marmiroli, and J. J. Miller, Cell division age dependency of meiosis in an apomictic variant of *Saccharomyces cerevisiae*, *Yeast*, 3:1–4 (1987).
80. Mortimer, R. K., and D. Schild, Genetic map of *Saccharomyces cerevisiae*, *Microbiol. Rev.*, 44:519–571 (1980).
81. Laner, G. O., J. M. Roberts, and L. C. Klutz, Determination of the nuclear DNA content of *Saccharomyces cerevisiae* and implications for the organization of DNA in chromosomes, *J. Mol. Biol.*, 114:507–526 (1977).
82. Strathern, J. N., C. S. Newlon, I. Herskowitz, and J. B. Hicks, Isolation of a circular derivative of yeast chromosome. III. Implications for the mechanism of mating type interconvesions, *Cell*, 18:309–319 (1979).
83. Schalit, P., K. Loughney, M. Olson, and B. D. Hall, Physical analysis of the CYCI-SUP4 interval in *Saccharomyces cerevisiae*, *Mol. Cell. Biol.*, 47:228–236 (1981).
84. Neigeborn, L., and M. Carlson, Genes affecting the regulation of SUC2 gene expression by glucose repression in *Saccharomyces cerevisiae*, *Genetics*, 108:845–858 (1984).
85. Neigeborn, L., and M. Carlson, Mutations causing constitutive invertase synthesis in yeast: Genetic interaction with *snf* mutations, *Genetics*, 115:247–253 (1987).
86. Mortimer, R. K., and D. C. Hawthorne, Genetic mapping in *Saccharomyces*. IV. Mapping of temperature sensitive genes and use of disomic strains in localizing genes, *Genetics*, 74:33–54 (1973).

87. Mortimer, R. K., and D. C. Hawthorne, Genetic mapping in yeast, in *Methods in Cell Biology. XI. Yeast Cells*, D. M. Prescott (Ed.), Academic Press, New York, pp. 221–223 (1975).

88. Mortimer, R. K., and D. C. Hawthorne, Yeast genetics, in *The Yeasts*, Vol. 1, A. H. Rose and J. S. Harrison (Eds.), Academic Press, New York, pp. 385–460 (1969).

89. Wood, J. S., Mitotic chromosome loss induced by methyl benzimmidiazole-2-yl-carbamate as a rapid mapping method in *Saccharomyces cerevisiae, Mol. Cell. Biol.*, 2:1080–1087 (1982).

90. Wickner, R. B., Mapping chromosomal genes of *Saccharomyces cerevisiae* using an improved genetic mapping method, *Genetics*, 92:803–821 (1979).

91. Klapholz, S., and R. E. Esposito, A new mapping method employing a meiotic *rec* mutant of yeast, *Genetics*, 100:387–412 (1982).

92. Sherman, F., G. R. Fink, and J. B. Hicks, *Methods in Yeast Genetics*, Cold Spring Harbor Laboratory, Cold Spring Harbor, NY (1983).

93. Pringle, J. R., Induction, selection and experimental uses of temperature-sensitive and other conditional mutants of yeast, in *Methods in Cell Biology*, D. M. Prescott (Ed), Vol. XII., Academic Press, New York, pp. 233–272 (1975).

94. Moat, A. G., N. Peters Jr., and A. M. Srb, Selection and isolation of auxotrophic yeast mutants with the aid of antibiotics, *J. Bacteriol.*, 77:673–677.

95. Snow, R., An enrichment method for auxotrophic yeast mutants using the antibiotic "nystatin," *Nature*, 211:206–207 (1966).

96. Bilinski, C. A., A. M. Sills, and G. G. Stewart, Morphological and genetic effects of benomyl on polyploid brewing yeasts: induction of auxotrophic mutants, *Appl. Environ. Microbiol.*, 48:813–817 (1984).

97. Littlewood, B. W., Methods for selecting auxotrophic and temperature-sensitive mutants in yeasts, in *Methods in Cell Biology XI, Yeast Cells*, D. M. Prescott (Ed.), Academic Press, New York, pp. 278–285 (1975).

98. McCammon, M., and L. W. Parks, Enrichment for auxotrophic mutants in *Saccharomyces cerevisiae* using the cell wall inhibitor echinicandin, *Mol. Gen. Genet.*, 186:295–297 (1982).

99. Young, J. D., J. W. Gorman, J. A. Gorman and R. M. Bock, Indirect selection for auxotrophic mutants of *Saccharomyces cerevisiae* using the antibiotic netropsin, *Mutat.Res.*, 35:423–428 (1976).

100. Henry, S. A., T. F. Donahue, and M. R. Culbertson, Selection of spontaneous mutants by inositol starvation in yeast, *Mol. Gen. Genet.*, 143:5–11 (1975).

101. Henry, S. A., and B. Horowitz, A new method for mutant selection in *Saccharomyces cerevisiae, Genetics*, 79:175–186 (1975).

102. Barclay, B. J., and J. G. Little, Selection of yeast auxotrophs by thymidylate starvation, *J. Bacteriol.*, 132:1036–1037 (1977).

103. Person, S., and H. L. Lewis, Effects of decay of incorporated [^{3}H]-thymidine on bacteria, *Biophys. J.*, 2:451–463 (1962).

104. Lubrin, M., Selection of auxotrophic bacterial mutants by tritium-labeled thymidine, *Science*, 129:838–839 (1959).

105. Littlewood, B. S., and J. E. Davies, Enrichment for temperature-sensitive and auxotrophic mutants in *Saccharomyces cerevisiae* by tritium suicide, *Mutat. Res.*, 17:315–322 (1973).

106. Littlewood, B. S., A method of obtaining antibiotic-sensitive mutants in *Saccharomyces cerevisiae, Genetics*, 77:305–308 (1972).
107. Hardie, I. D., and I. W. Dawes, Optimal condition for selecting specific auxotrophs of *Saccharomyces cerevisiae* using temperature-sensitive suicide mutants, *Mutat. Res.*, 42:215–222 (1977).
108. Letts, V. A., and I. W. Dawes, Mutations affecting lipid biosynthesis of *Saccharomyces cerevisiae*: Isolation of ethanolamine auxotrophs, *Biochem. Soc. Trans., Proc. 583rd Meeting, Cambridge*, 7:976–977 (1979).
109. Pearson, B. M., L. J. Fuller, D. A. Mackenzie, and M. H. J. Keenam, A red/white selection for *Saccharomyces cerevisiae* auxotrophs, *Lett. Appl. Microbiol.*, 3:89–91 (1986).
110. Hutchinson, H. T., and L. H. Hartwell, Macromolecular synthesis in yeast spheroplasts, *J. Bacteriol*, 94:1697–1705 (1967).
111. Deutch, C. E., and J. M. Parry, Spheroplast formation in yeast during transition from exponential phase to stationary phase, *J. Gen. Microbiol.*, 80:259–268 (1974).
112. Ferenczy, L., Enrichment of fungal mutants by selective cell wall lysis, *Nature*, 253:46–47 (1975).
113. Piedra, D., and L. Herrera, Selection of auxotrophic mutants in *Saccharomyces cerevisiae* by a snail enzyme digestion method, *Folia Microbiol.*, 21:337–341 (1976).
114. Walton, E. F., B. L. A. Carter, and J. R. Pringle, An enrichment method for temperature-sensitive and auxotrophic mutants of yeast, *Mol. Gen. Genet.*, 171:111–114 (1979).
115. Morita, T., and Y. Yamagihara, Osmotic-sensitive mutants of *Saccharomyces cerevisiae* as screening organisms for promutagens and procarcinogens, *Chem. Parm. Bull.*, 33:1576–1582 (1985).
116. Ono, B., and E. Sakamoto, *Saccharomyces cerevisiae* strains sensitive to inorganic mercury. I. Effect of tyrosine, *Curr. Genet.*, 10:179–185 (1985).
117. Ohya, Y., Y. Ohsumi, and Y. Anraku, Isolation and characterization of Ca-sensitive mutants of *Saccharomyces cerevisiae, J. Gen. Microbiol.*, 132:979–988 (1986).
118. Kunz, B., and R. H. Haynes, Phenomenology and genetic control of mitotic recombination in yeast, *Annu. Rev. Genet.*, 15:57–89 (1981).
119. Vaschishat, R. K., and S. N. Kakar, UV-sensitive mutants of *Saccharomyces cerevisiae*. 1;. Isolation and characterization, *Indian J. Exp. Biol.*, 17:28–32 (1979).
120. Prakash, L. and S. Prakash, Isolation and characterization of MMS-sensitive mutants of *Saccharomyces cerevisiae, Genetics*, 86:33–55 (1977).
121. Ruhland, A., E. Haase, W. Siede, and M. Brendel, Isolation of yeast mutants sensitive to the bifunctional alkylating agent nitrogen mustard, *Mol. Gen. Genet.*, 181:346–351 (1981).
122. Littlewood, B. S., A new method for obtaining antibiotic-sensitive mutants in *Saccharomyces cerevisiae, Genetics*, 71:305–308 (1972).
123. Gorenstein, C., K. D. Atkinson, and E. V. Fallce, Isolation and characterization of an actinomycin D-sensitive mutant of *Saccharomyces cerevisiae, J. Bacteriol.*, 136:142–147 (1978).
124. Chan, R. K., and C. A. Oate, Isolation and genetic analysis of *Saccharomyces cerevisiae* mutants supersensitive to G1 arrest by *a* factor and α factor pheromones, *Mol. Cell. Biol.*, 2:11–20 (1982).

125. Meade, J. H., and T. R. Maney, Sensitivity of tryptophan, tyrosine and phenylalanine mutants of *Saccharomyces cerevisiae* to phenethyl alcohol, *Genetics*, 104:235–240 (1983).

126. Ono, B., and E. Sakamoto, *Saccharomyces cerevisiae* strains sensitive to inorganic mercury. II. Effect of glucose, *Curr. Genet.*, 10:187–195 (1985).

127. Ohya, Y., S. Miyamoto, Y. Ohsumi, and Y. Anraku, Calcium-sensitive *cls4* mutant of *Saccharomyces cerevisiae* with a defect in bud formation, *J. Bacteriol.*, 165:28–33 (1986).

128. Middlekauf, J. E., S. Hino, S. P. Yang, C. C. Lindegren, and G. Lindegren, Gene control of resistance versus sensitivity to actidione in *Saccharomyces*, *Genetics*, 42:66–71 (1957).

129. Wilkie, D., and B. K. Lee, Genetic analysis of actidione resistance in *Saccharomyces cerevisiae*, *Genet. Res. Camb.*, 6:130–138 (1965).

130. Patel, P. V., and J. R. Johnston, Dominant mutation for nystatin resistance in yeast, *Appl. Microbiol.*, 16:164–165 (1968).

131. Wilkie, D., Analysis of mitochondrial drug resistance in *Saccharomyces cerevisiae*, In *Symposium of the Society for Experimental Biology XXIV, Control of Organelle Development VIII*, P. L. Miller (Ed.), Academic Press, New York, pp. 71–83 (1970).

132. Brusick, D. J., Induction of cycloheximide-resistant mutants in *Saccharomyces cerevisiae* with N-methyl-N'-nitro-N-nitrosoguanidine and ICR-170, *J. Bacteriol.*, 109:1134–1138 (1972).

133. Baranowska, H., R. Polakowska, and A. Putrament, Spontaneous and induced non-specific drug resistance in *Saccharomyces cerevisiae*, *Acta Microbiol. Pol.*, 28:181–201 (1973).

134. Ulaszewski, S., M. Grenson, and A. Goffeau, Modified plasma-membrane ATPase in mutants of *Saccharomyces cerevisiae*, *Eur. J. Biochem.*, 130:235–239 (1983).

135. Falco, S. C., and K. S. Dumas, Genetic analysis of mutants of *Saccharomyces cerevisiae* resistant to the herbicide sulfometuron methyl, *Genetics*, 109:21–35 (1985).

136. Thomas, J. H., N. F. Neff, and D. Botstein, Isolation and characterization of mutations in the β-tubulin gene of *Saccharomyces cerevisiae*, *Genetics*, 112:715–734 (1985).

137. Ross, I. S., and A. L. Walsh, Resistance to copper in *Saccharomyces cerevisiae*, *Trans. Br. Mycol. Soc.*, 77:27–32 (1981).

138. Fogel, S., and J. W. Welch, Tandem gene amplification mediates copper resistance in yeast, *Proc. Natl. Acad. Sci. U.S.A.*, 79:5342–5346 (1982).

139. Ono, B. I., and M. Weng, Chromium resistant mutants of the yeast *Saccharomyces cerevisiae*, *Curr. Genet.*, 6:71–77 (1982).

140. Ono, B. I., M. Weng, L. Chen, L. Yu, and K. Tong, Genetic analysis of chromium resistance in *Saccharomyces cerevisiae* and *Bacillus subtilis*, *Eisei Kagaku*, 28:66 (1982).

141. Gadd, G. M., A. Stewart, C. White, and J. H. Mowll, Copper uptake by whole cells and protoplasts of a wild-type and copper-resistant strain of *Saccharomyces cerevisiae*, *FEMS Microbiol. Lett.*, 24:231–234 (1984).

142. Roncero, C., M. H. Valdivieso, J. C. Ribas, and A. Durán, Isolation and characterization of *Saccharomyces cerevisiae* mutants resistant to calcofluor white,

     *J. Bacteriol.,*170:1950–1954 (1988).

143. Chen, S. Y., *C. R. Hebd. Seance. Acad. Sci., Paris,* 230:1887–1899 (1950).

144. Pomper, S., and P. R. Burkholder, Studies on the biochemical genetics of yeast, *Proc. Natl. Acad. Sci. U.S.A.,* 35:456–464 (1949).

145. Fowell, R. R., Hybridization of yeasts by Lindegren's technique, *J. Inst. Brew.,*57:180–195 (1957).

146. Conde, J., and G. R. Fink, A mutant of *Saccharomyces cerevisiae* defective for nuclear fusion, *Proc. Natl. Acad. Sci. U.S.A.,* 73:3651–3855 (1976).

147. Nilsson-Tillgren, T., J. G. L. Petersen, S. Holmberg, and M. C. Kielland-Brandt, Transfer of chromosome III during *Kar* mediated cytoduction in yeast, *Carlsberg Res. Commun.,* 45:113–117 (1980).

148. Dutcher, S. K., Internuclear transfer of genetic information in *Kar1-1/KAR1*heterokaryons in *Saccharomyces cerevisiae, Mol. Cell. Biol.,* 1:246–253 (1981).

149. Brown, A. J. P., A. R. Goodey, and R. S. Tubb, Interstrain transfer of the 2 μm DNA plasmid of *Saccharomyces cerevisiae* by cytoduction, *J. Inst. Brew.,* 87:234–238 (1981).

150. Young, T. W., and M. Yagui, A comparison of the killer character in different yeasts and its classification,*Antonie vanLeevwenhoekJ. Microbiol. Serol.,* 44:59–77(1978).

151. Young, T. W., Brewing yeast with anti-contaminant properties, in *European Brewery Convention, Proceedings of the 19th Congress, London,* IRL Press, Oxford, U.K., pp. 129–136 (1983).

152. Hammond, J. R. M., and K. W. Eckersley, Fermentation properties of brewing yeast with killer character, *J. Inst. Brew.,* 90:167–177 (1984).

153. Kielland-Brandt, M. C., T. Nillson-Tillgren, S. Holmberg, J. G. L. Petersen, and B. A. Svenningsen, Transformation of yeast without the use of foreign DNA, *Carlsberg Res. Commun.,* 44:77–87 (1979).

154. Klar, A. J. S., Mating-type functions for meiosis and sporulation in yeast act through cytoplasm, *Genetics,* 94:597–605 (1980).

155. Kau, K. N., and M. R. Michayluk, A method for high frequency intergeneric fusion of plant protoplasts, *Planta (Berl.),* 115:355–357 (1974).

156. Ferenczy, L., F. Kevei, M. Szegedi, A. Franko, and I. Rojik, Factors affecting high-frequency fungal protoplast fusion, *Experientia,* 32:1156–1158 (1976).

157. Stewart, G. G., and I. Russell, Current use of new genetics in research and development of brewer's yeast strains, in *European Brewery Convention, Proceedings of the 17th Congress, Berlin (West),* DSW, The Netherlands, pp. 475–490 (1979).

158. Ferenczy, L., Fungal protoplast fusion: basic and applied aspects, in *Cell Fusion: Gene Transfer and Transformation* R. F. Beers Jr. and E. G. Basset (Eds.), Raven Press, New York, pp. 145–168 (1984).

159. Bulder, C. J. E. A., Induction of petite mutations and inhibition of synthesis of respiratory enzymes in various yeasts. *Antonie van Leevwenhoek,* 30:1–9 (1964).

160. Bulder, C. J. E. A., Lethality of the petite mutation in petite negative yeasts, *Antonie van Leevwenhoek,* 30:442–454 (1964).

161. Spencer, J. F. T., and D. M. Spencer, Genetic improvement of industrial yeasts, *Annu. Rev. Microbiol.,* 37:121–142 (1983).

162. Spencer, J. F. T., P. Laud, and D. M. Spencer, The use of mitochondrial mutants in

the isolation of hybrids involving industrial strains. Use in isolation of hybrids obtained by protoplast fusion, *Mol. Gen. Genet.*, 178:651–654 (1980).

163. Gunge, N., and A. Tarmaru, Genetic analysis of products of protoplast fusion in *Saccharomyces cerevisiae, Japan. J. Genet.*, 53:41–49 (1978).

164. Ferenczy, L., and A. Maraz, Transfer of mitochondria by protoplast fusion in *Saccharomyces cerevisiae, Nature*, 268:524–525 (1977).

165. Sipiczky, M., and L. Ferenczy, Protoplast fusion of *Schizosaccharomyces pombe* auxotrophic mutants of identical mating type, *Mol. Gen. Genet.*, 151:77–81 (1977).

166. Van Solingen, P., and J. B. Van der Plaat, Fusion of yeast spheroplasts, *J. Bacteriol.* 130:946–947 (1977).

167. Sipiczky, M., Interspecific protoplast fusion in fission yeasts, *Curr. Microbiol,* 3:37–40 (1979).

168. Spencer, J. F. T., and D. M. Spencer, The use of mitochondrial mutants in hybridization of industrial yeasts. III. Restoration of mitochondrial function in petites of industrial yeast strains by fusion with respiratory-competent protoplasts of other yeast species, *Curr. Genet.*, 4:147–180 (1981)

169. De Van Broock, M. R., M. F. Siena, and L. I. Figueroa, Intergeneric fusion of yeast protoplasts, in *Proceedings of the Fifth International Symposium on Yeasts, Current Developments in Yeast Research*, G. G. Stewart and I. Russell (Eds.), Pergamon Press, Toronto, Canada, pp. 171–176 (1981).

170. Morgan, J. A., J. Heritage, and P. A. Wittaker, Protoplast fusion between petite and auxotroph mutants of the petite negative yeast, *Kluyveromyces lactis, Microbiol. Lett.*, 4:103–107 (1977).

4
Yeast Selection in Brewing

Gregory Paul Casey
Anheuser-Busch Companies
St. Louis, Missouri

I. INTRODUCTION

Just over a century ago the genetic improvement of strains of brewer's yeasts began at Carlsberg Laboratories with selection of pure yeast cultures in the laboratory of Emil C. Hansen. Of course improvements by the empirical method of trial and error had been occurring for millenia prior to Hansen's contributions, but these were done without the knowledge that yeast was the causative organism of fermentation.

Hansen's discoveries revolutionized the brewing industry, as the vast majority of breweries around the world, especially those producing lager beers, adopted the practice of pure yeast cultures by the end of the nineteenth century. Since that time, there has been an explosive increase in our understanding of yeast genetics, yeast physiology, and molecular biology. A person outside of the brewing industry would likely conclude, quite logically, that developments in these areas must be profoundly influencing the way with which modern-day yeast strains are selected for brewing. A global examination of today's brewing industry, however, would reveal that virtually all production strains of brewer's yeasts in use today have been selected based on the principles of Emil Hansen; i.e., screen and select pure cultures of wild-type yeast.

This chapter, which focuses almost entirely on strain selection in bottom-fermenting strains of lager yeasts (first, as this is the yeast of most familiarity to the author, and second, owing to our detailed knowledge concerning the genomic composition of lager yeasts), will explore the reasons behind the limited implementation of traditional genetics and many of the "new" genetic techniques in yeast-breeding programs. While research in these areas has greatly increased our basic understanding of the genetic complexity of brewer's yeasts (and provided employment to many a yeast geneticist!), one would be hard pressed, for example, to find a production lager strain produced by protoplast fusion or "rare mating."

Despite the negative nature of these statements, this author is of the opinion that the brewing industry is within 10 years of a revolution in terms of the way brewer's yeast strains are selected. This revolution, which will most probably erupt first in Europe, will be fueled by developments in the area of recombinant deoxyribonucleic acid (DNA) technology. While no recombinant brewer's yeast is used in production today, such yeasts have been successfully tested through pilot plant scale fermentations without exhibiting the shortcomings of yeast strains developed using pre-recombinant DNA genetic techniques. Before examining this exciting area of strain selection, however, the historical roots of lager yeast will be explored in order to better appreciate the developments of the past 2 decades.

II. HISTORICAL PERSPECTIVE

Yeasts belonging to the genus *Saccharomyces* have often been referred to as the oldest plants cultivated by man. Indeed, brewing itself may represent the world's oldest profession, if one accepts the premise that the first member of the "other" profession claiming that distinction consumated the first transaction only after sampling a fermented product! In any case, for most of the past few millenia, yeast strains present on raw materials and equipment were responsible for the "natural," but quite unpredictable, fermentation which we now call brewing. As brewers during those times were quite unaware of the existence of yeast (many a past brewer has viewed divine intervention as the most important ingredient for successful brewing!), little effort was expended on selecting yeast strains for brewing. As a result, variable product quality and frequent incidences of microbial spoilage have been the trademarks of brewing for most of its existence.

Up until the middle of the last century, most of the world was using top-cropping yeasts. Bottom-cropping lager yeasts were only utilized by Bavarian brewers, especially in Munich. In 1842, however, the yeast and fermentation techniques of lager beer production were literally smuggled to Czechoslovakia, followed by Copenhagen in 1845. Keeping in mind this was in a world devoid of automobiles and refrigeration, the modern-day brewer delights in imagining the difficulties endured to smuggle a viable yeast culture by stagecoach from Bavaria and Czechslovakia to Denmark! Four decades later, the lager yeast smuggled to Denmark became the focus of study for a scientist at Carlsberg Laboratory named Emil Christian Hansen. Before examining Hansen's contributions it is important to note the accomplishments of a contemporary of his, Louis Pasteur. In Pasteur's 1876 treatise *Etudes sur la Biere*, he presented convincing evidence that yeast used to produce beer contained contaminants in the form of acetic acid bacteria, lactics, molds, and wild yeast, and that several of these were directly responsible for particular types of beer spoilage (1). To reduce problems of yeast contamination one of Pasteur's recommendations was the implementation of cleaning procedures which favor the growth of the plant yeast (e.g., treatment with potassium bitartrate and ethanol to inhibit bacteria [1]). While assisting in reducing bacteria-related problems, this practice did little toward developing a pure culture of yeast as we think of it today.

It was at this point in Pasteur's treatise that Hansen wrote "How do you get that absolute pure culture . . . since not one but several yeast species survive the mentioned treatment?" (2). This question spurned Hansen to write his 1883 landmark paper on the subject of yeast strain selection: "Recherches sur la physiologie et la morphologie des ferments alcooliques II-IV"(3). In it he described how single yeast cells could be isolated for the first time by serial dilutions onto gelatin plates or in wort broth, and that these could be propagated to larger scale pure cultures through the use of aseptic techniques. In this manner he isolated pure cultures of bottom-fermenting yeasts from Tuborg and Carlsberg breweries, a top-fermenting yeast from an Edinburgh brewery, and lines of *Saccharomyces ellipsoideus* and *Saccharomyces pastorianus*. He proved the latter two yeasts were the causative organisms of beer ropiness and beer off-flavors, respectively, and demonstrated that pure cultures of the bottom-fermenting yeasts could be used for lager beer production. The first of these, called Carlsberg Yeast No. 1, was introduced into the Carlsberg brewery on a production scale on November 12, 1883 (2). By 1884, the entire production at Carlsberg was based on the use of clones of bottom-fermenting yeast, and within 10 years the advantages of the use of pure cultures became so apparent that Pabst, Schlitz, and Anheuser Busch breweries (as well as 50 smaller North American breweries) were also using pure yeast cultures (2).

Hansen's pure culture technique at last provided brewers with a way to select yeast clones and screen them for specific fermentation characteristics and flavor profiles. This ability alone resulted in improved consistency in plant fermentations and product quality, and undoubtedly accounted for the rapid acceptance of the technique in brewing. The major drawback of clonal screening however, is that it is an empirical selection process which does not employ a deliberate or systematic breeding strategy. To accomplish this required knowledge of the life cycle and genetics of yeast, and developments in these fields would have to wait until Europe was on the brink of becoming embroiled in World War II.

Once again the setting was Carlsberg Laboratory in Denmark. In 1935, Ojvind Winge and coworkers screened pure cultures initially isolated by Hansen and found among the nonproduction cultures several which sporulated and germinated regularly(4). Using *Saccharomyces ellipsoideus*, Winge elaborated the haploid and diploid phases of the yeast life cycle(5), and together with Lausten followed that accomplishment in 1937 with the demonstration of Mendelian segregation for the first time in yeast (this time with *Saccharomyces ludwigii*)(6). The second world war understandably resulted in a temporary halt in the exchange of yeast strains across the Atlantic, but by 1949 the Lindegrens, at Carbondale, Illinois, went on to isolate strains of *Saccharomyces cerevisiae* yeast of *a* and α mating types. Industrial yeast genetics aside, it is worthwhile to note that many of the laboratory strains widely used in the study of yeast genetics and molecular biology today can trace their origins to the laboratories of Lindegren and Winge (reviewed by Mortimer and Johnston[7]).

These discoveries in the 1930s and 1940s, on the surface at least, appeared to lay the groundwork for breeding specific new strains of yeast by classic hybridization programs. It soon became apparent, however, that brewer's yeasts were not like the easily manipulable haploid yeast now widely gaining use in academic laboratories around the world. Instead brewer's yeasts were found to be usually sterile polyploids and aneuploids. Worst of all, they rarely sporulated, and even when spores could be obtained, they usually were inviable. Yeast geneticists working for breweries therefore found themselves caught in the midst of a "catch 22"—the very traits which from a production standpoint made yeast strains more stable and less susceptible to mutational changes (and which likely led to their selection in the first place) prevented the development and implementation of Medelian genetics in yeast breeding programs.

Solutions to these obstacles would not come until the 1970s and 1980s with the discovery of "new" genetic techniques. Included among these are protoplast fusion, rare-mating, single chromosome transfers, transformation, and recombinant DNA technology. While advances in these areas did not originate from industrial laboratories, industrial geneticists can take some comfort from knowing that the haploid strains used in these areas originated from Hansen's effort in a Danish brewery some 100 years ago! What goes around comes around!

Unfortunately, many of the techniques have proven to be impractical for selecting new strains of brewer's yeasts and have largely been used for basic studies aimed at elaborating the genomic composition of brewer's yeasts (e.g., chromosome transfers). The remainder of this chapter (summarized in Table 1) will examine the principles behind each of these new techniques, with particular emphasis being placed on the potential for practical applications for each technique. Before any of these can be addressed, however, it is first necessary to examine our present understanding of the genome structure of lager yeasts. By doing so, the genetic idiosyncracies of different strain development programs will be better appreciated.

III. GENOME STRUCTURE OF LAGER YEAST

Lager yeasts are generally characterized by a number of traits which interfere with normal genetic analyses: homothallism, polyploidy, aneuploidy, poor sporulation, and low spore viability(8). While seemingly insurmountable obstacles, researchers

Table 1. A Historical Perspective of Brewer's Yeast Strain Selection

Year	Development	Comment
Pre 1880s	Yeast cultures are a mixture of lager yeast, *S.ellipsoideus* and other wild yeasts, lactics, and acetic acid bacteria	Variability is the norm
1880s	Emil Christian Hansen develops the first pure cultures of lager yeasts	Spoilage problems and wide variations in flavor are dramatically reduced
1880s-1930s	Use of pure cultures becomes lager industry standard. Extensive screening of clonal isolates to select best yeast	Selection programs involve random screening and selection. No deliberate breeding possible owing to ignorance of yeast genetics
1930s-1940s	Life cycle and mating types of *S.cerevisiae* yeast established (Winge, Lausten, Lindegren). First yeast hybrids produced by cross breeding of haploids	Production yeast sporulate poorly and do not mate
1950s-1960s	Attempts to produce new strains of brewer's yeasts by cross breeding prove fruitless	New genetic techniques needed to overcome mating and sporulation problems
1970s	Rare-mating, cytoduction, and protoplast fusion (which circumvent sterility problems) are utilized in brewer's yeast strain selection programs. Potential is at first thought to be considerable	No new real-world production strains produced through these technologies. Failure attributed to lack of specificity (too many traits transferred)
1978	First transformation in yeast with *LEU2*. Selective genetic engineering now possible	Potential for genetic engineering of brewers yeasts by recombinant DNA technology. Lack of dominant markers delays studies with prototrophic brewer's yeasts for 6-7 yrs
Early 1980	Researchers at Carlsberg study genome structure of production lager yeast. Results indicate a species hybrid origin of lager yeast (involving *S.cerevisiae, S.monacensis,* and *S.bayanus*)	Amphiploidy can also be considered a cause of mating and sporulation deficiencies in lager yeast
Mid-Late 1980	First reports of recombinant brewer's yeasts published. Efforts primarily focused on introducing β-glucanase, glucoamylase, or	

continued

Table 1 *(continued)*

Year	Development	Comment
	acetolactate decarboxylase into the genome. Many tested at pilot brewery stage	Most constructs contain bacterial DNA or are not integrated into the genome. Regulatory and stability concerns remain
1990s	Constructs will likely eliminate stability and regulatory concerns. First recombinant yeast to be used in the U.K. or northern Europe?	Regulatory decisions concerning use of recombinant food and beverage grade yeast will be made in late 1980s. Unfavorable rulings may delay implementation of recombinant strains indefinitely. Favorable rulings will open floodgates?

at Carlsberg Laboratory have combined the isolation of rare prototrophic *MAT*α and *MAT*a spore isolates (9) with the technique of single chromosome transfer (10) to overcome sterility problems and conduct detailed genetic analyses of the genome of one lager yeast; i.e., *Saccharomyces carlsbergensis* strain 244.

Single chromosome transfers are based on the *kar1* mutation which dramatically reduces the frequency of karyogamy after zygote formation (11), but still permits chromosome transfer between nuclei. Thus, in *kar1* X *KAR1* crosses, chromosomes can be transferred between nuclei during the transient heterokaryotic state, making it possible to isolate and analyze individual wild-type chromosomes from the lager yeast in the genetic background of a standard haploid strain of *S.cerevisiae*. In this manner, chromosomes III, V, X, XII, and XIII have been studied—with fascinating results (10, 12, 13, 14)! Observations specifically on chromosome X are described below to illustrate the principles involved in this type of analysis.

Prototropic spores of *S.carlsbergensis* strain 244 exhibiting a mating type (obtained in low frequency) were crossed to *MAT*a and *MAT*α haploid laboratory recipient strains carrying the linked chromosome X markers *arg3, met3, ilv3, cyc1-1, cdc11, hom6*; the unlinked auxotrophic markers *ade2-40* or *lys1*; the unlinked recessive cycloheximide-resistant *cyh2* marker; and *kar1-1*. Laboratory strains containing wild-type chromosome X from 244 were selected by plating hybrids on medium containing adenine, lysine, and cycloheximide. Expression of the recessive *cyh2* marker selects against diploids—this condition was confirmed by maintenance of the *ade2-40* (red) and *lys1* phenotypes. Mere isolation of clones showed that *S.carlsbergensis* chromosomes X carried the functional genetic information of *S.cerevisiae* chromosome X. However, tetrad analyses of four chromosome X transfer lines showed that two different *S.carlsbergensis* chromosomes X had been transferred to the *S.cerevisiae* haploid. Neither chromosome was

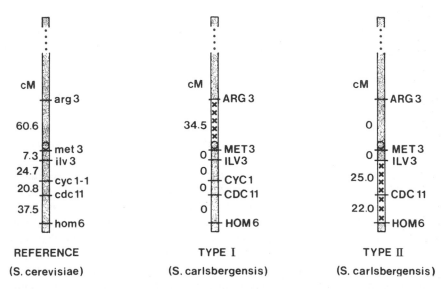

Figure 1. Map of relevant markers of chromosome X in *S.cerevisiae* and of type I and type II chromosomes X in *S.carlsbergensis*. The regions in chromosome X of *S.carlsbergensis* which recombine with the reference chromosome X of *S.cerevisiae* are indicated by crosses. (From Ref. 13, used with permission.)

completely homologous to the *S.cerevisiae* chromosome, since meiotic recombination was absent in different areas of type I and type II *S.carlsbergensis* chromosomes (Fig. 1)

Similar results were found with chromosomes III, V, XII, and XIII in that at least two structurally different versions of each chromosome exist in strain 244. However, in recombination studies with *S.cerevisiae*, these chromosomes were not all found to be like chromosome X. For example, homologues of chromosomes V (12), XII, and XIII (14) able to recombine with the *S.cerevisiae* chromosome at normal levels were found in addition to homeologues for chromosomes V (12) and XIII (14), unable to recombine at all with the *S.cerevisiae* chromosome. Mosaic lager chromosomes, like chromosome X (with recombination linked to parts of the chromosome), were also found for chromosomes III (10) and XII (14). Consistent with these findings are molecular hybridization studies which reveal the presence of two alleles differing in nucleotide sequence in strain 244 for the *CAN1, CYC7, HIS4, HML, HMR, ILV1, ILV2, ILV3, ILV5, LEU2, MAT, SUP-RL1,* and *URA3* loci (10, 12, 14-21). One allele strongly hybridizes to the *Saccharomyces cerevisiae* allele, whereas the other fails to hybridize or only does so under low Southern blotting stringency conditions. Similar observations of two alleles for the *ERG10, URA1,* and *ARG80* loci have been found in the French lager strain 0230 (22, 23).

Combined with the observations from recombination studies, the results clearly point to lager yeast being a species hybrid. However how many species are involved to produce this hybrid, and are these results typical of lager yeasts in general?

Answers to these questions have been provided in part by a powerful new tool in yeast genetics—pulsed field gel electrophoresis. Utilizing this technique, it is now possible to electrophoretically separate intact lager yeast chromosomes on a gel. While a variety of designs exists, the most frequently used include pulsed field gel electrophoresis (PFGE) (24), orthogonal-field-alternation-gel electrophoresis (OFAGE) (25), field inversion gel electrophoresis (FIGE) (26), contour-clamped homogeneous electric fields electrophoresis (CHEF) (27), and transverse alternating field electrophoresis (TAFE) (28). All of these designs share the use of two orientations of electric fields, with the orientation of the two fields and/or gel varying with the design of the apparatus. In the first systematic screening of lager yeasts by pulsed field electrophoresis, Mogens Pedersen (29) found that while slight chromosome length polymorphisms existed between six different European production strains of lager yeasts (including strain 244), the electrophoretic karyotype of these yeasts was very similar. While these results suggest a similar genome composition among production lager yeast strains, culture collection type strains of *S.carlsbergensis* had significantly different karyotypes (29).

When probed with cloned genes, specific homologous and homeologous chromosomes of lager yeasts can be visually identified in electrophoretic chromosome separations. For example, in separation gels used for chromosome X analyses (13), type I chromosome X migrated faster than the *S.cerevisiae* chromosome, whereas type II migrated slower (Fig. 2). A third *S.cerevisiae* chromosome X had similar electrophoretic characteristics as the *S.cerevisiae* chromosome (present in spore clone C6, 6S, but not C67), indicating strain 244 to be trisomic for at least one chromosome. In related studies with French lager strain 0230, chromosomes with differing mobilities were found with molecular probes for chromosomes XI and XIII (23).

What however is the combination of species which has resulted in the unique organism we now call lager yeast? Our best indication is that it is a species hybrid consisting of genomes originating from *S.cerevisiae* and either *S.bayanus* or *S.monacensis*. For example, DNA-DNA reassociation studies of Rosini et al. (30) indicate greater than 90% levels of sequence complementarity between the genomes of *S.uvarum* and *S.bayanus*. This chapter, however, will not explore the evidence behind this statement, but instead directs the reader to the convincing papers of Mogens Pedersen, who has provided most of the data on this subject via chromosome electrophoresis analyses, single chromosome transfer experiments, and molecular hybridization analyses on these yeasts (17- 21).

In summary, lager yeasts contain at least two genomes or types of chromosomes. One of these sets is very similar to the chromosomes of academic yeast—able to form normal bivalents (with subsequent recombination events) during meiosis, whereas the second set, at least in part, cannot form bivalents or recombine with academic standard chromosomes. This type of genotype organization in lager yeasts is like wheat or tobacco of the plant world—a permanent amphiploid hybrid. This composition explains why it has proven so difficult to breed brewer's yeasts by traditional genetic approaches (e.g., as recombination is a prerequisite for proper disjunction of chromosomes in meiosis [31], amphiploidy is undoubtably connected to poor sporulation, low spore viability, and aneuploidy in rare viable spores), and why advances in molecular biology have been a prerequisite for significant strain improvement of production lager yeasts.

IV. OBJECTIVES OF STRAIN DEVELOPMENT PROGRAMS

The need for strain development in brewing must go hand in hand with developments in brewing technology if brewers are to maximize raw material utilization and plant productivity and performance. For example, there is no dogma in brewing which states that brewer's yeasts are unaffected by changes in tank geometry or wort gravity—just ask any brewmaster with over 20 years of experience!

Regardless of the technology used to produce a beer, there are certain basic properties desirable in any strain of brewer's yeasts. These include:

1. Rapid fermentation rates
2. Efficient sugar utilization with maximum yield of ethanol, not cells
3. Tolerances to ethanol, osmotic pressure, oxygen, temperature, and carbon dioxide levels comfortably above those indigenous to each brewery
4. Appropriate flocculation and sedimentation properties
5. High culture viability for repitching purposes
6. Consistent production of flavor and aroma compounds
7. High genetic stability

Significant strain improvement in any or part of these properties have of course been accomplished by breweries through traditional methods of yeast screening programs. However, closer examination of these traits reveal two properties which makes deliberate breeding programs difficult. First, many of the above traits are clearly polygenic and in many cases the biochemistry (let alone genetics) relevant to the particular phenotype is not understood or fully elucidated. Second, brewers may not even be able to define the components involved in contributing to a specific property in a fermentation or finished beer. A classic example of this is defining all the compounds necessary to produce a beer of a characteristic taste, flavor, and aroma. While countless man years have been devoted to this area of research in brewing, countless more will have to be spent before we have a more complete grasp of this subject. Thus, before embarking on any strain improvement program it is well worth the effort to clearly define the objective and to ascertain if the available physiological and genetic information on the subject is adequate to carry out the job.

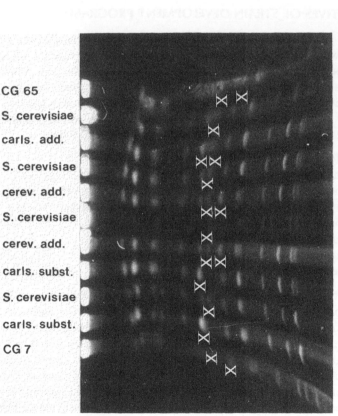

Figure 2. (A) Photograph of an OFAGE gel in which chromosomes of spore clones CG 7 and CG 65 (isolated from Carlsberg's lager strain 244), recipient *S.cerevisiae* yeasts for chromosome X analyses, and chromosome X addition and substitution lines are separated.

(B) Autoradiograph of the same gel probed with *ILV3* alleles that are unique to the lager yeast (*ILV3* carls) or shared with haploid *Saccharomyces cerevisiae* yeast (*ILV3* cerev) under low (60°C) and high (68°C) stringency conditions. (Taken from Ref. 13, used with permission.)

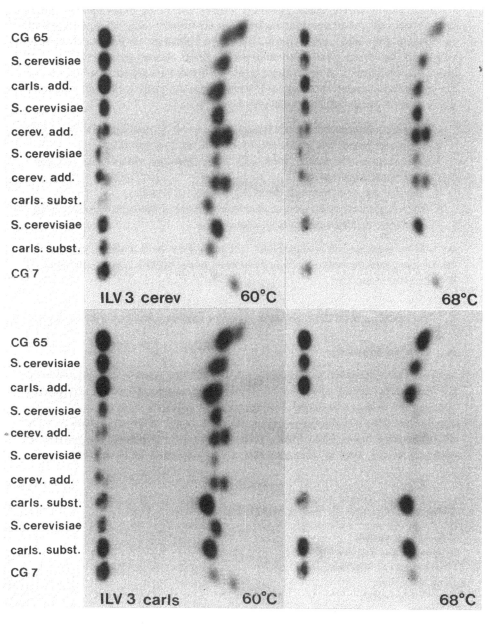

There are, however, many potential targets for strain improvement which involve traits of limited genetic complexity for which the biochemistry is often well understood. Previously, the major hurdle to the implementation of most of these properties has been the shortcomings of conventional genetic techniques. However, advances in genetics have removed these barriers, and the addition of the following enzymes to the existing complement in yeast is all well within the reach of modern-day strain improvement programs:

1. Amylases (starch and dextrin fermentation; production of light and dry beers)
2. Proteases (improved wort nitrogen utilization, chill proofing)
3. β-Glucanase (improved wort and beer filtration properties; increased brewhouse capacity)
4. Acetolactate decaboxylase (reduced diacetyl production)
5. Enzymes for the utilization of alternative carbohydrate/adjunct sources (e.g., cellulase, cellobiase, β-galactosidase)

As will be apparent, while signigicant advances have been made in some of the above categories of yeast selection programs, many hurdles (often nontechnical) remain to be overcome.

V. TECHNICAL STRATEGIES FOR STRAIN IMPROVEMENT PROGRAMS

A. General Remarks

In theory, all techniques developed for laboratory yeast genetics are potential tools for the genetic improvement of brewer's yeast cultures. However, the genetic peculiarities of these yeasts, as discussed in the previous sections, make certain approaches more attractive than others. As such, many of the techniques listed in Table 2 have proven to be of more value in studying the genetic composition of brewer's yeasts than in changing the actual inherited traits of such yeasts.

Table 2. Techniques for Genetic Manipulation of Brewer's Yeast

Classical approaches
   mutation and selection
   screening and selection
   cross breeding
Modern approaches
   rare-mating
   cytoduction
   protoplast fusion
   recombinant DNA technology

Therefore, each technique, whether classic or part of the "new genetics," will be analyzed not only in terms of the technical aspects of the approach, but also in terms of its potential for actually creating a new production strain of brewer's yeast (or to borrow a phrase in vogue today, conduct a "reality check" on each technique).

B. Strain Improvement Using Classic Genetic Techniques

1. Selection Programs

As discussed earlier, yeast selection was first employed by Emil Christian Hansen at Carlsberg Laboratory when he isolated the first pure cultures of lager yeast just over 100 years ago. The significance of this discovery can be measured by appreciating the relative speed with which brewing in general (a notoriously conservative industry) adopted this practice world wide. This represented a quantum leap in terms of producing more consistent fermentations (and beers) with greatly reduced frequencies of beer spoilage and off-flavors.

While flavor considerations are normally the bottom line in any selection program, there are a large number of yeast traits (with direct relevance to production parameters) which are also used as criteria in selection programs. Included among these are:

1. Yeast oxygen requirements (which dictate optimal trub and wort dissolved oxygen levels)
2. Yeast flocculation properties (top or bottom? influence of tank dimensions and geometry on yeast flocculation; centrifugation involvement?)
3. Speed and degree of fermentation (adjunct concentrations, nitrogen requirements)
4. Clarity of final product (filtering requirements)

No two brewer's yeasts are quite identical when measured against the above criteria (in some cases, the differences are very dramatic), and every brewery around the world has made extensive use of screening and selection to generate a production strain best suited for their particular system. While precise numbers are not available, this author is certain that the vast majority of production lager yeast strains in use today have been isolated using screening and selection procedures.

One obvious consideration in selection programs is how one conducts such analyses most efficiently. Obviously a brewery cannot run full plant scale trials on hundreds of potential yeast cultures and still remain in business. In addition to being practical and affordable, screening equipment must also permit testing under conditions as closely related to plant parameters as possible (i.e., temperature, wort gravity and composition, pitching rates, heat transfer, etc.). Many a promising yeast culture in pilot scale fermentations has failed in production testing due to a lack of attention to production level parameters.

Perhaps the most practical, but sophisticated method for screening yeast cultures is the Multiferm System developed at Carlsberg Laboratory by Sisgaard and Rasmussen (32). This automated and computerized fermentation system has a capacity for 60 2-liter fermentation vessels (at the core of each jacketed stainless steel fermentor is a tall EBC tube constructed according to EBC [European Brewery Convention] Analytica Microbiologia Method 2.5.2 [33]) with functions such as wort filling, temperature regulation, sampling, and cleaning, all under the control of a microcomputer. Any culture which performs well over five generations in the Multiferm System is progressively studied in 50-liter and 10-hl scale fermentations before production scale fermentations (Diter von Wettstein, personal communication). The staff at Carlsberg Laboratory will conduct tours of the pilot brewery to interested professionals, and this author strongly recommends making this part of any visit to Copenhagen. Before leaving this section it is imperative to stress the importance of preserving any yeast culture that has been selected (in genetics, the only thing that is constant is change!). The genetic homogeneity of a pure culture will not last indefinitely, as mitotic recombination and unequal chromosome segregation during mitotic division will eventually lead to the development of substrains within that culture (besides, in comparison to the time, the effort needed to preserve a yeast strain is minimal). This principle was clearly illustrated by Thorne (34) when he examined clones isolated from a "stable" lager yeast culture grown for 9 months under continuous culture conditions. Of 48 clones tested, 27 mutations affecting brewing-related phenotypes were measured in nearly 50% of the isolates. Preservation under liquid nitrogen is generally considered to be the method of choice, as outlined in Kirsop's recent extensive review on the subject of yeast preservation techniques (35).

2. Mutation and Selection

Induced or natural mutations are generally of little practical value with brewing strains owing to their polyploid genetic composition. Typical are the results of Aigle et al. (36) on two French production lager yeast cultures which, on the basis of DNA content/stationary phase cell, were found to be tetraploids. Therefore, while recessive mutations, whether spontaneous or induced, no doubt occur with the same frequency in brewer's yeasts as they do in haploid yeasts, they do not reveal themselves due to the presence of nonmutated alleles. This "backup" genetic safety cushion results in inherent stability against the expression and build-up of potentially undesirable gene mutations. As a result, while many tantilizing reports exist demonstrating the potential for inducing desired brewing properties into haploid yeasts by mutation (e.g., the construction of yeasts which do not produce diacetyl when upstream genes in the isoleucine-valine pathway are inactivated [37]), the results cannot be mimicked in brewer's yeasts.

Despite these drawbacks, mutation and selection has been attempted as a method for strain development. Mutagens such as ultraviolet light, ethylmethane sulfonate (EMS), and N-methyl-N-nitrosoguanidine (NTG) have all been applied to polyploid brewing strains in an attempt to increase the genetic variability within a yeast culture. Molzahn (38), for example, has used this approach to isolate mutants with increased flocculation or altered diacetyl/H_2S production. As these strains, however, were not amenable to genetic analysis, the exact mechanism giving rise to the altered phenotypes could not be precisely determined (as frequencies exceeded that expected from simultaneous inactivation of all alleles of a wild-type gene). A word of caution concerning mutation (even in wild-type brewer's yeasts). By their very nature, these mutations are uncontrolled and widespread. Thus, hidden undesirable mutations can become expressed well after the initial experiments by such events as chromosome loss, mitotic recombination, or mutation of other wild-type alleles. Analogous in some respects to computer viruses in computer software these could someday surface to spoil the efforts of a protracted strain improvement effort.

Reports also exist indicating the value of screening cultures for spontaneous mutants. For example, Jones et al. (39) have used brewer's yeasts resistant to 2-deoxyglucose to select for those with improved maltose utilization rates. These mutants, derepressed for glucose repression of maltose uptake, ferment the disaccharide without first requiring a 50-60% drop in wort glucose levels, and hence result in faster fermentation rates. Once again, the ease with which mutants were selected indicated the genetics was much less complicated than the simultaneous inactivation of all wild-type alleles. Heredia and Heredia (40) reached similar conclusions on studies with haploid strains when they found the rate of spontaneous mutation to 2-deoxyglucose resistance to be dramatically higher than rates for other phenotypes (e.g., auxotropic mutations). Long-term stability of these mutants remains unclear until the mutation mechanism itself is elucidated.

There is, of course, one class of mutation which can be easily detected even in the genetic background of a polyploid yeast—a dominant mutation. Perhaps the mutation with the most potential for selecting new production strains of lager yeast is dominant resistance to the herbicide sulfometuron methyl (SM). The target for SM is the enzyme acetohydroxyacid synthase (AHAS), which catalyzes the synthesis of acetolactic acid—the precursor of diacetyl. Researchers, once again at the Carlsberg Laboratory, have isolated resistant clones of their production lager strain which are resistant to SM (41). Resistance was determined to result from different point mutations at only the *ILV2* locus (which encodes for AHAS), indicating the dominant nature of this mutation. By screening numerous resistant clones with the Multiferm System, several were found with a reduced ability to produce diacetyl (a result no doubt from altered activity of at least one AHAS enzyme in the yeast). One low-diacetyl strain was selected for a full-scale 4500-hl test and was found to produce beers below the threshold level for diacetyl detection after only 8-9 days of lagering (compared to the traditional 20-day period) (41). It is not known if Carlsberg has actually adopted one of these SM resistant clones as a new production lager yeast strain.

3. Cross Breeding

The first new yeast ever created by hybridization was described by Winge and Lausten in the late 1930s (42). Rapid adoption of the technology did not become standard practice with brewer's yeasts owing to the already described problems associated with mating and sporulation. A major obstacle against the use of cross breeding is that extensive backcrossing is required (after the initial crosses have introduced the trait or traits of interest into the brewer's yeast strain) in order to eliminate the nondesirable traits simultaneously introduced during hybridization experiments. Cross breeding is *not* a razor-blade approach to strain alteration, as it is impossible to leave all other brewing traits unaffected. While backcrossing is standard practice in higher plant breeding programs, this can be extremely tedious (if not impossible) work with brewer's yeasts owing to low sporulation and poor spore viability. Thus, much like sequencing work, where the initial 90% is often sequenced very quickly, it is the finishing touches (necessary after the primary traits are introduced) which take up the bulk of the time. Further hampering these efforts is our frequent ignorance of the physiology and genetics of many of the "desirable" traits found in brewer's yeasts.

Despite these difficulties, there are intrepid geneticists who have attempted to produce yeasts with improved brewing characteristics by the use of sexual hybridization. An example is the work of Emeis (43), who in 1971 crossed a mating strain of lager yeast with a strain of *Saccharomyces diastaticus* of the opposite mating type. After repeated backcrossings, he was able to produce a new strain of yeast which could ferment starch and dextrins without imparting the characteristic phenolic off-flavor often associated with *S.diastaticus*. Such yeast strains, however, have not been adopted for use in brewing—primarily owing to flavor differences in beers produced by these yeasts.

The most complete study employing cross breeding has undoubtably been the patient and elegant 1981 work of Gjermansen and Sigsgaard with *S.carlsbergensis* lager strain 244 (9). This strain exhibits the same low mating ability analogous to diploid laboratory strains which are heterozygous for the mating type alleles *MATα* and *MATa*. Originally classified as nonsporulating when examined at standard temperatures used for laboratory strains (~ 30°C), they found that by lowering the temperature to 22°C and below, asci with 1, 2, or 3 spores could be formed. While most of the viable spores were found to be nonsporulating and nonmating, 11 and 30 of the 394 tested clones were found to be of the *a* and α mating types, respectively.

When five hybrids were prepared by matings between presumed haploid clones, one was eventually found to be equal in brewing performance to the parent strain in Mutiferm fermentations. In 1983, it was reported that in studies up to 575 hl in volume (over four generations of use), the same yeast hybrid produced beer of acceptable quality (although the yeast was found to be slightly more flocculent and produced less diacetyl than the parent strain[2]). Clearly, these results indicate the potential for recombining the genomes of meiotic segregants to produce new strains of brewer's yeasts (which often continue to remain poor sporulators—a positive aspect of strain stability). In addition, mutations can be more easily introduced into meiotic segregants to assist in easier monitoring/conducting of cross-breeding programs. To date, however, Carlsberg Breweries have not exploited these meiotic segregants for breeding purposes, but rather for detailed genetic analyses of the genome of strain 244 discussed earlier.

Bilinski et al. (44), in 1986, found that substitution of glucose with acetate as the presporulation carbon source, and inoculating sporulation medium with stationary rather than exponential phase cells, also increased the frequency of tetrad formation in brewer's yeasts. With meiotic segregants of ale and lager yeast produced in this manner, Bilinski and Casey (45) went on to form ale/lager hybrids. Several of these were found to have faster rates of attenuation and ethanol production in 16°P wort than either parent (owing to an increased rate of maltose uptake). Significantly, they produced beers of good palate, lacking the sulfury character of lager beers, but keeping the estery aroma of ales. It remains to be seen if such industry hybrids can be used for the production of new beer products in today's rapidly expanding marketplace for new categories of beer.

Before leaving this section on brewer's yeast cross breeding, it is worthwhile to examine the 1988 paper of Tsuboi and Takahashi (46) describing their genetic analysis of the nonsporulating phenotype in six strains of brewer's yeasts. When they crossed, by protoplast fusion, nonsporulating petites of brewer's yeast with homothallic diploid laboratory strains of *S.cerevisiae* carrying sporulation-deficient mutations, all hybrids sporulated at frequencies of 15-63%. This demonstrated both that the brewer's yeasts did not carry any dominant sporulation-deficient mutations and that they have wild-type sporulation genes. Perhaps the most interesting discovery in this paper was that crosses between brewer's yeasts with haploid or diploid strains *homozygous* for the mating-type locus had poor or no sporulation (in sharp contrast to the high frequency of sporulation involving crosses with strains heterozygous for the mating-type locus). This revealed that the polyploid, aneuploid, and/or amphiploid composition of brewer's yeasts does not solely account for the nonsporulating phenotype of brewer's yeasts, but may also involve a deficiency of the mating-type genes. One practical implication of these results is that brewers could perhaps increase the frequency of sporulation in their strains by first transforming their yeast with a plasmid containing both the *MATα* and *MATa* loci.

C. Strain Improvement Using Modern Genetic Techniques

1. General Remarks

All techniques covered in this section, i.e., protoplast fusion, cytoduction, rare-mating, and genetic engineering using recombinant DNA technology, share a number of common features. These include:

1. First reports on these techniques with yeast appeared in the late 1970s.
2. All operate independently of yeast ploidy and mating type.
3. In first light, each technique has appeared to hold great promise for constructing new strains of brewer's yeasts.

With the benefit of nearly 15 years of experience with these techniques it is now clear that all but recombinant DNA technology have the same fatal "good news/bad news" flaw which limits their use to basic studies; i.e., none of these are able to alter or introduce a specific trait of interest into a brewer's yeast without simultaneously disrupting many other desirable traits in that yeast. Numerous backcrosses would be needed to restore the disrupted traits (many of which have poor or no elucidation of the biochemistry or genetics involved), but the difficulty of accomplishing this has already been reviewed.

Recombinant DNA technology on the other hand is more of a "surgeon's knife" approach to strain improvement, and as such does not appear to be limited by the above obstacle. Regulatory issues aside, it therefore holds the greatest promise for being a practical approach to brewer's yeast strain development programs. For this reason most of the following discussions will deal with recombinant DNA (rDNA) technology, with the others being discussed only for the sake of completeness.

2. Rare-Mating

Rare-mating, also called forced-mating, is the technique in genetics which gives meaning to the expression never say never. First described with laboratory yeasts by Gunge in 1972 (47), and then by Spencer (48) with industrial yeasts in 1977, this technique involves mixing two cultures (at least one of which is nonmating) at high cell densities. At very low frequencies true hybrids with fused nuclei can be isolated using appropriate selection markers.

Perhaps the best known example of this technology is the brewer's yeast cross, described by Tubb et al. (49), between a respiratory-deficient yeast and an auxotrophic, nonbrewing, haploid dextrin-fermenting strain of *Saccharomyces*. This latter yeast was first manipulated by mutation and strain selection to eliminate the *POF1* gene, (which is responsible for the enzymatic decarboxylation of ferulic acid to 4-vinylguaicol that lends the characteristic herbal phenolic off-flavor typical of all haploid laboratory strains used in genetic studies). While brewing strains capable of producing low carbohydrate beers were isolated, the hybrids only produced beers of "acceptable" quality—not identical quality to that of the original parent. Neither Tubb et al. (49) nor the Brewing Research Foundation has continued to pursue strain construction by rare-mating.

3. Cytoduction

Cytoduction is a specialized form of rare-mating in which only the cytoplasmic components of the donor strain are transferred into a brewer's yeast. This approach has been used by Young in 1981 (50), and Hammond and Eckersley in 1984 (51) to construct "killer" strains of brewer's yeast (with anticontamination properties) by introducing cytoplasmic double-stranded RNA (dsRNA) (which encodes for the K1 "killer" zymocin and the associated immunity). In this case, the frequency of nuclear fusion was lowered even further by using a haploid killer strain containing the *kar1* mutation—shown by Conde and Fink (11) to impair nuclear fusion (see Fig. 3). While killer strains of brewer's yeasts could be isolated, many had signigicantly impaired fermentation characteristics. Hammond and Eckersley (51) found this was largely due to the nature of the mitochondria introduced by the killer haploid. Surprisingly, if haploids were used where the mitochondria originated from a brewer's yeast, not cultures derived from genetic stock centers, the constructs generally had enhanced brewing properties. Unfortunately, no molecular explanation linking mitochondrial DNA to brewing performance has been established, but it does drive home the fact that not all mitochondria from respiratory- competent yeast are the same. Importance must therefore be placed on nonnuclear DNA in brewer's yeast performance, further complicating the utility of these latter two techniques in strain development programs.

As a closing note on this subject, it remains to be determined how rare-mating or cytoduction occurs. The classic explanation is that mating ability arises at low frequencies via mitotic recombination or gene conversion events, producing homozygosity at the *MAT* locus. However, as cytoductants of brewer's yeast do *not* exhibit mating abilities once isolated in pure culture (K. Oldfield, unpublished data) the mechanism, whether with haploid or brewer's yeast, remains a mystery.

4. Protoplast Fusion

Protoplast fusion, first described with yeast in 1977 by van Solingen and van der Plaat (52), is conceptually similar to rare-mating and cytoduction in that it can be used to produce either hybrids or cytoductants. It is unlike these techniques, however, in that it can be used to conduct interspecific crosses, thereby greatly increasing the potential genetic pool available for strain development programs.

Since 1977 there has been a plethora of literature on this subject (one need only conduct a literature search using the name Spencer to rapidly find an impressive number of papers on this technique), but this section will be limited to crosses which specifically involve brewer's yeasts.

Protoplast fusion itself is a relatively efficient technique which can be even further facilitated by the use of electrofusion of protoplasts (where short electric field pulses of high intensity replace polyethylene glycol as the fusing agent) (53). The fusion of two genomes, however, is never a predictable procedure, with fusion products often being very different than either parent. Thus, once again it is very difficult to selectively introduce a single trait into a brewer's yeast using this technique. This obstacle has been revealed in publications dealing with protoplast fusion as a means of introducing into brewer's yeast:

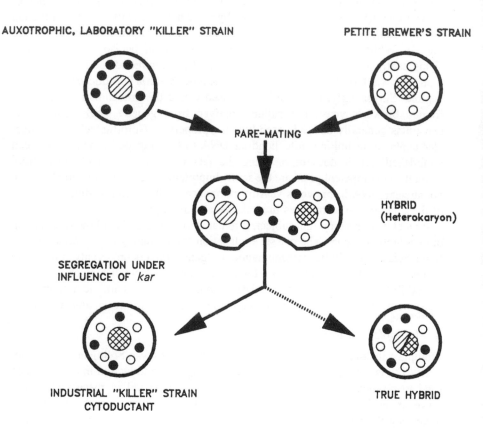

Figure 3. Rare-mating between brewer's and laboratory strains of yeast leading to the production of true hybrids (when nuclei fuse) or cytoductants (when only cytoplasmic elements are transferred).

1. Increased ethanol tolerance (54).
2. Dextrin fermenting ability (55, 56).
3. β-Galactosidase activity for lactose uptake (57).

Recently, it has been possible to use pulsed field electrophoresis to actually follow chromosomes during protoplast fusion experiments. Hoffman et al. (58) used OFAGE to assess the complete genetic composition of hybrids between two haploid *Saccharomyces* strains and found it to yield observations agreeing with conventional genetic analyses. More importantly, as hybrids can be analyzed directly, it was shown to provide definite genetic answers to situations where conventional genetics could not be used to reach firm conclusions (e.g., when hybrids are unable to sporulate, or one parent lacks adequate markers). In addition, it can be used to provide information on all chromosomes in the two parents, not just those with markers. While undoubtably a valuable tool for basic research papers, the ability to visually monitor chromosomes does not eliminate the previously discussed problems preventing the implementation of protoplast fusion as a tool to construct new strains of brewer's yeasts.

Protoplast fusion can, however, be a valuable technique when flavor is not the bottom line for judging success. Stewart et al. (59) demonstrated this when they fused a lager yeast with *S.diastaticus* and isolated hybrids with potential use in fuel alcohol production. In some cases, hybrids could be found having higher osmotic pressure resistance and alcohol yields (at higher temperatures) than all *Saccharomyces* yeasts in the Labatts Culture Collection. Should interest in gasohol be revived by future sharp rises in oil prices (high interest persists in some countries already; see Chapter 8), such strains could figure prominently in the alcohol-producing industry.

5. Recombinant DNA Technology

Even in a world where change is commonplace, it is still almost difficult to believe that genetic engineering of yeasts has been possible only since 1978. That year, Hinnen et al. (60) demonstrated the complementation of a *leu2* mutation in a haploid laboratory strain by transformation with a yeast *LEU2* gene. Since then, refinements in cloning procedures, vector construction, transformation procedures, and easy availability of the consumables necessary to conduct recombinant DNA research have all combined to make genetic engineering virtually routine in haploid laboratory strains of *Saccharomyces cerevisiae*.

Here at last then, from the perspective of an industrial yeast geneticist, was a technique that could selectively modify the genetic composition of a brewer's yeast without disrupting the countless desirable traits characteristic of every brewer's strain. Genes to be introduced could in theory be drawn from any source in nature (including the gene pool of all various raw materials used in brewing, e.g., amylases in barley), and by using in vitro mutagenesis genes could be manipulated to make the resulting proteins more productive under brewing parameters (e.g., a glucoamylase with increased activity at acidic brewing pHs).

Despite all this initial optimism, the first reports of genetic engineering of brewing strains did not emerge for a full 6-8 years after Hinnen's 1978 publication. Three reasons for this can be given. First, nearly all early vectors designed for use in haploid laboratory strains almost invariably took advantage of auxotrophic mutations as the basis for selection markers (principally *URA3, LEU2,* and *TRP1*). Such mutations in brewer's yeasts are neither available nor desirable. Second, integration strategies for use with wild-type yeasts, leading to acceptable levels of mitotic stability under nonselective conditions, did not become available until the 1980's. Third, and perhaps most importantly, all early vectors had bacterial sequences as part of their constitution. While government regulatory considerations alone would prevent use of a recombinant brewer's yeast for this last reason, no brewery sensitive to its public image would in the first place market a product produced by a yeast containing DNA from a fecal coliform (e.g., *Escherichia coli*).

As we approach the end of the 1980's, solutions to all three of the above problems have been found. The remainder of this section will be devoted to analyzing advances in these areas as they directly apply to brewer's yeasts. It will not be an overview of yeast genetics per se or recombinant DNA technology (e.g., sequences involved with gene regulation or protein secretion). For this, readers are directed to recent reviews on the subject (61-63) and to other chapters in this monograph.

Dominant Selection Markers. Table 3 provides a partial list of dominant selection markers which have been shown to be functional in *Saccharomyces* yeast (a more complete listing can be found in ref. 64). All of these have stemmed from research conducted in the 1980s, and all have particular relevance to vectors designed for use with brewer's yeasts as they circumvent the need for auxotrophic selection markers. This binding factor aside, not all of these markers have equal potential for use with brewer's yeasts, and it is the opinion of this author that the *SMR1- 410* gene offers the greatest number of advantages for recombinant DNA manipulations of brewer's yeasts. *SMR1-410* provides resistance to the herbicide sulfometuron methyl (SM), N- [(4, 6-dimethylpyrimidin-2-yl) aminocarbonyl-2-methoxycarbonyl-benzene-sulfonamide], manufactured by Dupont. The target site of SM in *Saccharomyces cerevisiae* is the enzyme acetolactate synthase, an enzyme in the biosynthetic pathway of isoleucine and valine, and which is encoded by *ILV2* (65). Both *SMR1-410* and *ILV2* have been sequenced, and it has been determined that the two alleles are identical excpt for a C to T transition mutation at nucleotide 574 of the open reading frame of *ILV2*(66). This results in a proline to serine change in the amino acid sequence of acetolactate synthase which confers resistance to inhibition by SM.

Casey et al. (67) have shown that all wild-type *Saccharomyces cerevisiae* yeast, not just brewer's yeasts, are amenable to transformation and selection with

Table 3. Dominant Selection Markers Shown To Be Expressed In *Saccharomyces cerevisiae*

Dominant marker	Source	Ref.
Resistance to chloramphenicol (chloramphenicol acetyl transferase)	*E.coli*	(73, 74)
Resistance to G418/geneticin (aminoglycosidase phosophotransferase)	*E.coli*	(69, 70)
Resistance to hygromycin B (hygromycin B phosphotransferase)	*E.coli*	(72)
Resistance to phleomycin (DNA scission)	*E.coli*	(76)
Resistance to methotrexate (dihydrofolate reductase)	Mouse	(71)
Copper resistance (*CUP1*, methallothionen)	Yeast	(68)
Sulfometuron methyl resistance (*SMR1*, acetolactate synthase)	Yeast	(67)
Dextrin fermenting ability (*STA1, STA3, DEX1* glucoamylases)	*S.diastaticus*	(78)
Melibiose utilization (*MEL1*, α-galactosidase)	*S.uvarum*	(77)
Killer and immunity phenotype (M1 dsRNA)	*S.cerevisiae*	(80–82)

Table 4. Transformation Frequencies of Industrial Strains of *Saccharomyces* with pCP2-4-10 DNA using *SMR1* as the Selection Marker

Yeast strain	Industrial application	No. of transformants/ 10 µg DNA
S.cerevisiae ATCC 18824	Ale yeast	800 (11)
S.cerevisiae ATCC 6037	Baker's yeast	3,340 (6)
S.cerevisiae ATCC 7754	Baker's yeast	810 (14)
S.cerevisiae ATCC 32120	Baker's yeast	540 (9)
S.cerevisiae ATCC560	Distiller's yeast	1,640 (48)
S.cerevisiae ATCC 287	Lager yeast	3,650 (18)
S. cerevisiae ATCC 2700	Lager yeast	1,440 (14)
S.carlsbergensis M244 (Carlsberg Breweries)	Lager yeast	920 (5)
S.cerevisiae var sake ATCC 26421	Sake yeast	6,800 (64)
S.cerevisiae ATCC 4098	Wine yeast	200 (28)
S.cerevisiae ATCC 4108	Wine yeast	580 (8)

Number in parentheses indicates the number of resistant colonies which arose on control plates in the absence of pCP2-4-10 DNA.
Source: Modified from Ref. 13, used with permission.

SMR1-410-based vectors (see Table 4). The phenotype of SM resistance is easily scored. Prototrophic *Saccharomyces* yeasts are completely inhibited by SM concentrations of 15 μg/ml or less, whereas transformants obtained using yeast-replicating or yeast-integrating vectors are resistant to at least 70 μg/ml sulfometuron methyl. The ease of detecting transformants using sulfometuron methyl resistance is in sharp contrast to several other dominant selection markers (e.g., copper resistance and G418 [69,70]), where the difference between the resistance of untransformed and transformed cells, especially during the initial screening of transformants, is not very substantial. On plates of SD + 20 μg/ml SM, untransformed cells do not grow at all, whereas transformants containing *SMR1-410* give rise to easily visible discrete colonies within 48 hr at 30°C (67).

A significant property of *SMR1-410* is that it allows for the selection of transformants when present in only a single copy. This is in contrast to markers conferring resistance to copper (68) and methotrexate (71), where multiple copies of the markers are required for maximum efficiency. As dominant selection markers intended for use with industrial strains of *Saccharomyces* yeasts must be stably inherited (i.e., via chromosomal integration) it is preferable that only a single copy of the gene conferring resistance be sufficient to score the phenotype.

Perhaps the greatest advantage of *SMR1-410* over other currently available dominant selection markers is that it is not derived from sources other than *Saccharomyces*, but instead is a mutation of the *ILV2* gene from *S.cerevisiae* which has only a single base pair transition. For example, the kanamycin-resistance gene of *Tn903* (69), the *hph* gene for hygromycin B phosophotransferase (72), and the CAT coding sequence from *Tn9* (73, 74) (which render *S.cerevisiae* resistant to G418, hygromycin B, and chloramphenicol, respectively [see Table 2]), all originate from *E.coli*. Methotrexate resistance in *S.cerevisiae* is mediated by a mouse-derived dihydrofolate reductase gene (71). The fact that these markers are not derived from *Saccharomyces* likely limits their application for use in genetically manipulating food- and beverage-grade strains of *Saccharomyces* yeasts by recombinant DNA technology. Government regulatory agencies will likely be more concerned about vectors employing non-yeast DNA than they will be with natural yeast mutations used as selection markers. While pWX509 contains *E.coli* DNA, Xiao and Rank (75) have gone on to modify this vector so that recombinant brewer's yeasts can be constructed free of bacterial sequences (see below).

Of the remaining yeast-derived markers, two have practical considerations which limit their potential. For example, utilization of melibiose requires the *MEL1* gene product, β-galactosidase, and transformants containing *MEL1*-based vectors can be conveniently detected on Tubb and Liljestrom's (77) X-α-gal-based medium. However, as all lager yeasts are already able to ferment melibiose (17), use of this marker is limited to ale yeasts only (which are melibiose negative). To date, the use of *MEL1* in ale yeast engineering has not yet been reported. Genes related to starch utilization (e.g., *STA1, STA3,* and *DEX1*) can in theory be both selection markers (e.g., haloes in starch—bromocresol medium) and desirable genetic traits for brewer's yeasts (e.g., light beers [78]). There are no reports, however, of the successful detection of transformants using these as markers with brewer's yeasts, despite the efforts of Tubb in 1987 (79) with *DEX1*.

The last yeast-based marker in Table 3, the M1 dsRNA-based killer and immunity phenotypes, is intriguing as it has the potential to perform "autoselection." M1 double-stranded RNA found in killer yeast strains codes for an extracellular toxin as well as an immunity factor that protects infected cells from the toxin; sensitive cells that do not contain dsRNA are killed by the killer toxin. In 1984, Lolle et al. (80) cloned cDNA of M1 dsRNA into YEp vector and selected for haploid transformants by complementation of *leu2*. Transformants produced authentic preprotoxin precursor and were also immune to the toxin; the toxin and immunity phenotypes were coded for by the same open reading frame. In 1985, Bussey and Meaden (81) developed a transformation protocol for expression of the complementary DNA (cDNA) copy based on medium-containing killer toxin. Two industrial yeasts transformed by this procedure were shown to have stable inheritance of the killer and immunity phenotype. The maintenance of plasmids containing killer cDNA was shown to result from an autoselection for cells maintaining the immunity factor coded for by the killer cDNA gene. Transformation of killer has been proposed as a means of conferring industrial strains resistant to contaminant yeasts (82). Despite the promise and potential of this marker, there are no reports yet on the utility of this in brewer's yeast strain development programs.

Construct Stability.   It is critically important that transferred genes are stably inherited. For commercial production, nonintegrating YEp and YRp vectors are inadequate in this respect, since they segregate at high frequency and it is too expensive, and generally unsound, to maintain vectors carrying resistance markers by addition of cellular inhibitors. Stability of autonomously replicating vectors in haploid strains has been improved by the construction of plasmids containing only yeast sequences (83). These constructs showed a dramatic increase in stability compared to standard YEp or YRp plasmids. Up to 99% of some haploid strains maintained some of these plasmids in the absence of selection pressure (84, 85). However, diploids (85) and other haploids lost these yeast plasmids at a higher rate. Thus, these plasmids are likely too unstable to be of use to industry and will remain so until they approach the stability of the endogenous 2-μm plasmids.

Presently, satisfactory levels of stability can be reached by the use of integrating vectors (YIp) which deposit the gene of interest into one or several chromosome locations. Lacking yeast origins of replication, these plasmids can only be stabilized by reciprocal recombination within an area of genetic homology. While integration occurs naturally at a low frequency, it can be enhanced, and targeted, by linearizing the vector at a restriction endonuclease site within the plasmid-integrating gene (86). The resulting integrant structure has a direct duplication of the integration gene sequence, flanking the other plasmid borne markers (which

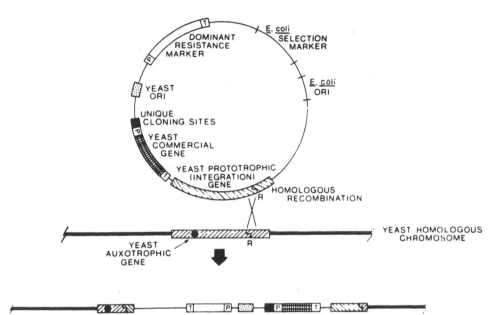

Figure 4. Classic integration of a generic plasmid indicating common sequences and chromosomal integration in an area of sequence homology. Autonomously replicating YEp and YRp plasmids contain a 2-μm or chromosomal *ORI* which allows replication in yeast. Most autonomous plasmids also carry an *E. coli ORI* which functions as a replication origin in *E. coli.* Other *E. coli* sequences (–) include a resistance selection marker, commonly Amp[R] or Tet[R], and nonspecific plasmid sequences. Yeast prototrophic genes, commonly *URA3, TRP1, LEU2,* or *HIS3,* are used to select for plasmid transformation in haploid cells containing a defective auxotrophic mutation (*) within an homologous chromosomal gene. Alternatively a dominant resistance selection marker (Table 3, column 1) can be used to provide for plasmid selection. Enhanced expression of resistance markers has been achieved by the use of control sequences from other genes–promoters (P) and terminators (T). The inclusion of unique cloning sites allows for the incorporation of commercial genes of interest (e.g., β-glucanase). Plasmid integration into a chromosomal area of homology is enhanced by restriction endonuclease digestion (R) within the plasmid homologous gene. Special integrating vectors (YIp) do not contain *ORI,* and thus can only be maintained by chromosomal integration. (From Ref. 64, used with permission.)

normally include nonyeast sequences from the original vector). Figure 4 is illustrative of the principles involved with YIp type vectors.

For use with brewer's yeast strain development two additional features have to be considered. First, the integrated structure produced in the manner depicted in Figure 4, while relatively stable, can be lost at low frequencies under nonselective conditions (0.001% per cell generation) by a reversal of the integration process following reciprocal recombination within the directly duplicated sequence (87). Secondly, the inclusion of bacterial DNA sequences during the integration would certainly raise objections from government regulatory agencies.

Papers by Yocum (87) and Xiao and Rank (75) demonstrate how these obstacles can be overcome in brewer's yeasts. The former author employed vectors based on resistance to antibiotic G418 and single integrated copies were selected readily in brewer's yeasts at the *HO* locus in the classic manner. After integration though, the bacterial G418 resistance gene and accompanying *E.coli* sequences were deleted or "jettisoned," leaving only the desired gene in the yeast chromosome. This was accomplished by including in the integrating plasmid the *E.coli lacZ*gene. This encodes for the enzyme β-galactosidase, which causes the original trans-formants to turn blue on X-β-gal plates. Also, the plasmid was constructed so that upon integration at the *HO* locus there existed a short stretch of *HO* DNA that is present as a direct repeat flanking the G418 resistance and *lacZ* ganes (but not flanking the gene of interest). Homologous recombination between the flanking repeats will result in looping out of the *E.coli* DNA sequences (and loss of G418 resistance and blue colony color), an event which can easily be detected on appropriate selection media. Repeating the process of integrating, and jettisoning at the remaining *HO* loci within polyploid brewer's yeasts results in constructs with effectively "permanent" stability of the introduced trait. To date, Yocum (87) has used this approach to produce lager strains containing the *Aspergillus niger*-glucoamylase enzyme for the production of low calorie beers, and is presently examining the potential of these yeasts in pilot scale fermentations at the Brewing Research Foundation.

Xiao and Rank (75) developed a one-step dominant selection procedure to construct recombinant lager yeast devoid of plasmid sequences. This strategy was based on Rothstein's (88) one-step gene-disruption procedure in which a linear chimeric construct with recombinogenic ends replaces the host-homologous segment by a double recombination event (eliminating concerns over the presence of direct repeats in integrant structures). In this case, the replacing sequences consisted of a region of a nonessential DNA from chromosome XIII in which internal sequences were replaced with *SMR1* and *MEL1* dominant selection markers as shown in Figure 5 (as Goebl and Petes (89) have estimated that up to 70% of the yeast genome is dispensible without noticeable phenotypic effects, a wide array of potential replacement sites exist). Integration by a double recombination event in Carlsberg's production lager strain 244 was easily detected by isolating sulfometuron-resistant clones exhibiting β-galactosidase activity (transformant colonies had a significantly deeper blue color on X-α-gal plates than wild-type colonies). Of course, the *MEL1* gene can simply be replaced by any gene of commercial interest (e.g., a glucoamylase gene), and there is no need to be concerned about stability problems as the replaced sequences lack the direct repeat produced by normal plasmid integration that could be reverted by homologous intrachromosomal recombination or gene conversion (75). While it remains to be seen whether this protocol will be exploited in brewer's yeasts, it is the opinion of this author that this is the most promising technique developed to date for modifying brewer's yeasts via recombinant DNA technology.

These latter two papers are only examples of how genes of interest to brewing can be introduced into a yeast without concerns over the stability of the construct or the involvement of bacterial sequences. Alternative strategies certainly are available, including the spontaneous amplification of recombinant genes by integration at the multicopy rDNA locus on chromosome XII (90), via integrating cassettes involving the yeast *Ty* transposable element (91), or by incorporation into native 2-μm plasmid DNA (92). The reader is encouraged to seek out these references to stimulate ideas on this subject as the only limit on this type of research is perhaps only man's imagination itself.

With these thoughts in mind, it is now appropriate to review efforts aimed at constructing new strains of brewer's yeasts via recombinant DNA technology. Emphasis will be placed not only on the technical considerations of the constructs, but also on the likelihood of each construct being successfully introduced into the brewing industry. Examination of Table 5 reveals that except for three papers

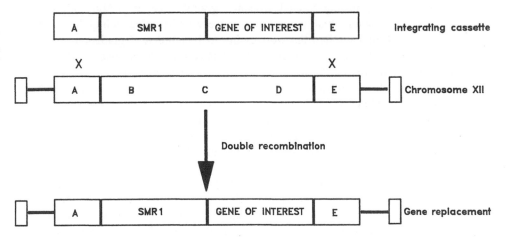

Figure 5. Principles of gene replacement leading to the insertion of recombinant constructs without the addition of nonyeast sequences.

Table 5. Published Reports On Recombinant Wild-Type Brewer's Yeast Strains

Year	Gene activity	Source	Ref.
1984	Glucoamylase	*Aspergillus niger*	(87)
1985	*CUP1*—copper resistance	*S. cerevisiae*	(68)
1985	*DEX1*—Glucoamylase	*S. diastaticus*	(104)
1985-86	β—Glucanase	*Bacillus subtilis*	(93,94,98–101)
1986	Human serum albumin	Human	(128)
1986	β—Glucanase	Barley	(95)
1987	β—Glucanase	*Trichoderma reesii*	(96)
1987-88	α—ALDC	*Enterobacter aerogenes*	(121–124)
1987	Extracellular protease	Wild yeast	(126)
1988	*SMR1*—sulfometuron methyl	*S. cerevisiae*	(67)
1988	*DEX1*—glucamylase	*S. diastaticus*	(105)
1988	*MEL1*—melibiase	*S. cerevisiae*	(75)

dealing with an examination of different selection marker, virtually all recombinant brewer's yeasts constructed to date involve one of three different enzymes: β-glucanase, amylases, or acetolactate decarboxylase. The rationale behind choosing each of the enzymes as well as the technical and practical accomplishments seen to date will now be examined.

Breeding of β-Glucanase-Positive Yeast: During malting the endosperm cell walls are degraded by the enzyme β-glucanase ([1-3,1-4]-β-D-glucan 4-glucanohydrolase), making the enclosed starch granules subject to degradation by α-amylases. Frequently, malt is used in a brewery which releases high levels of water-soluble β-glucans into the wort owing to incomplete degradation of the endosperm wall materials. In the brewhouse this can lead to prolonged filtration times, thereby decreasing brewhouse capacity and efficiency, and when passed onto the beer, also result in filtration and haze problems. Problems at least with beer filtration could be solved by using lager yeast strains capable of degrading the β-glucans during the primary fermentation. As brewing yeast do not contain a β-glucanase gene, accomplishing this requires introducing a cloned β-glucanase gene from a non-yeast source via recombinant DNA technology. To date β-glucanase has been cloned from three sources: species of *Bacillus* (93,94), barley (95) and the fungus *Trichoderma reesii* (96). The rationale for choosing *B. subtilis* stemmed from its β-glucanase being an approved food-grade enzyme (95), and as it acts in a similar manner to barley β-glucanase (producing trimeric and tetrameric oligosaccharides from high molecular weight β-glucans (96)). Hinchliffe (98) and Hinchliffe and Daubney (99) introduced the *Bacillus* β-glucanase into NCYC 74, NCYC 240, NCYC 1026, and various Bass brewing production yeasts on a yeast episomal-based vector with CUP1 as the selection marker. While β-glucanase transformants were isolated and pilot scale beers were indistinguishable from the control beers on the basis of organoleptic analysis, two problems exist:

1. The vector contain *E.coli* sequences, thereby making regulatory approval unlikely.
2. Insufficient levels of β-glucanase were produced to enable successful commercial implementation.

Cantwell et al. (100) increased the levels of β-glucanase produced by placing the *Bacillus subtilis* β-glucanase gene under the control of the highly expressed yeast *ADH1* promoter. Lager transformants were able to degrade five times as much β-glucans as ale transformants (100), but secreted levels were still too low for production requirements. In an attempt to increase secretion levels, the β-glucanase gene was fused to the prepro-leader region of the yeast *MAT* mating factor gene (100). While increased efficiency of secretion (over the natural *Bacillus* signal sequences) was found in trials with laboratory strains of *Saccharomyces,* similar trials with production brewer's yeasts were not reported. One year later, however, Lancashire and Wilde (101) described the transformation of ale strain BRG 6003 with a yeast replicating vector containing the *Bacillus subtilis* β-glucanase gene and a dominant selection marker lending chloramphenicol resistance. In this case secretion was also accomplished by the use of the MFα prepro secretion signal sequence. In tall-tube ale fermentations approximately one-third of the wort β-glucan was degraded during the fermentation. Beers had moderately reduced viscosity values and significantly increased filterability index values (101). No detailed fermentation or flavor data were presented in the paper, in which the authors noted that the results were those of a first-generation construct which would have stability and regulatory issues "cleaned up" before use in production applications (101).

The barley β-glucanase gene was cloned by researchers at Carlsberg Laboratory in 1986 (95) by splicing together two partial cDNA clones derived from barley aleurone. While a brewing strain was successfully shown to degrade significant amounts of β-glucans during primary fermentation, the constructs could only be used for basic studies, and not be considered for production purposes as:

1. Regulatory concerns remain due to the use of an *E.coli*-derived selection marker and a mouse derived signal factor directing β-glucanase secretion.
2. β- Glucanase activity was not stabilized by integration, but was introduced on inherently unstable yeast episomal vectors.

Efforts at Carlsberg Laboratory are presently focused on in vitro site directed mutagenesis of the cloned enzyme to increase its heat stability (102). The intent is not to introduce this into yeast, but rather to reintroduce it back into barley whenever an efficient transformation system for barley becomes available. For this purpose yeast is only serving as a host for the production of β-glucanases to test in mashing experiments. Similar research is also in progress at Carlsberg with a hybrid *B. subtilis/B. amylofaciens* β-glucanase (103) to produce a heat stable enzyme for use in mashing.

The most sophisticated β-glucanase construct reported to date is that of Enari et al. (96), where the β-glucanase gene from *Trichoderma reesii* was integrated into the genome of a lager yeast tested over three generations in 50-liter pilot-scale fermentations. Integration in this case was accomplished by the cotransformation method, where a yeast episomal plasmid containing the β-glucanase gene under the control of a PGK promoter and secretion signal sequences (source not stated) was cotransformed with a *CUP1*-containing plasmid (the β-glucanase-containing gene had first been linearized in the *LEU2* gene to direct integration to that site in the lager yeast genome). Copper-resistant transformants were subsequently screened to select for those exhibiting β-glucanase activity and a cotransformed yeast was then cured of the selection plasmid by growth under nonselective conditions. The pilot-scale fermentations demonstrated that the integrant yeast produced a beer with nearly three times the filtrability values of the control beer and virtually no β-glucans (compared to 200–300 mg/liter in the control beers). The fermentation profiles of integrant and control strains were very similar and no significant differences in flavor were noted in taste testing of beers from the two yeasts (96). At this time the only drawback to the construct is that the integrant contains bacterial sequences, although this should be straightforward to rectify.

Breeding of Amylolytic Yeasts: Introduction of genes for extracellular amylolytic enzymes into brewer's yeasts (including glucoamylase, α- and β-amylases, pullulanase, or other debranching enzymes) would dramatically improve the efficiency of utilization of wort carbohydrates. Such yeast would have great utility in the production of low- carbohydrate "light" beers (as well as the new "dry beer" products) by replacing or reducing the levels of malt or enzymes needed to convert starch to fermentable sugars (which in wort are limited to fructose, glucose, maltose, and maltotriose [57]). In theory, at least, geneticists argue that amylolytic yeast could ferment malt substrates directly without prior modification or processing in the brewhouse. In reality, such a dramatic development is unlikely in the near future, although some involvement through a recombinant amylolytic yeast construct does seem a distinct possibility.

Like publications dealing with protoplast fusion, there is a plethora of papers dealing with the construction of amylolytic haploid laboratory strains of *S.cerevisiae* yeast, but a scarcity of papers dealing exclusively with recombinant brewer's yeast strains. Certainly the first report involving a brewer's yeast was that of Yocum (87). While the technical considerations of this have already been reviewed earlier (including the exclusion of *E.coli* sequences and stabilization by integration at several HO loci), initial studies with a lager yeast constructed at the

Brewing Research Foundation revealed that insufficient quantities of an *Aspergillus niger* glucoamylase were secreted by the yeast to produce a sufficiently attenuated beer (R. Yocum, personal communication). At last report, an attempt to increase the yield of glucoamylase was done by replacing the *TPI* promoter with the more efficiently expressed glucose inducible *EN02* promoter for continued BRF pilot scale fermentation (R. Yocum, personal communication). The author is unaware of the results of these tests (presumably conducted in late 1988 or early 1989). Should sufficient levels of enzyme be produced, it will be interesting to see if such a construct is approved for use in the United Kingdom (where the BRF has petitioned the Ministry of Agriculture, Fisheries and Food for permission to use a recombinant amylolytic brewer's yeast in brewing).

In 1985, Meaden and Tubb (104) also reported the construction of a recombinant amylolytic brewer's yeast. In this case, the cloned glucoamylase was the *DEX1* gene from *Saccharomyces diastaticus* in a yeast episomal type vector (which unlike the *Aspergillus* enzyme lacks α-1-6 debranching activity, thereby limiting its ability to fully hydrolyze starch). Three years later Perry and Meaden (105) reported results obtained in 271-scale pilot fermentations at the Brewing Research Foundation with lager strain NCYC 1324 transformed with the vector pLHCD6, also used in 1985. They found that a transformant could utilize about 30% of the dextrins in a 1.040 O.G. wort (superattenuating to a specific gravity of 1.0023 compared to 1.0046 for the parental strain). However, as the transformant had an adversely affected growth rate, fermentation times were significantly enhanced. A solution to this was found by using a mixed culture of transformant:parental yeasts to maintain traditional fermentation rates (105). The authors did note that the negative effect on growth rate may be strain specific, as a transformant of ale yeast NCYC 240 exhibited no significant effect on the rate of growth or fermentation (105). However, as both the 1985 and 1988 publications reported use of the yeast episomal type vector (during the pilot scale fermentations the proportion of $CUP^R DEX^+$ cells fell to 52% in the studies with NCYC 240) containing *E.coli* sequences, the constructs could not be considered for practical applications owing to stability and regulatory concerns. Presumably researchers at the BRF are presently designing second-generation vectors to overcome these problems.

If the BRF is engaged in designing a new generation of vectors for engineering amylolytic yeasts, it is highly possible that they have substituted the *DEX1* gene from *S.diastaticus* with a glucoamylase gene from another source. One attractive possibility is yeast from the genus *Schwanniomyces* (e.g., *S.castelli* and *S.occidentalis*), as not only do they possess a glucoamylase having both α-1-4 and α-1-6 bond-cleaving activities (enabling a greater utilization of wort dextrins than that possible with *DEX1*), the enzyme is completely inactivated by pasteurization temperatures, thereby minimizing concerns about enzyme activity causing flavor and spoilage problems in bottled products (106). This is in sharp contrast to the *Aspergillus* glucoamylase enzyme employed by Yocum (87), which is con-

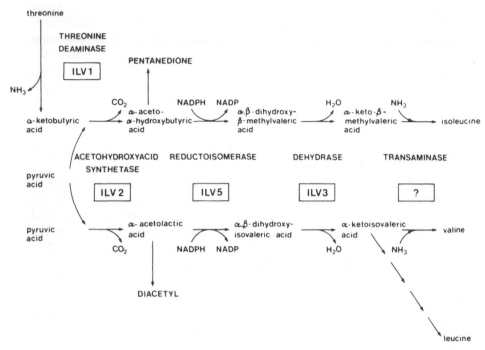

Figure 6. The biosynthetic pathways for isoleucine and valine in *Saccharomyces cerevisiae*. (From Ref. 15, used with permission.)

siderably more heat resistant. These differences between glucoamylases from varying sources illustrates the relevance of the expression "consider the source" in any strain improvement programs; i.e., before any genetic research is initiated, it is of considerable value to compare the biochemistry of the alternatives (e.g., pH, temperature, substrate range) before deciding which gene(s) should be cloned for introduction into a particular brewing strain. Such time would be well spent as the genetic phase of strain construction can easily require a 3- to 5-year period.

Breeding of a Diacetyl-Free Yeast: The production of vicinal diketone aroma compounds during wort fermentation, especially diacetyl and pentanedione, is of considerable importance to the brewing industry as consumers prefer low levels of these in beer (37). Diacetyl and pentanedione are not actually produced by yeast, but are side products of the isoleucine-valine pathway, shown in Figure 6. Excess acetohydroxy acids formed by acetohydroxyacid synthase are excreted during the primary fermentation into the wort. A nonenzymatic decarboxylation follows with acetolactate being converted into diacetyl and acetohydroxybutyrate to pentanedione. A major reason for costly secondary fermentations in brewing is to reduce the levels of vicinal diketones during lagering. A genetically engineered brewing yeast which produces low levels of vicinal diketones would therefore be of great value to the brewing industry.

It is the personal opinion of this author that the first recombinant yeast to be used in brewing will be one which produces decreased levels of diacetyl. Four reasons lend to this opinion:

1. Both the biochemistry and genetics of diacetyl formation are well evaluated.
2. All genes involved in the isoleucine-valine pathway have been cloned.
3. A wide array of recombinant-based strategies are available to construct diacetyl-free/reduced yeasts.
4. Extracellular enzymes are not required to decrease diacetyl levels, eliminating concerns over postfermentation enzymes in the finished product.

As shown in Figure 6, five enzymes catalyze the biosynthesis of isoleucine and valine. The first enzyme, threonine deaminase (encoded by *ILV1*), converts threonine to α-ketobutyrate. The remaining four enzymes are shared with the biosynthetic pathway for valine, converting α-ketobutyrate to isoleucine and pyruvate to valine. The acetohydroxyacid synthetase is coded for by *ILV2*, acetohydroxyacid reductoisomerase by *ILV5*, and dihydroxy acid dehydrase by *ILV3*. All enzymes appear to be located in the yeast mitochondria in an enzyme complex (107). Two alleles for each of these loci have been established to exist in Carlsberg lager strain 244. *ILV1* (108), *ILV2* (66), *ILV3* (109), and *ILV5* (109) from *S.cerevisiae* have been cloned as well as the alleles of *ILV1* (110), *ILV2* (41), and *ILV3* (15) unique to the lager yeast. These clones are available for genetic engineering purposes and possible strategies include:

1. Changes in the regulatory sequences in front of *ILV1* and *ILV2* genes can be introduced by gene replacement to reduce the levels of threonine deaminase and acetohydroxyacid synthase.
2. The activity of the acetohydroxyacid synthase enzyme can be decreased by in vitro mutagenesis of the coding regions of the two *ILV2* genes for in vivo gene replacement, thereby reducing the amount of acetolactate formed.
3. The levels of reductoisomerase or dehydrase may be enhanced by mutations in front of *ILV5* or *ILV3*, respectively, or by inserting genes allowing amplification of the copy number of the genes. Both strategies would enhance the flow of the acetohydroxyacid intermediates into isoleucine and valine.
4. Screening for clones with altered acetohydroxyacid synthase activity, selected on the basis of sulfometuron methyl resistance.
5. Regulation of *ILV2* gene expression by the use of antisense RNA to one or both alleles of *ILV2*.

Of the above strategies, those involving gene replacement of regulatory sequences are perhaps the most far off. While the alleles of *ILV1* (111), *ILV2* (66), and

ILV5 (112) shared between lager yeast 244 and *S.cerevisiae* haploids have been sequenced and mapped for transcripts, of the alleles unique to the lager yeast only *ILV1* has been sequenced (113). Complete sequencing data of regulatory sequences in front of all alleles of *ILV2* and *ILV5* will be needed before a comprehensive attempt at strain engineering by this approach can be initiated.

Carlsberg is apparently taking a three-pronged approach toward the diacetyl question. One involves the construction of in vitro deletion mutations of both *ILV2*alleles for substitution into the lager yeast by gene replacement (41). No reports describing the progress of this approach have been published, although gene replacement of the *S.cerevisiae* allele has been accomplished (C. Gjermansen, personal communication). The second involves the amplification of *ILV5* copy number (41). To date, this approach has only been studied using 2-μm based YEp vectors with laboratory strains of *S.cerevisiae*, where diacetyl production was cut in half (41). Dillemans et al. (114) also saw similar levels of reduction in diacetyl in haploid yeast transformed with a multicopy plasmid vector. Presumably simple modification of the vector to make it amenable to the 2-μm based approach of Hinchliffe et al. (115) (where *ILV5* would be deposited and amplified into the native 2-μm plasmid without the incorporation of bacterial sequences) could result in a "natural" lager yeast producing significantly lower levels of diacetyl. The third approach is to select for clones with spontaneous dominant mutations resistant to the herbicide sulfometuron methyl, as these have altered AHAS structure and activity (65). While this promising approach has been discussed earlier, it is necessary to mention that Galvan et al. (116) found no significant reduction in diacetyl production in sulfometuron methyl-resistant clones.

Regulation of *ILV2* gene expression by the use of antisense RNA has been tested by Xiao and Rank (117) in haploid laboratory yeasts. Results were promising as the antisense RNA to *ILV2* under *GAL10* control was able to repress cellular growth and produce a bradytrophic auxotroph fully revertable by the addition of isoleucine and valine. Inhibition of ALS enzyme levels reached up to 60%, clearly indicating the potential of this approach for engineering brewer's yeasts producing reduced levels of diacetyl (although it would be necessary to replace *GAL10* control with that of promoter from a highly expressed constitutive yeast gene, e.g., *ADH1* or *PGK*).

Before leaving this section it must be pointed out that perhaps the most practical approach to constructing diacetyl-free yeast is not to focus on the biochemistry and/or genetics of the pathway itself, but instead to introduce into a yeast the enzyme α-acetolactate decarboxylase, α-ALDC (absent in yeast but present in *Enterobacter* and *Lactobacillus* bacteria), which converts free acetolactate directly into acetoin. As acetoin has a much higher taste threshold than diacetyl, lagering times could be significantly reduced.

During the early 1980s, studies with externally added α-ALDC from *Enterobacter aerogenes* and *Lactobacillus casei*, researchers at Carlsberg (118, 119) found that beer lagering could be replaced by a 24-hr incubation of primary beer with α-ALDC (also proving that diacetyl production is the main function of lagering). In 1987, Jensen et al. (120), of Novo, found that 0.02 mg/liter of pure α-ALDC from *B.brevis*, added at the time of pitching, could reduce diacetyl levels to below the taste threshold (0.15 ppm) by the end of the primary fermentation. At last check, however, there is no commercial supply of α-ALDC produced by Novo.

With the philosophy that whatever a chemical supplier cannot provide, a geneticist can clone, researchers at VTT in Finland (121) and Kirin Breweries in Japan (122-124) went on to clone the α- ALDC gene from *Enterobacter aerogenes*. In the latter case, Sone et al. (123, 124) demonstrated in 11°P wort fermentations that α-ALDC on a YEp-type vector resulted in a brewer's yeast which produced 140 ppm diacetyl after the primary fermentation (compared to 1,040 ppm in the control) (124). Importantly, no significant differences in fermentation properties (i.e., attenuation, rate of fermentation, or volatiles profile) was seen between the transformant and the control yeast. While illustrating the potential, attempts to integrate only the α-ALDC gene into the yeast genome proved unsuccessful (90).

Penttila et al. (121), however, from VTT's Biotechnical Laboratory in Finland, were able to integrate an α-ALDC expression cassette (under yeast *PGK* or *ADC1*promoters and terminators) into the genomes of commercial brewing strains (using selection for copper resistance). In 50 l pilot-scale fermentations several yeast transformants produced no detectable amount of diacetyl during the primary fermentation. When bottled after the primary fermentations, and compared to control beers, the quality and flavor was judged to be indistinguishable to that of the control. In theory, this construct could completely circumvent the need for lagering cellars. At this time, it is not clear where the expression cassette was integrated and whether *E.coli* sequences were included. Also, it is not certain how regulatory agencies or brewery owners will view the use of an *Enterobacter aerogenes* gene in a food- and beverage-grade yeast (although this species can be found in Dutch lambic fermentations). Perhaps a better source of α-ALDC is that from *Streptococcus lactis*, already used in the cheese-making industry. Goelling and Stahl (125) have recently cloned the α-ALDC gene from the organism with the ultimate intention of introducing into brewer's yeast.

Proteolytic Brewer's Yeast. In 1987 Young and Hosford (126) described the introduction of an extracellular protease gene, cloned from a wild yeast, into an ale yeast. The intent was to produce a yeast with chill-proofing activity, thereby eliminating the need for adding papain to treat chill-haze (for breweries using this practice). The protease gene was incorporated into mini-chromosome plasmid YCp50, with the G418 resistance gene added as the dominant selection marker. In 1.25 liter EBC tall-tube fermentations of control and transformant yeasts, fermentation profiles were found to be essentially the same. Unexpectedly, however, the transformant yeast had much higher levels of diacetyl at the end of primary fermentation than the control yeast. This, combined with the presence of bacterial sequences of YCp50, severely limits the practical application of this construct.

Future Prospects for Recombinant Brewers Yeast. The perspective of a North American brewer's yeast geneticist toward recombinant DNA technology is much different than that of his or her peers elsewhere in the world. Close examination of the last several sections reveals that aggressive attempts to construct recombinant brewer's yeasts are underway in Denmark, Finland, Great Britain, and Japan. Conspicuously absent from this list are the United States and Canada, where (as far as this author is aware) no major brewery is conducting recombinant DNA research. What is the reason for this and why, even in countries where breweries have a strong genetics program, are there no recombinant brewer's yeasts in use today?

Possible answers to this question are listed in Table 6. Perhaps the most important reason (and this is difficult to say as an industrial geneticist) is the relatively poor track record of brewing yeast genetics in producing "real-world" production yeast strains. Certainly it has taught us much concerning the genetic composition of brewer's yeasts, but by and large it has not generated an obvious return on investment from a head-office viewpoint. As considerable investments of financial and time resources are necessary to initiate and complete the construction of a recombinant yeast, the reluctance of brewery management to do so can therefore be appreciated.

Second, there is the very newness of the technology itself. Lack of suitable dominant selection markers delayed serious attempts at strain construction until 1984-1985, and we are only now seeing the first generation of recombinant brewing strains. That serious stability and regulatory concerns exist in many of these constructs cannot be disputed. However, it also cannot be disputed that rapid advances in recombinant DNA technology are providing practical solutions to all of these concerns. The first generation of some constructs have already proven the

Table 6. Factors Hindering The Introduction Of Recombinant DNA Technology In Brewer's Yeast Strain Selection Programs

1. Poor track record of industrial yeast genetics
2. Newness of rDNA technology with brewer's yeasts
3. Uncertain government regulatory approval
4. Availability of alternative traditional solutions
5. Inadequate knowledge concerning critical areas of yeast biochemistry and brewing science
6. Patent complications
7. Concerns over consumer acceptance

feasibility of the technology. The second generation will prove that recombinant DNA can produce real-world production strains.

Government regulatory approval will of course have to be a prerequisite for introducing recombinant yeast into brewing. While no petitions are under consideration by the Food and Drug Administration (FDA) in the United States, the Brewing Research Foundation in the United Kingdom has petitioned the Ministry of Agriculture, Fisheries and Food for permission to test the use of a recombinant amylolytic yeast (B. Lanchashire, personal communication). In many countries, regulatory guidelines for recombinant organisms are just now evolving and will likely be in place when the next generation of recombinant constructs are ready to be tested. It is the belief of this author that approval will eventually be given for the use of recombinant yeast in brewing, as government-approved field trials are already underway in many parts of the world with plant species modified by recombinant DNA technology. In the United Kingdom there is no reason in principle why approval would not be given to using a genetically modified yeast (127). This assumes, of course, that the recombinant yeast contains "natural" DNA sequences, not cloned human DNA as has been recently proposed (128).

Fourth, there are some instances in which the technology itself may not be appropriate. For example, while β-glucanase and glucoamylase-positive strains of brewer's yeasts are within reach, perhaps cheaper traditional brewing solutions already exist. The modern brewmaster has proven to be extremely versatile over the past 10 years, using the brewhouse very effectively to produce the diverse range of beers now commonly made by all breweries. The onus then is clearly on the geneticist to prove why recombinant approaches to improved product quality or plant efficiency should be considered over others.

Fifth, and a complication which should not be underestimated, is our lack of basic knowledge concerning many aspects of yeast biochemistry which influence beer flavor and yeast phenotypes relevant to brewing. A classic example is the phenomenon of yeast flocculation. While the *FLO1* gene has recently been cloned by Watari et al. (129) at Sapporo Breweries in Japan, use of this gene for genetic engineering purposes is severely handicapped by our lack of understanding of the chemistry and biochemistry of flocculation (although availability of the *FLO1* clone should assist efforts to elucidate the exact mechanism of yeast flocculation). Exactly how gene disruptions or replacement events influence the subtle chemistry of flavor compound formation is also not clear. It is difficult to accept that gene alterations only affect the locus and pathway in question, given the obvious complexity and interconnection of metabolic pathways. For lighter lager beers in particular, subtle differences in taste and/or flavor might prevent the implementation of an otherwise acceptable recombinant yeast. For this reason, recombinant strategies should be as focused as possible (e.g., reducing diacetyl production by the use of α-ALDC rather than altering genes within the isoleucine-valine pathway) in order to have the greatest chance of practical success.

Concerns over patent rights and intellectual property are also likely limiting research progress. Many of the techniques, genes, and/or vectors used in recombinant work are patented or have patent applications on file. For example, Biotechnica International of Cambridge, Massachusetts, holds a European patent on the use of any recombinant amylolytic brewer's yeast (R. Yocum, personal communication). This certainly clouds the picture from a management viewpoint, especially if patents stipulate a royalty on production volume, not a flat rate fee.

Last, and perhaps the bottom line concerning the use of a recombinant yeast, is uncertainty over how the consumer would react to the use of a recombinant yeast. Certainly brewing by nature is a very conservative industry and consumers have become accustomed to the wholesome, traditional image of beer. While they may be willing to ingest pharmaceuticals produced by a recombinant yeast, the consumption of a recombinant-made beer may be another matter (especially if effectively exploited by negative marketing on the part of competitors). For this reason, most breweries will likely adopt a wait and see approach to recombinant yeast even after all technical and regulatory concerns have been eliminated. However, should a major brewery use a recombinant yeast in the 1990s, without consumer backlash, to gain a significant global advantage in the economics of brewing, the impact on strain selection could be equal to that caused by Hansen some 100 years earlier.

REFERENCES

1. Pasteur, M.L., Etudes sur la biere, Gauthier-Villars, Paris, pp. 1-383 (1876).
2. Wettstein, D. von, Emil Christian Hansen Centennial Lecture: From pure yeast culture to genetic engineering of brewers yeast. Proc. Eur. Brew. Conv., pp. 97-119 (1983).
3. Hansen, E.C., Recherches sur la physiologie et al morphologie des ferments alcooliques II-IV. Meddelelser fra Carlsberg Laboratoriet, 2, 29-102 (1883).
4. Winge, O., and A. Hjort, On some *Saccharomycetes* and other fungi still alive in the pure cultures of Emil Chr. Hansen and Alb. Klocker. Comp. Rend. Lab. Carlsberg, Ser. Physiol., 21:51-58 (1935).
5. Winge, O., On haplophase and diplophase in some *Saccharomycetes*. Comp. Rend. Lab. Carlsberg, Ser. Physiol. 21:77-111 (1935).
6. Winge, O., and O. Lausten, On two types of spore germination, and on genetic segregations in *Saccharomyces* demonstrated through single-spore cultures. Comp. Rend. Lab. Carlsberg, Ser. Physiol., 22:99-116 (1937).
7. Mortimer, R.K., and J.R. Johnston, Geneology of principal strains of the yeast genetic stock center. Genetics, 113:35-43 (1986).
8. Spencer, J.F.T., and D.M. Spencer, Genetic improvement of industrial yeasts. Ann. Rev. Microbiol., 37:121-142 (1983).
9. Gjermansen, C., and P. Sigsgaard, Construction of a hybrid brewing strain of *Saccharomyces carlsbergensis* by mating of meiotic segregants. Carlsberg Res. Commun., 46:1-11 (1981).
10. Nilsson - Tillgren, T., C. Gjermansen, M.C. Kielland - Brandt, J.G.L. Petersen, and

S. Holmberg, Genetic differences between *Saccharomyces carlsbergensis* and *S.cerevisiae*. Analysis of chromosome III by single chromosome transfer. Carlsberg Res. Commun., 46:65-76 (1981).

11. Conde, J., and G.R. Fink, A mutant of *Saccharomyces cerevisiae* defective for nuclear fusion. Proc. Natl. Acad. Sci. U.S.A., 73:3651-3655 (1976).

12. Nilsson - Tillgren, T., C. Gjermansen, S. Holmberg, J.G.L. Petersen, and M.C. Kielland- Brandt, Analysis of chromosome V and the *ILV1* gene from *Saccharomyces carlsbergensis*. Carlsberg Res. Commun., 51:309-326 (1986).

13. Casey, G.P., Molecular and genetic analysis of chromosome X in *Saccharomyces carlsbergensis*. Carlsberg Res. Commun., 51:327-341 (1986).

14. Petersen, J.G.L., T. Nilsson-Tillgren, M.C. Kielland-Brandt, C. Gjermansen, and S. Holmberg, Structural heterozygosis at genes *ILV2* and *ILV5* in *Saccharomyces carlsbergensis*. Curr. Genet., 12:161-174 (1987).

15. Casey, G.P., Cloning and analysis of two alleles of the *ILV3* gene from *Saccharomyces carlsbergensis*. Carlsberg Res. Commun., 51:327-341 (1986).

16. Holmberg, S., Genetic differences between *Saccharomyces carlsbergensis* and *S.cerevisiae* II. Restriction endonuclease analysis of genes in chromosome III. Carlsberg Res. Commun., 47:233-244 (1982).

17. Pedersen, M.B., DNA sequence polymorphisms in the genus *Saccharomyces*. I Comparison of the *HIS4* and ribosomal RNA genes in lager strains, ale strains and various species. Carlsberg Res. Commun., 48:485-503 (1983).

18. Pedersen, M.B., DNA sequence polymorphisms in the genus *Saccharomyces*. II Analysis of the genes *RDN1, HIS4, LEU2*, and *Ty* transposable elements in Carlsberg, Tuborg and 22 Bavarian brewing strains. Carlsberg Res. Commun., 50:263- 272 (1985).

19. Pedersen, M.B., DNA sequence polymorphisms in the genus *Saccharomyces*. III Restriction endonuclease fragment patterns of chromosomal regions in brewing and other yeast strains. Carlsberg Res. Commun., 51:163-183 (1986).

20. Pedersen, M.B., DNA sequence polymorphisms in the genus *Saccharomyces*. IV. Homeologous chromosomes III of *Saccharomyces bayanus, S.carlsbergensis* and *S.uvarum*. Carlsberg Res. Commun., 51:185-202 (1986).

21. Casey, G.P., and M.B. Pedersen, DNA sequence polymorphisms in the genus *Saccharomyces*. V. Cloning and characterization of a *LEU2* gene from *S.carlsbergensis*. Carlsberg Res. Commun., 53:209-219 (1988).

22. Deguin, S., R. Gloeckler, C.J. Herbert, and F. Boutelet, Cloning, sequencing and analysis of the yeast *S.uvarum ERG10* gene encoding acetoacetyl CoA thiolase. Curr. Genet., 13:471-478 (1988).

23. L'Hote, H., S. Deguin, and F. Boutelet, Identification of some chromosomes of the brewing yeast *Saccharomyces uvarum*. FEMS Microbiol.Lett., 52:219-224 (1988).

24. Schwartz, D.C., and C.B. Cantor, Separation of yeast chromosome-size DNA's by pulsed field gradient gel electrophoresis. Cell, 37:67-75 (1984).

25. Carle, G.F., and M.V. Olson, Separation of chromosome DNA molecules from yeast by orthogonal-field alternation gel electrophoresis. Nucl. Acids Res., 12:5647-5664 (1984).

26. Carle, G.F., M. Frank, and M.V. Olson, Electrophoretic separations of large DNA molecules by periodic inversion of the electric field. Science, 232:65-68 (1986).

27. Chu, G., D. Vollrath, and R.W. Davis, Separation of large DNA molecules by contour-clamped homogenous electric fields. Science, 234:1582-1585 (1986).

28. Stewart, G., A. Furst, and N. Avdalovic, Transverse alternating field electrophoresis (TAFE). Biotechniques, 6:68-75 (1988).

29. Pedersen, M.B., Practical use of electro-karyotypes for brewing yeast identification. Proc. Eur. Brew. Conv., pp. 489-496 (1987).

30. Rosini, G., F. Federico, A.E. Vaughan, and A. Martini, Systematics of the species of the yeast genus *Saccharomyces* associated with the fermentation industry. European J. Appl. Microbiol. Biotechnol., 15:188-193 (1987).

31. Nilsson-Tillgren, T., M.C. Kielland - Brandt, S. Holmberg, J.G.L. Petersen and C. Gjermansen, Is lager yeast a species hybrid? Utilization of intrinsic genetic variation in breeding. Proc. IVth Int. Symp. Genet. Indust. Microorganisms, pp. 143-148 (1982).

32. Sigsgaard, P., and J.N. Rasmussen, Screening of the brewing performance of new yeast strains. Am. Soc. Brew. Chem. J., 43:104-108 (1985).

33. Enari, T.M., Method 2.5.4 Tubes E.B.C., E.B.C. Analytica Microbiologica, J. Inst. Brew, 83:117-118 (1977).

34. Thorne, R.S., Some observations on yeast mutation during continuous fermentation. J. Inst. Brew., 74:516-524 (1968).

35. Kirsop, B.E., Culture and preservation In: Living Resources For Biotechnology: Yeasts. Kirsop, B.F., and C.P. Kurtzman (Eds.). Cambridge University Press, pp. 74-98 (1988).

36. Aigle, M., Erbs, D. and M. Moll, Some molecular structures in the genome of lager brewing yeasts. Am. Soc. Brew. Chem. J., 42:1-6 (1984).

37. Cabane, B.C., C. Ramos-Jeunehomme, N. Lapage, and C.A. Masschelein, Vicinal diketones—the problem and prospective solutions. Am. Soc. Brew. Chem. Proc., pp. 94-99 (1974).

38. Molzahn, S.W., A new approach to the application of genetics to brewing yeast. Am. Soc. Brew. Chem. Proc., pp. 54-59 (1977).

39. Jones, R.M., I. Russell, and G.G. Stewart, The use of catabolite derepression as a means of improving the fermentation rate of brewing yeast strains. Am. Soc. Brew. Chem. J., 44:161-166 (1986).

40. Heredia, M.F., and C.F. Heredia, *Saccharomyces cerevisiae* acquires resistance to 2-deoxyglucose at a very high frequency. J. Bact., 170:2870-2872 (1988).

41. Kielland-Brandt, M.C., C. Gjermansen, T. Nilsson-Tillgren, S. Holmberg, and J.G.L. Petersen, Diacetyl and brewer's yeast. Yeast, 4:Special Issue, p. 470 (1988).

42. Winge, O., and Lausten, O. On 14 new yeast types, produced by hybridization. Compt. Rend. Lab. Carlsberg, Ser. Physiol., 22:337-352 (1939).

43. Emeis, C.C., A new hybrid yeast for the fermentation of wort dextrins. Am. Soc. Brew. Chem. Proc., 29:58-62 (1971).

44. Bilinski, C.A., I. Russell, and G.G. Stewart, Analysis of sporulation in brewers yeast:induction of tetrad formation. J. Inst. Brew., 92:594-598 (1986).

45. Bilinski, C.A., and G.P. Casey, Developments in sporulation and breeding of brewers yeast. Yeast, 5:429-438 (1989).

46. Tsuboi, M., and T. Takahashi, Genetic analysis of the non-sporulating phenotype of brewers yeasts. J. Ferment. Technol. 66:605-613 (1988).

47. Gunge, N., and Y. Nakatomi, Genetic mechanisms of rare matings of the yeast *Saccharomyces cerevisiae* heterozygous for mating type. Genetics, 70:41-58 (1972).
48. Spencer, J.F., and Spencer, D.M., Hybridization of non-sporulating and weakly sporulating strains of brewers and distillers yeasts. J. Inst. Brew., 83:287-289 (1977).
49. Tubb, R.S., B.A. Searle, A.R. Goodey, and A.J. Brown, Rare mating and transformations for construction of novel brewing yeasts. Proc. Eur. Brew. Conv., pp. 487-496 (1981).
50. Young, T.W., The genetic manipulation of killer character into brewing yeast. J. Inst. Brew., 87:292-295 (1981).
51. Hammond, J.R., and K.W. Eckersley, Fermentation properties of brewing yeasts with killer character. J. Inst. Brew., 90:167-177 (1984).
52. Van Solingen, P., and J.B. van der Plaat, Fusion of yeast spheroplasts. J. Bacteriol., 130:946-947 (1977).
53. Halfmann, H.J., W. Rocken, C.C. Emeis, and U. Zimmermann, Transfer of mitochondrial function into a cytoplasmic respiratory-deficient mutant of *Saccharomyces* by electro-fusion. Curr. Genet., 6:25-28 (1982).
54. Seki, T., S. Myoga, S. Limtong, S. Uedono, J. Kumnuanta, and H. Taguchi, Genetic construction of yeast strains for high ethanol production. Biotechnol. Lett. 5:351-356 (1983).
55. Freeman, R.F., Construction of brewing yeasts for production of low carbohydrate beers. Proc. Eur. Brew. Conv., pp. 497-504 (1981).
56. Russell, I., I. Hancock, and G.G. Stewart, Construction of dextrin fermenting yeast strains that do not produce phenolic off-flavors in beer. J. Amr. Soc. Brew. Chem., 41:45-51 (1983).
57. Stewart, G.G., The genetic manipulation of industrial yeast strains. Can. J. Microbiol., 27:973-990 (1981).
58. Hoffmann, M., M. Zimmermann, and C.C. Emeis, OFAGE as a means for the analysis of somatic hybrids obtained by protoplast fusion of different *Saccharomyces* strains. Curr. Genet., 11:599-603 (1987).
59. Stewart, G.G., I. Russell, and C.J. Panchal, Genetically stable allopolyploid somatic fusion product useful for the production of fuel alcohols. Canadian Patent No. 1, 199, 593 (1986).
60. Hinnen, A., J.B. Hicks, and G.R. Fink, Transformation of yeast. Proc. Natl. Acad. Sci. U.S.A., 75:1929-1933 (1978).
61. Sturley, S.L., and T.W. Young, Genetic manipulation of commercial yeast strains. Biotechnol. Genet. Eng. Rev., 4:1-31 (1986).
62. Schwab, H., Strain improvement in industrial microorganisms by recombinant DNA techniques. Adv. Biochem. Eng. Biotechnol., 37:129-161 (1988).
63. Fincham, J.R., Transformation in fungi. Microbiol. Rev., 53:148-170 (1989).
64. Rank, G.H., G.P. Casey, and W. Xiao, Gene transfer in industrial *Saccharomyces* yeasts. Food Biotechnol., 2:1-41 (1988).
65. Falco, S.C., and K.S. Dumas, Genetic analysis of mutants of *Saccharomyces cerevisiae* resistant to the herbicide sulfometuron methyl. Genetics, 109:21-35 (1985).
66. Falco, S.C., K.S. Dumas, and K.J. Livak, Nucleotide sequence of the yeast *ILV2* gene which encodes acetolactate synthase. Nucl. Acids Res., 13:4011-4027 (1985).
67. Casey, G.P., W. Xiao, and G.H. Rank, A convenient dominant selection marker for

gene transfer in industrial strains of *Saccharomyces* yeast: *SMR1* encoded resistance to the herbicide sulfometuron methyl. J. Inst. Brew., 94:93-97 (1988).

68. Fogel, S., and J.W. Welch, Tandem gene amplification mediates copper resistance in yeast. Proc. Natl. Acad. Sci. U.S.A., 79:5342-5346 (1982).

69. Jimenez, A., and J. Davies, Expression of a transposable antibiotic resistance element in *Saccharomyces*. Nature, 287:869-871 (1980).

70. Sakai, K., and M. Yamamoto, Transformation of the yeast *Saccharomyces carlsbergensis*, using an antibiotic resistance marker. Agr. Biol. Chem., 50:1177-1182 (1986).

71. Zhu, J., R. Contreras, and W. Fiers, Construction of stable laboratory and industrial yeast strains expressing a foreign gene by integrative transformation using a dominant selection system. Gene, 50:225-237 (1986).

72. Gritz, L., and J. Davies, Plasmid-encoded hygromycin B resistance: the sequence of hygromycin B phosphotransferase gene and its expression in *Escherichia coli* and *Saccharomyces cerevisiae*. Gene, 25:179-188 (1983).

73. Hadfield, C., J.A. Kirkham, and W.E. Lanchashire, Dominant marker replacement: a general method for introducing new genetic material into yeast chromosomes without unwanted accompanying sequences. Yeast, 2:Special Issue, p. S142 (1987).

74. Hadfield, C., A.M. Cashmore, and P.A. Meacock, Sequence and expression characteristics of a shuttle chloramphenicol-resistance marker for *Saccharomyces cerevisiae* and *Escherichia coli*. Gene, 52:59-70 (1987).

75. Xiao, W. and G.H. Rank, The construction of recombinant industrial yeast free of bacterial sequences by directed gene replacement into a non-essential region of the genome. Gene, 76:99-107 (1989).

76. Gatignol, A., M. Baron, and G. Tiraby, Phleomycin resistance encoded by the *kle* gene transposon Tn5 as a dominant selectable marker in *Saccharomyces cerevisiae*, Mol. Gen. Genet., 207:342-348 (1987).

77. Tubb, R.S., and P.L. Liljestrom, A colony- colour method which differentiates α-galactosidase - positive strains of yeast. J. Inst. Brew., 92:588-590 (1986).

78. Meaden, P., K. Ogden, H. Bussey, and R.S. Tubb, a *DEX* gene conferring production of extracellular amyloglucosidase on yeast. Gene, 34:325-334 (1985).

79. Tubb, R.S., Gene Technology for industrial yeasts. J. Inst. Brew., 93:91-96 (1987).

80. Lolle, S., N. Skipper, H. Bussey, and D.Y. Thomas, The expression of cDNA clones of yeast M1 double-stranded RNA in yeast confers both killer and immunity phenotypes. EMBO J., 3:1383-1387 (1984).

81. Bussey, H., and P. Meaden, Selection and stability of yeast transformants expressing cDNA of a M1 killer toxin-immunity gene. Curr. Genet. 9:285-291 (1985).

82. Thomas, D.Y., N. Skipper, P.C.K. Lou, S. Lolle, and H. Bussey, Production and secretion of proteins and polypeptides in yeast. Canadian Patent 479062 (1987).

83. Toh-e, A., P. Guerry-Kopecko, and R.B. Wickner, A stable plasmid carrying the yeast *LEU2* gene and containing only yeast deoxyribonucleic acid. J. Bacteriol., 141:413-416 (1980).

84. Fagan, M.C., and J.F. Scott, New vectors for construction of recombinant high-copy-number yeast acentric-ring plasmids. Gene, 40:217-229 (1985).

85. Panchal, C.J., L. Bast, T. Dowhanick, J. Johnstone, and G.G. Stewart, Studies on stability of miniplasmids comprised of only yeast DNA. Curr. Genet., 12:15-20 (1987).

86. Orr-Weaver, T.L., J.W. Szostak, and R.J. Rothstein, Yeast transformation: a model system for the study of recombination. Proc. Natl. Acad. Sci. U.S.A., 78:6354-6358 (1981).

87. Yocum, R.R., Genetic engineering of industrial yeasts. In: Proceedings of Biology Exposition 1986, Butterworth, Stoneham, MA, pp. 171-180.

88. Rothstein, R.J., One step gene disruption in yeasts. Methods Enzymol., 101:202-211 (1983).

89. Goebl, M.G., and T.D. Petes, Most of yeast genomic sequences are not essential for cell growth and division. Cell, 46:983-992 (1986).

90. Fujii, T., K. Kondo, H. Sone, J. Tanaka, and T. Inoue, Expression of α-acetolactate decarboxylase in brewers yeast. Yeast, 4:Special Issue, p. 5466 (1988).

91. Boeke, S.D., Xu, H., and G.R. Fink, A general method for the chromosomal amplification of genes in yeast. Science, 239:280-282 (1988).

92. Chinery, S.A., and E. Hinchliffe, A novel class of vector for yeast transformation. Curr. Genet., 16:21-25 (1989).

93. Cantwell, B.A., and D.J. McConnell, Molecular cloning and expression of Bacillus subtilis β-glucanase gene in *Escherichia coli*. Gene, 23:211-219 (1983).

94. Borriss, R., H. Baeumlein, and J. Hofemeister, Expression in *Escherichia coli* of a cloned β-glucanase gene from *Bacillus amyloliquefacians*. Appl. Microbiol. Biotechnol., 22:63-71 (1985).

95. Jackson, E.A., G.M. Ballance, and K.K. Thomsen, Construction of a yeast vector directing the synthesis and release of barley (1-3, 1-4) - β-glucanase. Carlsberg Res. Commun., 51:445-458 (1986).

96. Enari, T.-M., J. Knowles, U. Lehtinen, M. Nikkola, M. Pentilla, M.L. Suikho, S. Hime and A. Vipola, Glucanolytic brewer's yeast. Proc. Eur. Brew. Conv., pp. 529-536 (1987).

97. Forage, A., Application of genetic engineering. Monat. fur Brau., 36:172 (1983).

98. Hinchliffe, E., β-Glucanase: the successful application of genetic engineering. J. Inst. Brew., 91:384-389 (1985).

99. Hinchliffe, E., and C.J. Daubney, The genetic modification of brewing yeast with recombinant DNA. J. Am. Soc. Brew. Chem., 44:98-101 (1986).

100. Cantwell, B.A., T. Ryan, J.C. Hurley, M. Doherty, and D.J. McConnell, Regulation of barley β-glucan by brewers yeast. Eur. Brew. Conv. Monograph - XII, pp. 186-198 (1986).

101. Lancashire, W.E., and R.J. Wilde, Secretion of foreign proteins by brewing yeasts. Proc. Eur. Brew. Conv., pp. 513-520 (1987).

102. Olsen, O., and K.K. Thomsen, Processing and secretion of barley (1-3, 1-4) - β-glucanase in yeast. Carlsberg Res. Commun., 54:29-40 (1989).

103. Borriss, R., O. Olsen, K.K. Thomsen, and D. von Wettstein, Hybrid *Bacillus* endo-(1-3, 1-4) - β-glucanases: construction of recombinant genes and molecular properties of the gene products. Carlsberg Res. Commun., 54:41-54 (1989).

104. Meaden, P.G., and R.S. Tubb, A plasmid vector system for the genetic manipulation of brewing strains. Proc. Eur. Brew. Conv., pp. 219-226 (1985).

105. Perry, C., and P. Meaden, Properties of a genetically-engineered dextrin-fermenting strain of brewers yeast. J. Inst. Brew., 94:64-67 (1988).

106. Howard, J.J., R.J. Wilde, A.T. Carter, J.A. Kirkham, D.J. Best, and W.E. Lancashire,

Amylolytic activities secreted by yeast of the genus *Schwanniomyces*. Yeast, 4:Special Issue, p. S467 (1988).

107. Ryan, E.D., and G.B. Kohlhaw, Subcellular localization of isoleucine-valine biosynthetic enzymes in yeast. J. Bacteriol., 120:631-637 (1974).

108. Petersen, J.G.L., S. Holmberg, T. Nilsson - Tillgren, and M.C. Kielland - Brandt, Molecular cloning and characterization of the threonine deaminase (*ILV1*) gene of *Saccharomyces cerevisiae*, Carlsberg Res. Commun., 48:149-159 (1983).

109. Polaina, J., Cloning of the *ILV2*, *ILV3* and *ILV5* genes of *Saccharomyces cerevisiae*. Carlsberg Res. Commun., 49:577-584 (1984).

110. Petersen, J.G.L., Molecular genetics of diacetyl formation in brewers yeast. Proc. Eur. Brew. Conv., pp. 275-282 (1985).

111. Kielland-Brandt, M.C., S. Holmberg, T. Nilsson-Tillgren, and J.G.L. Petersen, Nucleotide sequence of the gene for threonine deaminase (*ILV1*) of *Saccharomyces cerevisiae*. Carlsberg Res. Commun., 49:567-575 (1984).

112. Petersen, J.G.L., and S. Holmberg, The *ILV5* gene of *Saccharomyces cerevisiae* is highly expressed. Nucleic Acids Res., 14:9631-9651 (1986).

113. Nilsson-Tillgren, T.C., S. Holmberg, J.G.L. Petersen and M.C. Kielland-Brandt, Analysis of chromosome V and the *ILV1* gene from *Saccharomyces carlsbergensis*. Carlsberg Res. Commun., 51:309-326 (1986).

114. Dillemans, M., E. Goossens, O. Goffin, and C.A. Masschelein, The amplification effect of the *ILV5* gene on the production of vicinal diketones in *Saccharomyces cerevisiae*. J. Am. Soc. Brew. Chem., 45:81-84 (1987).

115. Hinchliffe, E., C.J. Fleming, and D. Vakeria, A novel "stable" gene maintenance system for brewing yeast. Proc. Eur. Brew. Conv., pp. 505-512 (1987).

116. Galvan, L., A. Perez, M. Delgado, and J. Conde, Diacetyl production by sulfometuron resistant mutants of brewing yeast. Proc. Eur. Brew. Conv., pp. 385-392 (1987).

117. Xiao, W., and G.H. Rank, Generation of an *ilv* bradytrophic phenocopy in yeast by antisense RNA. Curr. Genet., 13:283-289 (1988).

118. Godtfredsen, S.E., and M. Ottesen, Maturation of beer with α-acetolactate decarboxylase. Carlsberg Res. Commun., 47:93-102 (1982).

119. Godtfredsen, S.E., H. Lorck, and P. Sigsgaard, On the occurrence of α-acetolactate decarboxylases among microorganisms. Carlsberg Res. Commun., 48:239-247 (1983).

120. Jensen, B.R., I. Svendsen, and M. Ottesen, Isolation and characterization of an α-acetolactate decarboxylase useful for accelerated beer maturation. Proc. Eur. Brew. Conv., pp. 393-400 (1987).

121. Penttila, M., M.L. Suihko, K. Blomqvist, M. Nikkola, J.K.C. Knowles, and T-M Enari, Construction of brewers yeast strains expressing bacterial α-ALDC genes: a revolution in the brewing industry. Yeast, 4:Special Issue, p. S473 (1988).

122. Sone, H., T. Fujii, K. Kondo, and J. Tanaka, Molecular cloning of the gene encoding α-acetolactate decarboxylase from *Enterbacter aerogenes*. J. Bacteriol., 5:87-91 (1987).

123. Sone, H., K. Kondo, T. Fujii, F. Shimuzu, J. Tanaka, and T. Inoue, Fermentation properties of brewers yeast having α-acetolactate decarboxylase gene. Proc. Eur. Brew. Conv. pp. 545-552 (1987).

124. Sone, H., T. Fujii, K. Kondo, F. Shimizu, J. Tanaka, and T. Inoue, Nucleotide

sequence and expression of the *Enterobacter aerogenes* α-acetolactate decarboxylase gene in brewers yeast. Appl. Env. Microbiol., 54:38-42 (1988).

125. Goelling, D., and U. Stahl, Cloning and expression of an α-acetolactate decarboxylase gene from *Streptococcus lactis sbsp. diacetylicus* in *Escherichia coli*. Appl. Env. Microbiol., 54:1889-1891 (1988).

126. Young, T.W., and E.A. Hosford, Genetic manipulation of *Saccharomyces cerevisiae* to produce extracellular protease. Proc. Eur. Brew. Conv., pp. 521-528 (1987).

127. Dunn, A.J., Food safety approval for genetically manipulated yeast strains. Proc. Eur. Brew. Conv., Monograph XII, pp. 223-234 (1986).

128. Hinchcliffe, E., E. Kenny, and A. Leaker, Novel products from surplus yeasts via recombinant DNA technology. Proc. Eur. Brew. Conv., Monograph XII, pp. 139-154 (1986).

129. Watari, J., Y. Takata, M. Ogawa, N. Nishikawa, and M. Kaminura, Molecular cloning of a flocculation gene in *Saccharomyces cerevisiae*. Agr. Biol. Chem., 53:901-903 (1989).

5

Wine Yeast: Selection and Modification

Ronald Ernest Subden
University of Guelph
Guelph, Ontario, Canada

I. INTRODUCTION

A. Historical

This chapter provides a historical perspective on the use of pure starter culture yeasts for wine-making, parameters used for selection and the manipulatory techniques for genetically modifying existing wine yeast strains. More comprehensive treatments of specific aspects of the selection and manipulation of industrial yeasts and bacteria other reviews are available (1–8).

A provocative article by R. E. Kunkee (9) argues that notwithstanding 4,000 years of wine-making history, wine yeast selection began only 100 years ago, and attempts to modify wine yeast strains have a history of only two decades. He was referring to the fact that in 1866 Louis Pasteur published his "Etude sur Vin" (10) and 20 years later Hansen (11, 12) used selected yeast starter culture to pitch a wort while Mueller-Thurgau (13, 14) used a selected yeast starter culture to inoculate a must. The practice rapidly enjoyed widespread acceptance, then for vintners, widespread rejection. It was thought that the natural microflora, though less efficient, gave the wine greater complexity. The majority of wine fermentations in Europe and elsewhere now rely on selected inocula. The use of selected starter cultures was reintroduced primarily via the newer wine regions in North America, Australia, and South Africa. The rising sales of active dry wine yeast indicates that the situation in Europe, and Germany in particular, is still changing.

Selection in the early years was simple and empirical. Yeasts from wines with good fermentation profiles were retrieved and added to fresh must. The practice was often repeated, and eventually one or more phenotypes from certain species (usually *Saccharomyces cerevisiae*) would become dominant and persistent residents on winery equipment. It was thought that later when the pommace was ploughed back into the vineyard soil, the wine yeasts became a frequent constituent of the grape berry bloom to initiate the next cycle of selection. Initial selection procedures were based primarily on fermentation performance, and it was some years before there was yeast selection based on aroma or flavor criteria.

Lafon-Lafourcade (2) divided wine aroma into three classes: (a) the primary aroma of the grape, (b) the secondary aroma of fermentation, and (c) the tertiary bouquet that develops during aging. The components of the must dictate the potential for wine quality. The yeast can only metabolize those components present in the must, and each grape variety has a unique chemical composition. The process of aging and the elaboration of a barrel or bottle bouquet are a complex process, part of which involves acids and alcohols esterifying to produce new aromas not produced by either grape or yeast. The acids and alcohols present are directly attributable to the genetic constitution of the grape and yeast variety. Yeast metabolism is a greater determinant of aroma in young wines and "vin ordinaire" than it is in aged wines. The yeast therefore does play a role in the elaboration of aromas, but it is not a dominant one.

B. The Natural Microflora of Musts and Wines

Before discussing wine yeast selection protocols and genetic manipulation schemes, it is constructive to consider the inoculum and natural selection of wine fermentations without pure culture yeast starters. A number of references (2, 15–18) support the following description of grape berry, must, and wine microflora.

There are approximately 10^{11} microbes on the surface of a grape berry, of which about 10^3 are molds, 10^5 bacteria, and 10^{3-6} yeasts (15). The range in the frequency of yeasts is a function of climatic conditions, viticultural practices (use of sprays), general sanitation, and geography.

Up to 95% of the yeasts present on the bloom are *Hanseniaspora uvarm* (or its imperfect stage *Kloeckera apiculata*) and up to 40 different species make up the remaining 5%. Hanseniaspora and Kloeckera are lemon-shaped cells (apiculate) that practice bipolar budding, and their growth is not inhibited by 250 g sugar per liter of medium. In fresh unsulfited must they undergo short lag and log phases, followed by a very brief stationary phase before the number of live cells start to decline as a result of ethanol inhibition that occurs around 40 g/liter. As brief as their activity is, they produce significant quantities of metabolites other than ethanol—viz., acetic acid (approximately 1 g/liter), ethyl acetate (approximately 250 mg/liter) (19), amyl acetate, and glycerol (20). There has been a long, protracted debate over the value of these and other metabolites. Some claim it is the major source of complexity in the bouquet of natural fermentations, while others view them as undesirables and practice rigorous debourbage and must sulfiting to remove and inhibit the species.

Aerobic species of yeasts such as *Candida* spp. manage only one or two divisions before the redox potential of the wine decreases below the level required for their growth. They are not killed during fermentation, and species such as *C. vini, C. krusei, C. valida, C. plcherima, C. colliculoa, C. stellata,* and *C. melini* may reappear as a surface film if the SO_2 level is less than 150 mg/liter and oxygen is available. They may impart a "damp basement" smell to the wine and are not beneficial contributors to complexity. All are considered spoilage yeasts.

Perfect forms of some *Candida* spp. like *Pichia membanefaciens* are similar to *Candida* in their formation of films but produce considerable amounts of acetic acid, acetaldehyde, and a variety of acetate esters (20, 21). They are a greater spoilage problem than *Candida* because of their greater tolerance to SO_2 (250–500 mg/liter) and ethanol (up to 110 g/liter) (22, 23).

Brettanomyces is an insidious spoilage yeast that sporadically wreaks havoc, mostly in wines undergoing barrel aging. A particular severe outbreak occurred in the Napa Valley, California, in the mid 1970s. The yeast has a wide spectrum of substrates and though sensitive to SO_2 (100 mg/liter) it can tolerate 120 g ethanol per liter (21). It is one of a few wine yeasts naturally resistant to actidione (up to 20 mg/liter). Spoilage problems include the production of caproic, isobutyric, isovaleric acid, and ethyl acetate (24), giving the wine a peculiar fruity smell.

Hansenula anomala is a commonly found spoilage film yeast producing ethyl acetate and acetic acid (25). The film has a characteristic corduroy surface, and the ascospores are hat-shaped.

Another sporadic problem, *Saccharomycodes ludwigii*, is a large (25 µm long) apiculate yeast that is found in extremely low frequency on grapes. In unsanitary cellars the yeast can proliferate and become a severe problem, as it can ferment up to 16.8% v/v ethanol and is highly resistant to SO_2 (up to 600 mg/liter) (2). In bottles, the yeast has a tendency to flocculate and form compact sedimented particles (25).

Zygosaccharamyces bailii (syn. *Saccharomyces baillii, S. acidifiens, S. elegans*) is a bloom constituent (16, 17) that has a high tolerance for SO_2 (up to 500 mg/liter), ethanol (up to 110 g/liter), and sugar. It is a bothersome contaminant in valves, stop cocks, outlets, and other perturbations of wine tanks and pumping equipment. Though not producing particularly malodorous metabolites, it is a major cause of refermentation in bottles of sweet wines.

The fission yeasts *Schizosaccharomyces pombe* and *Sch. malidevorans* are infrequent bloom constituents (18). They are not competitive with *Saccharomyces* strains, as they have a long generation time and the optimal growth temperature is about 30°C (25). Because *Schizosaccharomyces* is extremely tolerant of SO_2 (up to 150 mg free SO_2/liter), it will grow in oversulfited musts where other yeasts are inhibited. In addition to pyruvic acid, acetaldehyde, and alpha-ketoglutarate, *Schizosaccharomyces* produces certain undesirable flavor metabolites (26). After fermentation with *Schizosaccharomyces* the wine may taste flat as a result of its ability to perform a complete malo-ethanolic fermentation on malate concentrations normally encountered in wine.

The genus *Saccharomyces* has several species of importance to wine makers. *S. cerevisiae* (syn. *S. elipsoideus, S. vini*) is the consummate wine yeast. It is found in low numbers on the grape bloom but proliferates rapidly to dominate the main fermentation. It is estimated that 98% of the yeasts present at the end of a fermentation are *S. cerevisiae* (27). The qualities that make *S. cerevisiae* such a good wine yeast will be discussed in the next section.

S. bayanus (syn. *S. oviformis, S. beticus, S. cheriensis, S. rouxii*) differs from *S. cerevisiae* in its inability to utilize glactose, its ability to form a film, and its tolerance to SO_2 and ethanol (up to 19% v/v). The numbers of *S. bayanus* rise toward the end of the fermentation. It is often used to restart stuck fermentations. It is a common cause of refermentation in bottles of sweet wines.

A review of the major fermentation and spoilage yeasts can be found elsewhere (2).

C. Natural Selection During Fermentation

Glucose and fructose are the primary carbon sources for any yeasts growing in must. The relationship between sugar concentration and the rate of ethanol production by yeasts can be expressed as a Monod-type equation:

$$v = V_{max} \frac{Cs}{(K_s + C_s)}$$

where v is the specific ethanol productivity (g ethanol/g cells/hr), C_s is the sugar concentration (g/liter), and K_s is a saturation constant. The K_s for most wine yeasts is 0.2–0.4 g glucose per liter (28, 29). Fresh must may be saturated with 9 g O_2 per liter, but the Crabtree effect represses respiration. Depending on the strain, hexose concentrations of 3–30 g/liter repress the production of the oxidative enzymes (catabolite repression) (28, 30), so only fermentation is possible in musts. At hexose concentrations > 150 g/liter the fermentative enzymes start to be inhibited and the conversion of sugar to ethanol is slowed (31, 32). Oxygen in must is usually depleted within hours as several active grape enzymes such as polyphenyloxidase require oxygen. Yeasts use O_2 for the synthesis of certain polyunsaturated lipids required as membrane constituents (33). Under these conditions there is selection against poor fermenters such as *Candida*.

During fermentation, there is a constant selection for more ethanol-tolerant yeasts. Ethanol inhibits yeast growth by denaturing some glycolytic enzymes and disrupting certain membrane structures (34, 35). Most yeasts are not affected by ethanol levels below 20 g/liter (36). *Hansenula anomala* will ferment slowly to about 35 g/liter and *Kloeckera apiculata* to 40 g/liter before growth and ethanol productivity are significantly reduced (37). Most wine strains of *S. cerevisiae* will ferment to about 110 g ethanol per liter at which time they may be succeeded by more ethanol-tolerant strains of *S. bayanus*, *Saccharomycodes ludwigii*, *Brettanomyces intermedius*, or *Zygosaccharomyces bailii*.

Acclimation or nongenetic factors may also play a role in selection. Starter yeast cultures grown aerobically have greater ethanol tolerance than those grown under anaerobic or partially anaerobic conditions. Certain "survival factors" are made during aerobic growth which promote ethanol tolerance under anaerobic conditions. The survival factors are believed to be ergosterol and related sterols possibly complexed with proteins (38–40).

Selection during fermentation may also be based on tolerance to various toxins released by yeasts during fermentation. Yeasts have been reported to release fatty acids which inhibit growth and fermentation (41). Reversal of the inhibition can be accomplished if one removes the fatty acids by binding with yeast ghosts. In addition to fatty acid toxins, wine yeasts and wild yeasts may cytoplasmically harbor a number of viruslike dsRNA particles that are responsible for the production and resistance to yeast protein exotoxins called the "killer toxins." It is uncertain how much selection is based on the production or sensitivity to these toxins as most wild yeasts also have viruslike dsRNA particles so are resistant and the pH of wine is considerably below the pH optimum of most toxins.

D. Desirable Phenotypes for Wine Yeasts

Most of the wine yeast strains selected to date have been *Saccharomyces cerevisiae* with a few *S. bayanus*. Strains of *S. beticus* and *S. cheriensis* have been selected for sherry production but are otherwise regarded as undesirable contaminants.

Naturally occurring wine yeast phenotypes have been selected for (a) reproductive fitness, (b) high specific ethanol productivity, (c) ethanol tolerance, (d) osmotolerance, (e) tolerance to low pH, (f) a tendency to sediment into a compact mass after fermentation, (g) low production of SO_2 and H_2S, (h) low foaming, and (i) tolerance to low temperatures.

II. SELECTION

A. Pure Cultures and Population Yeasts

Virtually all commercial yeast strains are naturally occurring variants that have been empirically selected because they have the fermentation characteristics described in the previous section. Often single cells are isolated and propagated to yield a pure culture starter yeast. Pure culture yeasts have the disadvantage that they may not be competitive at one or another stage during fermentation or that they are good fermenters but do not yield the desired spectrum of flavor metabolites. To circumvent this problem several pure cultures may be combined to produce a "population yeast starter culture." Because the yeasts in a population culture have different growth rates and different conditions wherein they dominate a culture, new population yeast inocula must be used for each fermentation batch.

Some commonly used commercial pure culture yeast strains are presented in Table 1.

B.   Fitness

If one assumes a bloom population of 10^7 yeasts per grape berry, a doubling during harvesting and crushing, and a 20-fold reduction during debourbage and sulfiting, then one is left with a naturally occurring must inoculum of 10^6 yeast cells/ml. Of this number, the wine yeasts will total about 0.1–1% or about 10^3 wine yeast cells/ml. At its peak the wine yeast population in a fermenting must will reach $2–4 \times 10^8$ cells/ml, requiring 17 generations in an uninoculated must. There will be several species and many genotypes within each species, and there will be competition to dominate the must numerically and metabolically. The ability to dominate depends on the strain's doubling time or fitness. This fitness can be expressed as follows:

$$N_{(t)} = \frac{1}{1 + ((1/N_o)-1 \ e^{-t \ 1N(F)}}$$

where $N_{(t)}$ is the proportion (0 to 1) of wine yeasts at generation "t," N_o is the original proportion of wine yeasts in the yeast population, and F is the relative fitness of the wine yeast. Given a strain with fitness of 1.01 (it divides 1% faster than its competitors) and a must frequency of $N_o = 0.01$, it will take 462 generations for the wine yeast to dominate the population to greater than 50% if the competing yeasts are also ethanol-tolerant. Fortunately, most competing yeasts are not. In a winery situation a 1% or 2% inoculum of a 10^8 cells/ml selected yeast starter culture is used to give a starting concentration of $1–2 \times 10^6$ cells/ml, requiring only seven doublings during fermentation. Dominance of the starter culture (42, 43) increases the probability of a regulated fermentation with consistent quality. Starter strain dominance, however, is not always assured, as experiments using isozyme strain identification have shown that sometimes the strain present at the end of a fermentation is not the same as the starter culture used (44).

Table 1. Some Commercial Selections from Naturally Occurring Yeasts

Strain	Species

Saccharomyces cerevisiae

Montrachet (UCD 522)	Isolated by Pacottet in Burgundy and brought to U. C. Davis by F. T. Boletti. Starts to ferment quickly and ferments cleanly.
Waedenswill 27	Isolated at the Federal Agricultural Research Station by K. Mayer at Waedenswill, Switzerland. A cold-tolerant yeast.
S_2(EG8)	From Federal Agricultural Research Institute, Colmar, France.
WE^{-14} and WE 372	From the Research Institute for Viticulture and Oenology, Stellenbosch, South Africa.
V1116 (K1)	From P. Barre at the Federal Agricultural Research Institute at Montpellier. Possesses the killer factor.
71B	From J. Maugenet at the Federal Agricultural Research Inst. at Narbonne. Produces a fruity character due to high levels of esters and higher alcohols. Also capable of utilizing 25–34% of the malic acid present in musts.
LW-128-91	From the Agricultural Research Station at Geisenheim, W. Germany.
Schazhofberger	Isolated by G. Wuerdig at the Regional Agriculture Institute at Trier, W. Germany. This strain produces very low levels of SO_2.
Steinberg (DGI 228)	A popular yeast strain of the Rhine Valley from the Agricultural Research Station in Geisenheim, W. Germany.
Epernay 2 (UVA CEG)	Selected at the Agricultural Research Institute, Geisenheim, W. Germany, from the Epernay Champagne strain.
Siha 1	Isolated by J. Maugenet at the Federal Agr. Inst. at Montepellier.
Siha 3 (WET136)	Isolated by F. Zimmerman, Technical University, Darmstadt, W. Germany.
Hefix (2000 and 3000)	Strains widely used by large wineries in the Rhine Valley.

Saccharomyces bayanus

Prise de Mousse (EC 1118)	Isolated at the Enology Institute of Champagne, Epernay, France. Widely employed for still and sparkling wine production.
SB1	From the Federal Agricultural Research Station in Montpellier, France. Used extensively to restart stuck fermentations.

Table 1. *(Continued)*

Strain	Species
Champagne (UCD 505)	Brought to U. C. Davis by F. T. Boletti from Champagne.
R-2	Isolated in Sauternes, France. Ethanol-tolerant and grows more rapidly than most *S. bayannus*. Widely used in Australia.
SW-185-25	From the Agricultural Research Station at Geisenheim, W. Germany.
Pasteur champagne (UCD 595)	From the Pasteur Institute in Paris. This strain does well in Charmat process sparkling wine production.
Siha-2 (Fermechamp)	Isolated by J. Maugenet at the Fed. Agr. Inst. at Montpellier.
	Saccharomyces fermentati
Flor sherry (UCD 519)	From the University of California, Davis.

At low ethanol concentration the value for *S. bayanus* F is less than that for *S. cerevisiae*, which dominates the early fermentation. At higher ethanol concentrations the *S. bayanus* F value is greater than that of *S. cerevisiae*, so *S. bayanus* may dominate. Clearly fitness at various alcohol concentrations is a major consideration when selecting wine yeast strains.

C. Associative Fermentations

In many cooler parts of the grape-growing world, acid levels in musts are excessive and must be reduced. Dilution (amelioration) and titration (with $CaCO_3$) often have adverse effects on quality. *Schizosaccharomyces pombe* is a fermenting yeast that can metabolize malic acid, but it has a low fitness and imparts undesirable flavors to wine. Various methods for partial fermentations with *Sch. pombe* followed by finishing fermentations with *S. cerevisiae* or *S. bayanus* have been reported (45–47) but have met with very little commercial acceptance. Fermentations requiring two or more species with specific metabolic roles are referred to as "associative fermentations." The most widely practiced associative fermentation in wine making is *S. cerevisiae* performing an ethanolic fermentation followed closely by a bacterial (usually Lactobacilliaceae) performing a malolactic fermentation.

D. Fermentation Efficiency

As early as 1789 Lavoisier was studying the stoichiometry of the glycolytic conversion of hexose to ethanol, and shortly thereafter Gay-Lussac published his famous equation:

$$C_6H_{12}O_6 \rightarrow 2C_2H_5OH + 2CO_2 + \text{energy}$$

hexose	ethanol	carbon dioxide	ATP
180 g	92 g	88 g	56 K cal

In wine fermentations a gram of glucose produces about 0.45–0.48 g EtOH or 88–94% of the theoretical maximum of 0.511 g EtOH. The remaining 6–12% is used for biomass, maintenance, and secondary products. The efficiency of the fermentation varies according to aeration, fermentation temperature, must composition yeast strain, and size of inoculum used. In warm regions, the specific ethanol productivity is not a concern. In some cold-climate regions, however, the amount of chaptalization is tightly regulated, and it is important to approach the theoretical conversion limit in order to produce the desired ethanol concentration. Most vintners empirically rely on yeasts to perform a 92% conversion, and they chaptalize with 17 g sugar per liter to produce an ethanol increment of 1% v/v.

A number of genetic and regulatory studies on the glycolytic and fermentative pathways of yeasts have been performed (48–55), and it appears that constructing and selecting strains with fermentation efficiencies approaching the theoretical maximum is a reasonable goal.

E. Organic Acid Metabolism

Concern over organic acid levels in musts is a function of the climate of the viticultural area. In hot regions, organic acid concentrations are low and must be retained or possibly augmented during fermentation while musts from cold regions often have to reduce acidity by chemical means or by associative fermentations. Yeast strains have been selected that either produce or utilize acids. Wine yeasts produce succinic acid, pyruvic acid, 2-oxoglutarate, and (most importantly) 10–20 mg of malic acid (56–58), depending in part on the concentration of certain nitrogenous compounds in the medium. (59). Clonal selections have been described that are suitable for low-acid musts (60).

Several of the bloom yeast species can utilize malate as a substrate (61). *S. bayanus* strains utilize very little if any malate while *S. cerevisiae* uses about 18–23% of the malate normally present in musts (26, 62–64). The 71B strain of Maugenet (Table 1) is a commercially available strain capable of degrading up to 34% of the malic acid normally present in musts (65).

Acetic acid production by wine yeasts ranges from 0.12 to 0.54 mg/liter and is an undesirable feature of some yeast strains. On a synthetic medium with a pH less than 3 or greater than 9, one commercial strain was able to produce 1.30 g/liter acetic acid. At present there are few reports on genetic mechanisms or control of acetic acid production by various yeast strains. General reviews on the organic acid metabolism of yeasts can be found in the publications of Whiting (66) and Radler (8).

F. Sulfur Metabolism

All wine yeast strains have been selected for low H_2S production, but only a few for low SO_2 production. Inquiries into sulfur metabolism in wine yeast began in the 1960s in several countries (67–71), and a number of reports since that time (72–82) have contributed to an understanding of the putative metabolic pathway shown in Figure 1. The pathway omits the influence of copper, tin, zinc, manganese, initial pH, cofactor and amino acid concentration, and other substrate elements that are reported to influence the pathway (83).

H_2S and mercaptan are of concern to vintners as they impart the odor of rotten eggs or skunk oil, respectively, even when present in concentrations of parts per billion. Wine yeasts produce H_2S from sulfate, sulfite, or elemental sulfur (79). Sulfate and sulfite are normal must constituents. Selection of low H_2S-producing yeast strains can be performed on acid bismuth yeast agar color test medium (84). Ironically many selected strains that are low H_2S producers are high SO_2 producers, and vice versa.

Elemental sulfur in musts is usually residue from antifungal sprays that can be converted both enzymatically and nonenzymatically to H_2S (76, 79).

SO_2 is used extensively in wine making (85, 86), to inhibit browning, to inhibit proliferation of microbial contaminants (87, 88), as a disinfectant of winery equipment, as a color extraction agent, and as a general antioxidant. Recent concern over SO_2 levels in wines arises from infrequent but traumatic allergic reactions in SO_2-hypersensitive individuals. In Europe there are multinational working groups reporting regularly on yeast production of SO_2 during fermentation (71, 89) and technologies for reducing the requirement for SO_2 in wines (90–94).

Schazhofberger is a low SO_2-producing yeast strain that is widely used in the Mosel Valley. To isolate this strain Wuerdig assayed 50 clones of the original strain for SO_2-producing ability. He took the lowest producer, propagated it, and again obtained clones of single-cell isolates which he assayed for SO_2 production. After four similar selection cycles he obtained a genetically stable strain that produced extremely low levels of SO_2.

Production of SO_2 in yeast is believed to be a function of the sensitivity of ATP sulfurylase (see Fig. 1) to its product adenosine-5 phosphosulfate and sulfide (96–98).

G. Flavor Metabolites

There are vintners who believe that the best one can hope for in a wine yeast is that it ferments sugars effectively and then sediments rapidly without leaving any

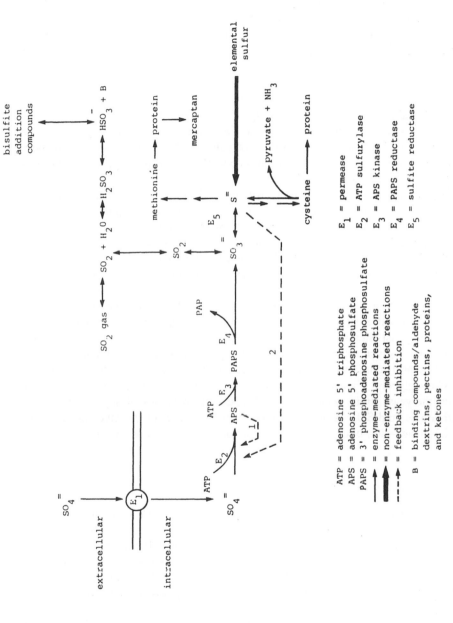

Figure 1. Sulfur metabolic pathway in yeast.

malodorous metabolites in the wine. For some time research efforts have been directed toward isolating yeasts making a positive contribution to wine flavor, especially yeasts for sparkling and "nouveau"-type wines. Clonal selection of flavor producing yeasts occurring in natural populations, has been based empirically on organoleptic evaluations. To date, chemical analyses have been used only to identify the products and not as a screening device for selection.

Numerous studies on the properties of wine yeast strains (99–103) and the analysis of wine flavor (104–106) have elucidated only a few correlations between yeast strains and the production of specific flavorful metabolites.

Studies on the chemical basis of wine flavor are complicated by the knowledge of synergistic effects of certain pairs of compounds and that human sensitivity thresholds vary so greatly that obvious concentration/sensory effect relationships cannot be readily drawn.

Yeast strains vary in their propensity to produce fusel oils which contribute to flavor complexity at concentrations around the sensory perception threshold but detract from quality at higher concentrations (107). Fusel oils or higher alcohols are produced when an alpha keto acid decarboxylase (probably pyruvate decarboxylase) and alcohol dehydrogenase metabolize the alpha keto acid precursor of amino acids. Yeasts excrete much of these higher alcohol products.

Quantitative but not qualitative differences in ester production are noted when comparing different wine yeast strains (102, 108). Though the principal isozymes of wine yeast esterases have been studied (109, 110), no correlation has been made between a given esterase allele and any organoleptic response ellicited.

Glycerol contributes "mouth feel" and, to a lesser extent, sweetness to wine. In wines, glycerol is usually 1/10 to 1/15 the concentration of ethanol (8), depending on the yeast strain, temperature, pH, and concentrations of sugar, amino acid, thiamine, and SO_2. Wine yeasts produce 0.1–3 g/liter of the glycerol derivative 2,3-butane-diol, but its precise contribution to flavor is not known (8, 11).

H. Foaming and Sedimentation

Yeasts that foam require the vintner to set aside 5–10% of fermentation tank capacity to prevent spillage. Foam may also deposit on tank walls, where after partial drying it may form a putrifying scum which can fall back into the wine with undesirable results. To select nonfoaming wine yeast strains, air is bubbled through the growth medium and the froth is routinely skimmed off. Foaming yeasts adhere to bubbles and can be selectively removed, leaving mostly nonfoaming yeasts in the medium (112).

Yeast strains range in their sedimentation character from those that sediment slowly forming fine sediments that are readily resuspended to those that form dense granular clumps forming a compact sediment not easily perturbed. The granular sediments are desirable because the wines require fewer clarification treatments (which tend to strip wines of complexity) and reduce production costs.

I. Cold Tolerance

Important volatiles such as ethanol, esters, aldehydes, ketones, and terpenes are often entrained by CO_2 during fermentation. Many of these compounds are responsible for the floral and fruity character of wines. Entrainment is directly related to temperature, bubble size, and concentration. For light floral wines it is necessary to ferment at low temperatures (102, 108). Low temperatures, however, raise the risk of premature cessation of fermentation. Chemostat and other methods have been developed for clonal selections of cold-tolerant yeasts (113).

J. Harmful Yeast Metabolites

In addition to SO_2, yeasts may produce allergens that can be reduced by selecting the appropriate yeast strain. At one time it was thought that red wine allergy (114) was the result of histamine synthesized by Lactobacilliaceae decarboxylation of must histidine during malolactic fermentation. Buteau et al. (115) found no relationship between must histidine, wine histamine, and malolactic fermentation but did suggest that other wine biogenic amines (mostly ethanolamine) were yeast secretory products. Recent evidence suggests that red wine syndrome is probably not due to histamines (114).

There is good evidence to conclude that most mutagens are carcinogens (116). A number of naturally occurring plant products are mutagenic or can be activated to mutagens by degradary enzymes of the enteric bacteria (117) in the liver (118) or other tissues (119). Wines in general (120, 121), wines from specific grape varieties (122), and wines fermented with various yeasts including contaminants (123) have been assayed for mutagen and promutagen content, and for the most part have been found to contain insignificant or low levels of mutagen. Mutagens that have been found arise more from certain grape varieties (Concord) than from the yeast strain used (122).

K. Active Dry Yeasts

Much of the resurgent interest in pure culture inocula and consequent discussion of the comparative virtues of different wine yeast strains in Europe and elsewhere are attributable to the availability of active dry yeast (ADY) inocula. For most producers, wine yeasts are a secondary market for their main commercial interest of supplying fresh yeast to bakeries. Active dry yeasts relieve the vintner of maintaining pure cultures throughout the year and the preparation of a "pied de cuve." They are convenient and inexpensive, and they offer a selection that now exceeds 150 wine yeast strains.

During the first 20 years of production of wine, ADY has risen to 200 tons in 1983 (124). They are produced by firms in several countries including Anchor Yeast (South Africa), Universal Foods (United States), Danske Spiritfabrikka Gaering Industri (Denmark), Gist Brocades (Holland), Erbsloeh (West Germany), Lallemand (Canada), Société Le Saffre (France), and Deutsche Hefe Werk (West Germany). In most cases cultures are grown aerobically on readily available carbon sources such as molasses supplemented with gaseous NH_3, phosphate, and other growth factors. The method of drying dictates whether the yeasts will be formed into pellets, granules, or small noodles (125, 126) before vacuum packaging. Initial live cell concentrations of $1-4\times10^{10}$ cells per gram ADY can be well maintained at 4–7°C but may drop off to less than 10^3 cells per gram if stored at 40°C for 6 months (127, 128).

During rehydration, ADY yeasts have impaired membranes that leak essential nutrients, resulting in the occasional cell death (128). Several producers have ongoing programs selecting clones of commercial strains with superior storage and rehydration survival. Most commercial ADY starters can survive rehydration well enough to dominate natural populations under most industrial conditions (129).

III. MODIFICATION

A. Mutagenesis

Occasionally a naturally occurring yeast population used in a selection program will not contain sufficient genetic diversity to contain yeast with a desired phenotype. It has been known for some time that genetic diversity of wine yeasts can be increased through mutagenesis (130–134). In situations such as those for cold tolerance, fitness, or sedimentation, the number and structure of the target genes involved in expression and regulation are not known, so all types of mutations are sought, including base-pair substitutions, frame-shift mutations, deletions, and duplications. The relationship between mutagen and mutations (new alleles) produced has been reviewed elsewhere (5).

More difficult than the induction of the mutation is the screening protocol by which a single mutant clone with a desired phenotype can be distinguished or isolated from among approximately 10^6 of its parental-type sibs. Empirical observation or analyses are limited to 10^4-10^5 clones, depending on the analysis. Mutants involving sulfur metabolism (135, 136) and malate utilization (137, 138) have been isolated using color indicators in solid support media while amino acid auxotrophy tests were used to isolate fusel oil production mutant strains (139, 140).

Killer yeasts have been identified and isolated by their characteristic halo on a lawn of sensitive yeasts (141), and mutant enrichment media have been used to isolate alcohol- or cold-tolerant phenotypes in batch or chemostat cultures (130, 133). Mutant phenotypes for recessive traits in wine yeasts occur at a lower frequency than that published for the heterothallic haploid research yeast strains used in the laboratory for research. Wine yeasts are homothallic and possibly polyploid or aneupolyploid. Methods have been published to mutagenize and disperse wine yeast ascospores during their brief haploid phase (140). This treatment increases the probability of isolating homozygous mutant alleles in the daughter cells.

B. Recombination

1. Sexual

To combine the desirable traits of two wine yeast strains or incorporate a trait from a nonwine yeast strain into a wine yeast strain, it may be necessary to sexually recombine the strains involved. Strains can be mated even if both are homothallic or if one is homothallic and the other a genetically homologous heterothallic laboratory strain (141–143). Using classical mating techniques, Thornton (144) was able to introduce the flocculating trait into a homothallic wine yeast.

Many wine yeast strains sporulate poorly. Most protocols for mating require ascospores from each parental strain to be situated adjacent to each other either on agar blocks or in liquid medium depending on the availability of complementary auxotrophic markers in the parental strains. A comprehensive review of yeast mating is given elsewhere (7).

During pure culture expansion and fermentation, selected strains undergo many spontaneous genomic modifications, including mutation, mitotic recombination, chromosomal rearrangements, polyploidy, and aneuploidy. Dominant lethal events are rapidly excluded from the population, but deleterious recessive alleles and genomic rearrangements accumulate, lowering the genetic fitness of the culture. It is necessary to periodically reisolate isogenic clones from single spores. Such clonal selections should be evaluated for growth and fermentation performance. Spore viability of known strains is often used as an indicator of acquired deleterious mutations and genomic anomalies.

Sexual recombination has been used to transfer the killer trait from a sake yeast strain into a wine yeast strain (141). Wild yeasts and some wine yeasts contain one or more of 10 different double-stranded RNA (146) or double- stranded DNA (147) viruslike particles (VLP) that are associated with the production of proteinaceous exotoxins (148). Certain VLPs bestow upon the host yeast resistance to specific killer toxins from wild yeasts. Wild yeast killer toxin problems have been documented in breweries (1) but not in wineries. The use of pure culture wine yeast strains is questionable if one considers the toxin dose-response stoichiometrically in wines with a pH range of 3.0–3.4. The pH range for killer toxin activity is 4–6. Hara et al. (141) avoided the problem of having the killer sake yeast parent kill the sensitive wine yeast parent simply by using fresh medium immediately before mating.

The *Karl- 1* allele (150) delays nuclear fusion after mating, so a hybrid is formed with the genotype of one parent and the cytoplasm of th other in a process called cytoduction. Cytoduction has successfully been used to transfer a killer viruslike particle from one strain into the Montrachet wine yeast (151). Occasionally during cytoduction, one (rarely two) chromosome is transferred (152). This *Karl-1-*mediated single chromosome transfer has been used effectively to map functions on brewing yeast chromosomes (153), but to date has had no application for wine yeasts.

2. Asexual

Mitotic recombination (149) has not been as well studied in yeast as it has in filamentous fungi. Cell fusion is an alternative where sexual recombination is not possible owing to gross phylogenetic or histocompatibility differences between species. Carrau et al. (154) fused *Sch. pombe* and *S. cerevisiae* to select fusants with malate degrading ability of the former and fermenting character of the latter. Fusants were obtained that could degrade more malate than *S. cerevisiae* but less than *Sch. pombe* and resembled one or other parent.

C. Transformation

Transformations with naked DNA has been used in brewing strains (155) using a maltotriose fermentation marker but has never been used successfully in wine yeasts.

Transformation involving a recombinant DNA plasmid has been employed to produce a yeast strain capable of a simultaneous ethanolic and malolactic fermentation (156). The malolactic gene from *Lactobacillus delbruekii* was first cloned in plasmid pBR322 in *E. coli* then transformed into a wine yeast strain. Unfortunately, expression of the malolactic activity was poor, and further work is required to determine the reasons for this problem. In a similar experiment the malolactic gene from *Leuconostoc enos* was cloned in *E. coli*, but again expression was poor (157).

S. cerevisiae wine yeast strains can utilize about 20% of the malate present in must via a malic enzyme, pyruvate decarboxylase, alcohol dehydrogenase-mediated maloethanolic pathway (8). *Sch. pombe* uses the same pathway but utilizes all the malate present in musts. It can do this because it has an active malate permease and malic enzyme present (137, 138). *S. cerevisiae* is reported not to have an active malate permease (8). It would be interesting to construct a *S. cerevisiae* wine yeast transformed with recombinant plasmids containing *Sch. pombe* malate permease and malic enzyme alleles that would be capable of complete malo-ethanolic fermentation.

IV. CONCLUSION

Unlike brewer's or distiller's wort, must has a very large number of competing species, each with an ecological niche during fermentation. The particular succession of species is responsible for many of a wine's faults as well as a mysterious complexity that defies simple definitions. The use of pure culture yeasts decreases the frequency of faulty fermentations, increases the consistency of wines, and if used properly can result in elegant complex wines. At present, pure culture yeasts are necessary for efficient production of everyday wines and desirable for the production of fine wines. Both the classic methods for genetic manipulation and the latter-day molecular or genetic engineering of wine yeast strains have held great promise in the construction of strains with hitherto unknown metabolic virtuosity, but the commercial success of such strains has been limited.

The advent of active dry yeast two decades ago has made pure culture inocula much more reliable and efficient than in past years. It has also focused the attention of the vintner on the relative merits of specific wine yeast strains and, in turn, put pressure on ADY producers to construct and select strains with superior fermentation performance. It is possible that ADY sales will be the motive force behind the construction of a variety of strains custom-made to the vintner's demands.

REFERENCES

1. Spencer, J. F. T., and D. M. Spencer, Genetic improvement of industrial yeast. Annu. Rev. Microbiol., 37:121-142 (1982).
2. Lafon- Lafourcade, S., Wine and brandy in Biotechnology V: Food and Feed Production with Micro-organisms, G. Reed (Ed.), Verlag Chemie, Deerfield Beach, FL, pp. 82-163 (1983).
3. Snow, R., Genetic improvement of wine yeast in *Yeast Genetics*, J. F. T. Spencer, D. M. Spencer, and A. R. W. Smith (Eds.), Springer- Verlag, New York, pp. 439–459 (1983).
4. Subden, R. E., Wine yeasts, *Dev. Indust. Microbiol.*, 24:221–229 (1983).
5. Subden, R. E., Current developments in wine yeasts, in *CRC Biotechnology*, CRC Press, Boca Raton, FL, 5:49-65 (1987).
6. Stewart, G. G., C. R. Murray, C. J. Panchal, I. Russell, and Sills, A. M., The selection and modification of brewers yeast strains, *Food Microbiol.* 1:289–302 (1984).
7. Beckerich, J. M., P. Fournier, C. Gaillardin, H. Heslot, M. Rochet, and B. Treton,

Yeasts in *Genetics and Breeding Industrial Organisms*, C. Ball (Ed.), CRC Press, Boco Raton, FL, pp. 115-157 (1984).

8. Radler, F., Microbial biochemistry. *Experientia*, 42:884–892 (1986).
9. Kunkee, R. E., Four thousand years of winemaking with an emphasis on the last one hundred years, in *Grape and Wine Centennial Symposium Proceedings*, A. D. Webb (Ed.), University of California Press, Davis, CA, pp. 76–78 (1982).
10. Pasteur, L., *Estudes sur le vin*, Imprimerie Impériale Victor Masson et Fils, Paris, p. 56 (1886).
11. Hansen, E. C., Undersogelser over alkohols jaersvanpenes fysiologi og morgolivi. V. Methoder til fremstilling af renkulturer af saccharomyceter of ligende. Mikro organismer, *Meddr. Carlsberg Lab.*, 2:152–167 (1886).
12. Hansen, E. C., Undersogelser fra gjaeringsindustriens praxis, *Meddr. Carlsberg Lab.*, 2:257–283 (1888).
13. Mueller-Thurgau, H., Ueber die Vergaehrung des Traubenmostes durch zugesetzte Hefe, *Weinbau Weinhandel*, 7:477–478 (1889).
14. Mueller-Thurgau, H., Ergebnisse neuer Untersuchungen auf dem Gebiete der Weinbereitung, in *Ber d. XII Dtsch. Weinbaukongr. in Worms*, p. 14 (1891).
15. Barnett, J. A., M. A. Delaney, E. Jones, A. B. Magnon, and B. Winch, The number of yeasts associated with wine grapes of Bordeaux, *Arch. Microbiol.*, 83:52–55 (1972).
16. Domercq, S., Etude et classification des levures de vin de la Gironde, *Ann. Tech. Agric.*, 6:139–183 (1957).
17. Minarik, E., *Ecology of Natural Wine Yeast Species in Czechoslovakia*, J. Kolek Publ., Bratislava (1966).
18. Florenzano, G., W. Balloni, and R. Materassi, Contributo alla ecologia dei lieviti *Schizosaccharomyces sulla. Vitis*, 16:38 (1977).
19. Ribereau-Gayon, J., and E. Peynaud, *Traite d'Oenologie*, Vol. I. Beranger, Paris (1960).
20. Sponholz, W. R., and H. H. Dittrich, Die Bildung von SO_2–bindenden Gaerungs–Nebenprodukten, hoeheren Alkoholen und Estern bei einigen Reinzuchthefestaemmen und bei einigen fuer die Weinbereitung wichtigen wilden Hefen, *Wein Wiss.*, 29:301–314 (1974).
21. Van Zyl, J. A., J. M. DeVries, and A. S. Zeeman, The microbiology of S. African winemaking. III. The effect of different yeasts on the composition of fermented musts, *Sci. Bull. Dep. Agric. Techn. Serv. Pretoria*, 6:165–180 (1963).
22. Scheffer, W. R., and E. M. Mrak, Characteristics of yeast causing clouding of dry white wine, *Mycopathol. Mycol. Appl.*, 5:236–249 (1951).
23. Schanderl, H., *Die Mikrobiologie des Mostes und Weines*, 2d Ed. Verlag Eugen Ulmer., Stuttgart, West Germany, p. 17 (1951).
24. Wenzel, K. W. O., M.Sc. thesis, University Stellenbosch, South Africa (1966).
25. Ribéreau-Gayon, and P. Sudraud, *Traité d'Enologie, Sciences et Techniques du Vin*, Vol. 2, Dunod Imp., Paris (1976).
26. Rankine, B. C. Decomposition of L-malic acid by wine yeasts, *J. Sci. Food Agric.*, 17:312–316 (1966).
27. Soufleros, E., Les levures de la région viticole de Naoussa (Greece). Ph.D. thesis, University of Bordeaux, France.

28. Moss, F. J., A. D. Rickard, F. E. Bush, and P. Caiger, The response by micro-organisms to steady-state growth in controlled concentrations of oxygen and glucose. II.*Saccharomyces carlsbergensis, Biotechnol. Bioeng.*, 13:63–75 (1971).

29. Maiorella, B. L., Ethanol, in *Comprehensive Biotechnology*, M. Moo-Young (Ed.), Pergamon Press, Oxford, U. K., pp 861–914 (1985).

30. Dedenken, R. H., The crabtree effect: A regulatory system in yeast,*J. Gen. Microbiol.*, 44:149–156 (1966).

31. Holzer, H., Catabolite inactivation in yeast, *TIBS*, 8:178–181 (1976).

32. Wang, D. I. C., C. L. Cooney, A. L. Demain, P. Dunnil, A. E. Humphrey, and M. D. Lilly, *Fermentation and Enzyme Technology*, Wiley, New York (1979).

33. Haukeli, A. D., and S. Lie, Controlled supply of trace amounts of oxygen in laboratory scale fermentations, *Biotechnol. Bioeng.*, 13:619–628 (1971).

34. Rose, A. H., M. J. Beavan, and C. Charpentier, Physiological basis for enhanced ethanol production by *Saccharomyces cerevisiae*, in *Overproduction of Microbial Products*, V. Krumphanzl, B. Sikyta, and Z. Vanek (Eds.), Academic Press, New York, pp. 211–219 (1982).

35. Millar, D. G., K. Griffiths-Smith, E. Algar, and R. K. Scopes, Activity and stability of glycolytic enzymes in the presence of ethanol, *Biotechnol. Lett.*, 4:601–606 (1982).

36. Bazua, C. D., and C. R. Wilke, Ethanol effects on the kinetics of a continuous fermentation with *S. cerevisiae, Biotechnol. Bioeng. Symp.*, 7:105 (1977).

37. Karuwanna, P., Wine fermentations with mixed cultures, M.Sc. thesis, University of California, Davis (1976).

38. Traverso-Rueda, S., and R. E. Kunkee, The role of sterols on growth and fermentation of wine yeasts under vinification conditions, *Dev. Indust. Microbiol.*,23:131–143 (1982).

39. Larue, F., S. Lafon-Lafourcade, and P. Ribéreau-Gayon, Relationship between the sterol content of yeast cells and their fermentation activity in grape must, *Appl. Environ. Microbiol.*, 39:808–822 (1980).

40. Ohta, H., and S. Hayashida, Role of Tween 80 and monoolein in a lipid-sterol-protein complex which enhances ethanol tolerance of sake yeast, *Appl. Environ. Microbiol.*, 46:821–825 (1983).

41. Lafon-Lafourcade, S., C. Geneix, and P. Ribéreau-Gayon, Inhibition of alcoholic fermentation of grape must by fatty acids produced by yeast and their elimination by yeast ghosts, *Appl. Environ. Microbiol.*, 47:1246–1249 (1984).

42. Rosini, G., Assessment of dominance of added yeast in wine fermentation and origin of *Saccharomyces cerevisiae* in wine making, *J. Gen. Appl. Microbiol.*, 30:249–256 (1984).

43. Heard, G., and G. H. Fleet, Growth of natural yeast flora during the fermentation of inoculated wines, *Appl. Env. Microbiol.*, 50:727–728 (1985).

44. Bouix, M., J. Y. Leveau, and C. Cuinier, Applications de l'électrophorèse des fractions exocellulaires de levures au control de l'efficacité d'un levurage en vinification, in *Current Developments in Yeast Research*, G. G. Stewart and I. Russell (Eds.), Pergamon Press, Toronto, pp. 87–92 (1981).

45. Gallander, J., Deacidification of eastern table wines with *Schizosaccharomyces pombe, Am. J. Enol. Vitic.*, 28:65–69 (1977).

46. Gandini, K. A., and D. Marendo, Prove semi industrial d'impiego di lieviti attivie di *Schizosaccharomyceti* nella vinificazione di mosti piemontesi, *Enotecnico*, 8:3–11 (1978).

47. Prida, I. A., S. A. Kishkovskays, V. S. Razuvaev, N. I. Bur'yan, and A. G. Reva, Method for biologically reducing acids in wines (in Russian), U.S.S.R. Pat. SU 941421.

48. Ciracy, M., and I. Breitenbach, Psysiological effects of seven different blocks in glycolysis of *S. cerevisiae.*, *J. Bacteriol.*, 139:152–160 (1979).

49. Sprague, G. F., Isolation and charterization of a *S. cerevisiae* mutatant deficient in pyruvate kinase activity, *J. Bacteriol.*, 130:232–241 (1977).

50. Maitra, P. K., Glucose and fructose metabolism in a phosphogluco isomerase less mutant of *S. cerevisiae, J. Bacteriol.*, 107:759–769 (1971).

51. Lobo, Z., and P. K. Maitra, Genetics of yeast hexokinase, *Genetics*, 86:727–744 (1977).

52. Clifton, D., S. B. Weinstock, and D. G. Fraenkel, Glycolysis mutants in *S. cerevisiae, Genetics*, 88:1–11 (1978).

53. Brown, S. W., and S. G. Oliver, Isolation of ethanol tolerant mutants by continuous selection, in *Tenth International Conference on Yeast Genetics and Molecular Biology, Louvain-la-Neuve*, France (1980).

54. Brown, S. W., S. G. Oliver, D. E. F. Harrison, and R. C. Righelato, Ethanol inhibition of yeast, growth and fermentation: Differences in the magnitude and complexity of the effect, *Eur. J. Appl. Microbiol. Biotechnol.*, 11:151–155 (1981).

55. Hoppe, A. K., and A. S. Hansford, Ethanol inhibition of continuous anaerobic yeast growth, *Biotechnol. Lett.*, 4:39–44 (1982).

56. Dakin, H. D., The formation of L-malic acid as a product of alcoholic fermentation by yeast, *J. Biol. Chem.*, 61:139–142 (1924).

57. Drawert, F., A. Rapp, and W. Ulrich, Bildung von Aepfelsaeure, Weinsaeure and Bernsteinsaeure durch verschiedene Hefen, *Naturwissenschaften*, 52:306–311 (1965).

58. Radler, F., and E. Lang, Malatbildung bei Hefen, *Wein Wiss.*, 37:391–399 (1982).

59. Ponader, W., Einfluss von Stickstoffverbindungen auf die Bildung von Carbonylverbindungen durch Hefen. Dissertation, Johannes Gutenberg University, Mainz, West Germany (1973).

60. Fatichenti, F., G. A. Farris, P. Deina, S. Ceccarelli, and M. Serra, Commercial trial of winemaking using two selected starters of *S. cerevisiae* which do not reduce malic acid content, *Am. J. Enol. Vitic.*, 32:236–240 (1981).

61. Goto, S., M. Yamazaki, Y. Yamakava, and J. Yokosutka, Decomposition of malic acid in grape must by wine and wild yeasts, *Hakkokogaku (J. Ferment. Technol.)* 56:133–135 (1978).

62. Fuck, E., and F. Radler, Aepfelsaurestoffwechsel bei *Saccharomyces* I. Der anaerobe Aepfelsaureabbau von *Saccharomyces cerevisiae*, *Arch. Microbiol.*, 87:149–164 (1972).

63. Fuck, E., G. Stark, and F. Radler, Aepfelsaurestoffwechsel bei *Saccharomyces*. II. Anreicherung und Eigenschaften eines Malatenzyms, *Arch. Microbiol.*, 89:223–231 (1972).

64. Wenzel, K., H. Dittrich, and B. Pietonka, Untersuechungen zur Beteiligung von Hefen am Aepfelsaureabbau bie der Weinbereitung, *Wein Wiss.*, 37:133–138 (1982).

65. Cunier, C., Fermentation sur milieu synthétique, in *Technical Data Form 71B (1122): Lalvin*, Lallemand Inc., Montreal (1983).

66. Witing, G. C., Organic acid metabolism of yeasts during fermentation of alcoholic beverages: A review, *J. Inst. Brew.* 82:84–92 (1975).

67. Kielhofer, E., Die Wirkung von schwefliger Saeure und Ascorbinsaeure auf die organoleptischen. Eigenschaften des Weines, *Weinberg Keller*, 5:573–587 (1958).

68. Rankine, B. C., Hydrogen sulphide production by yeasts, *J. Sci. Food Agric.*, 15:872–877 (1964).

69. Zambonelli, C., Richerche biometriche, sulla produzione di idiogeno solforato da solfati e solfiti in *Saccharomyces cerevisiae* var ellipsoideus, *Ann. Microbiol.*, 14:129–141 (1964).

70. Wuerdig, G., and H. A. Schlotter., SO$_2$-Bildung in gaerenden Traubenmosten, *Z. Lebensm. Unters.-Forsch.*, 134:7–13 (1967).

71. Wuerdig, G., Formation de SO$_2$ par reduction de sulfate pendant la fermentation, in *Rapport du Groupe Travail: "Microbiologie du Vin"*, Bull. O. I. V., April (1972).

72. Wuerdig, G., and H. A. Schlotter, SO$_2$-Bildung durch Sulfatreduktion waihrend der Gaerung I. Mitteilung. Versuche und Beobachtungen in der Praxis, *Wein Wiss.*, 23:356–371 (1968).

73. Wuerdig, G., and H. A. Schlotter, SO$_2$-Bildung durch Sulfatreduktion wahrend der Gaerung. II. Mitteilung. Beeinfluessung durch das Substrat und die Gaerungsbedingungen, *Wein Wiss.*, 25:283–297 (1970).

74. Eschenbruch, R., and J. deVilliers, On the metabolism of sulphate and sulphite during the fermentation of grape must by *Saccharomyces cerevisiae*, *Arch. Microbiol.*, 93:259–266 (1973).

75. Eschenbruch, R., and P. Bonish. The influence of pH on sulphite formation by yeasts, *Arch. Microbiol.*, 107:229–231 (1976).

76. Eschenbruch, R., P. Bonish, and B. M. Fisher, The production of H$_2$S by pure culture wine yeasts, *Vitis*, 17:67–74 (1978).

77. Heinzel, M., and H. G. Trueper, Sulfite formation by wine yeasts. II. Properties of ATP- sulfurylase, *Arch. Microbiol.*, 107:293–297 (1976).

78. Heinzel, M., and H. G. Trueper, Sulfite formation by wine yeasts. V. Regulation of biosynthesis of ATP and ADP sulfurylase by sulfur and selenium compounds, *Arch. Microbiol.*, 118:243–247 (1978).

79. Schütz, M., and Kunkee, R. E., Formation of hydrogen sulfide from elemental sulfur during fermentation by wine yeast, *Am. J. Enol. Vitic.*, 28:137–144 (1977).

80. Minarik, E., Formation of sulfurous acid through reduction of sulfate during fermentation of must, *Kvasny. Prumysl.*, 18:104–107 (1972).

81. Premuzic, D., T. Lovrics, O. Sofar, and V. Jovic, Production of SO$_2$ during fermentation of must as a result of the metabolism of some yeast strains and their effect on the colour of white wines, *Kem. Indust.* 21:9 (*Chem Abstr.* 76:152021g) (1973).

82. Weeks, C., Production of sulfur dioxide by two Saccharomyces yeasts, *Am. J. Enol. Vitic.*, 20:32–39 (1969).

83. Eschenbruch, R., Sulphite and sulphide formation during winemaking: A review, *Am. J. Enol. Vitic.*, 25:157–161 (1974).

84. Nickerson, W. J., Reduction of inorganic substances by yeast. I. Extracellular reduction of sulfate by species of *Candida, J. Infect. Dis.* 83:43–48 (1953).

85. Taylor, S. L., and R. K. Bush, Sulfites as food ingredients, *Food Technol.*, 40:47-52 (1986).

86. Wedzicha, B. L., Chemistry of Sulphur Dioxide in Foods, *Elsevier*, London (1984).

87. Macris, B. J., and P. Markakis, Transport and toxicity of sulfur dioxide in *Saccharomyces cerevisiae* var. *elippsodieus. J. Sci. Food Agric.*, 25:21–29 (1974).

88. King, A. D., J. D. Pointing, D. W. Sanshuck, R. Jackson, and K. Mihara. Factors affecting death of yeasts by sulfur dioxide, *J. Food Prot.*, 44:92–97 (1981)

89. Mayer, K. and A. Dufor, Etude sur la production de SO_2 par les levures au cours de la fermentation, Rapport du group travail "Microbiologie du Vin." *Bull. O.I.V.*, May (1973).

90. Gomes, M. J. V., Limitation de l'emploi de l'anhydride sulfureux compte tenu des besoins en matieres nutritives deslevures pendent la fermentation, *Bull. O. I. V.*, 46:316 (1973).

91. Gomes, M. J. V., and M. F. daSilva Babo, Les technologies de vinification permettant de dimineur les doses de SO_2, *Bull. O. I. V.*, 58:621 (1985).

92. Ruiz Hernandez, M., Les technologies de vinification permettant de dimineur les doses de SO_2, *Bull. O. I. V.*, 58:617–620 (1985).

93. Usseglio-Tomasset, L., Les technologies de vinification permettant de diminuer les doses de SO_2, *Bull. O. I. V.*, 58:606–616 (1985).

94. Valouyko, G. G., N. M. Pavlenko, and S. T. Ogoroduik, Les technologies de vinification permettant de dimineur les doses de SO_2, *Bull. O. I. V.*, 58:637–644 (1985).

95. Asvany, A. L., Les technologies de vinification permettant de diminuer les doses de SO_2, *Bull. O. I. V.*, 58:621–623 (1985).

96. Dott, W., M. Heinzel, and H. G. Trueper, Sulfite formation by wine yeasts. I. Relationships between growth fermentation and sulfite formation, *Arch. Microbiol.*, 107:289–292 (1976).

97. Dott, W., and H. G. Trueper, Sulfite formation by wine yeasts. II. Properties of ATP-sulfurylase, *Arch. Microbiol.*, 107:293–296 (1976).

98. Dott, W., M. Heinzel, and H. G. Trueper, Sulfite formation by wine yeasts. IV. Active uptake of sulfate by low and high sulfite producting wine yeasts, *Arch. Microbiol.*, 112:283–285 (1977).

99. Nykaenen, L. and I. Nykaenen, Production of esters by different yeast strains in sugar fermentations, *J. Inst. Brew.*, 83:30–31 (1977).

100. Kundu, B. S., S. Shuresh, and M. C. Bhardiya, Effect of different wine yeast strains on quality of white table wine from Perlette grapes, *Harv. Agric. U. J. Res.*, 10:65–69 (1980).

101. Benda, I., Reinzuchthefen in der Kellerwirtschaft, *Weinwirt*, 118:700–705 (1982).

102. Soles, R. M., C. S. Ough, and R. E. Kunkee, Ester concentration differences in wine fermented by various species and strains of wine yeast, in *Yeast Genetics Fundamentals and Applied Aspects*, J. F. T. Spencer, D. M. Spencer, and A. R. W. Smith (Eds.), Springer Verlag, N. Y., pp. 439–459 (1982).

103. Soles, R. M., C. S. Ough, and R. E. Kunkee, Ester concentration differences in wine fermented by various species and strains of yeasts, *Am. J. Enol. Vitic.*, 33:94–98 (1982b).

104. Nykaenen, L., and H. Suomolainen, *Aroma of Beer, Wine and Distilled Alcoholic Beverages*, Reidel, Dordrecht, Holland (1983).

105. Schreier, P., and J. Paroschy, Volatile constituents from Concord Niagara (*V. labruscaL.*) and Elvira (*V. labrusca* × *V. riparia*) grapes, *Can. Inst. Food Sci. Technol.*, 14:112–116 (1981).

106. Rapp, A., H. Hastrich, L. Engel, and W. Knipser, Possibilities of characterizing wine quality and wine varieties by means of capillary chromatography, in *Flavour of Food and Beverages*. G. Charlambous and G. E. Inglett (Eds.), Academic Press, New York, pp. 391–417 (1980).

107. Kunkee, R. E., and M. A. Amerine, Yeasts in wine making, in *The Yeasts*, Vol. III, J. A. Harrison and A. H. Rose (Eds.), Academic Press, New York, pp. 5–71 (1970).

108. Akhtar, M., R. E. Subden, J. D. Cunningham, C. Fyfe, and A. G. Meiering, Production of volatile yeast metabolites in fermenting grape musts, *Can. Inst. Food Sci. Technol. J.*, 4:280–283 (1985).

109. Woehrmann, K., and P. Lange, The polymorphism of esterases in yeast (*Saccharomyces cerevisiae*), *J. Inst. Brew.*, 86:174 (1980).

110. Subden, R. E., D. Irwin, J. D. Cunningham, and A. G. Meiering, Wine yeast isozymes. I. Genetic differences in eighteen stock cultures, *Can. J. Micro.*, 28:1047–1050 (1982).

111. Guymon, J. F., and E. A. Crowell, Direct gas chromatographic determination of levo and meso 2,3-butane-diols in wines and factors affecting their formation, *Am. J. Enol. Vitic.*, 18:200–209 (1967).

112. Eschenbruch, R., and J. M. Rassel, The development of non-foaming yeast strains for wine making, *Vitis*, 14:43–47 (1975).

113. Meyer, J. P., Les levures adaptées aux basses températures et leur sélection, *Rev. Fr. Oenol*, 76:45–49 (1979).

114. Masyczek, R., and C. S. Ough, The red wine reaction syndrome, *Am. J. Enol. Vitic.*, 34:260–264 (1983).

115. Buteau, C., C. L. Duitschaever, and G. C. Ashton, A study of the biogenesis of amines in a Villard Noir wine, *Am J. Enol. Vitic.*, 35:228–234 (1981).

116. Ames, B. N., Identifying environmental chemicals causing mutations and cancer, *Science*, 204:587–593 (1979).

117. Tamura, G., C. Gold, A. Ferro-Luzzi, and B. N. Ames, Fecalase: A model for the activation of dietary glycosides by intestinal flora, *Proc. Natl. Acad. Sci. U.S.A.*, 77:4961–4965 (1980).

118. Ames, B. N., J. McCann, and E. Yamasaki, Methods for detecting carcinogens and mutagens with the *Salmonella*/mammalian-microsome test, *Mutat. Res.*, 31:347–364 (1975).

119. Josephy, P. D., and R. E. Subden, Hydrogen peroxide-dependent acitivation of benzidine to mutagenic species, *Mutat. Res.*, 141:23–28 (1984).

120. Stoltz, D. R., B. Stavric, D. Krewski, R. Klassen, and B. Junkins, Mutagenicity screening of foods. I. Results with beverages, *Environ. Mutagen.*, 4:477–485 (1982).

121. Subden, R.E., T. Haffie, and D. Rancourt, Mutagen content of homemade Canadian wines, *Can. J. Food Sci.*, 16:45–51 (1983).
122. Subden, R. E., A. Krizus, and D. Rancourt, Mutagen content of table wines made from various grape species and hybrid cultivars, *Food Chem. Toxic.*, 22:309–313 (1983).
123. Subden, R. E., A. Krizus, and D. Josephy, Mutagen content of wines contaminated with common yeasts, *Food Chem. Toxic.*, 23:343–347 (1984).
124. Mur, A., Methods de controle utilisées pour les levures comèrcialisées dans les differèrents pays, *Bull. O. I. V.*, 56:268–278 (1983).
125. Chen, S. L., E. J. Cooper, and F. Gatmanis, Active dry yeast. I. Protection against oxidative deterioration during storage, *Food Technol.*, 20:1585–1589 (1966).
126. Reed, G., and S. L. Chen, Evaluating commercial active dry wine yeasts by fermentation activity, *Am. J. Enol. Vitic.*, 29 (3):165–168 (1978).
127. Beckers, M. E., I. J. Krause, E. J. Ventina, J. G. Kontakevichl, and H. Damberga, Factors of wine yeast resistance at their dehydration, in *Fifth International Symposium on Yeasts,* London, Ontario, Canada (1980).
128. Kraus, J. K., R. Scopp, and S. L. Chen, Effect of rehydration on dry wine yeast activity, *Am. J. Enol. Vitic.*, 32:132–134 (1981).
129. Lemperle, E., and E. Kerner, Vergleichende Pruefung von Trocken-Reinzuechthefen. 2. Mitteilung: Ergebnisse aus dem Jahre 1980, *Weinwirtschaft*, 30:899–902 (1981).
130. Kusewicz, D., and J. Johnston, Genetic analysis of cryophilic and mesophilic wine yeasts, *J. Inst. Brew.*, 86:25–27 (1980).
131. Alikhanyam, S. I., and G. M. Nalbandyan, Selection of wine yeasts with the use of mutagens. I. Selection of *S. cerevisiae* used in the production of strong table wines, *Sov. Genet.*, 7:1200–1205 (1971).
132. Alikhanyan, S. I., G. M. Nalbandyan, and B. P. Avakyan, Selection of wine yeasts using mutagens. II. Selection of highly active alcohol resistant strains of *S. oviformis*for the production of Sherry wines, *Sov. Genet.*, 7:1287–1293 (1971).
133. Alikhanyan, S. I., G. M. Nalbandyan, and B. P. Avakyan, Selection of wine yeasts using mutagens. III. Selection of new strains of *S. vini* used in the production of champagne wines, *Sov. Genet.*, 7:1452–1457 (1971).
134. Avakyan, B. P., and N. A. Ter-Balyan, The selection of wine yeast with mutagens for the production of red wines, *Biol. Zh. Arm.*, 29:89–93 (1976).
135. Zambonelli, C., M. E. Guerzoni, M. E. Nanni, and G. Gianstefani, Selezione genetica nei lieviti della fermentazione vinaria. 2. Il potre stabilizzante del colore, *Rev. Vitic. Enol.*, 25:111–129 (1972).
136. Zambonelli, C., M. E. Guerzoni, and M. E. Nanni, Selezione genetica nei lieviti della fermentazione vinaria. 5. La modalita di sviluppo, *Rev. Vitic. Enol.*, 26:104–112 (1972).
137. Osothsilp, C., and R. E. Subden, Isolation and characterization of *Schizzosaccharomyces pombe* mutants with defective NAD dependent malic enzyme, *Can. J. Micro.*, 32:481–486 (1986).
138. Osothsilp, C., and R. E. Subden, Isolation and characterization of *Schizzosaccharomyces pombe* with defective malic acid permease, *J. Bacteriol.*, 168:1439–1443 (1986).
139. Rous, C. V., R. E. Kunkee, and R. Snow, Wine fermentations with amino acidless mutant strains of Montrachet resulting in lowered production of higher alcohols, *Am.*

Soc. Enol., abstract in *Proc. A. G. M.*, Los Angeles, CA (1980).

140. Rous, C. V., R. Snow, and R. E. Kunkee, Reduction of higher alcohols by fermentation with a leucine auxotrophic mutant of wine yeast, *J. Inst. Brew.*, 89:274–278 (1983).

141. Hara, S., Y. Iimura, and K. Otsuka, Breeding of useful killer wine yeasts, *Am. J. Enol. Vitic.*, 31 (1):28–33 (1980).

142. Thornton, R. J., and R. Eschenbruch, Homothallism in wine yeasts, *Antonie van Leeuwenhoek J. Microb. Serol.*, 42:503–509 (1976).

143. Cummings, J., and S. Fogel, Genetic homology of wine yeasts with *S. cerevisiae, J. Inst. Brew.*, 84:267–270 (1978).

144. Thorton, R. J., The introduction of flocculation into a homothallic wine yeast. A practical example of the modification of wine making properties by the use of genetic techniques, *Am. J. Enol. Vitic.*, 36:47–51 (1985).

145. Snow, R., Towards genetic improvement of wine yeast, *Am. J. Enol. Vitic.*, 30:33–37 (1979).

146. Young, T. W., and M. Yagiu, A comparison of killer character in different yeasts and its classification, *Antonie van Leeuwenhoek. J. Microbiol. Serol.*, 44:59 (1978).

147. DeLouvencourt, L., H. Fukuhara, H. Heslot, and M. Wesolowski, Transformation of *Kluyveromyces lactics* by killer plasmid deoxyribonucleic acid, *J. Bacteriol.*, 154:737–742 (1983).

148. Pfeiffer, P., and F. Radler, Purification and characterization of extracellular and intracellular killer toxin of *Saccharomyces cerevisiae* strain 28, *J. Gen. Microbiol.*, 128:2699–2706 (1982).

149. Zimmerman, F. K., Procedures used in the induction of mitotic recombination and mutation in the yeast *S. cerevisiae, Mutat. Res.*, 31:71–77 (1975).

150. Conde, J., and G. R. Fink, A mutant of *S. cerevisiae* defective for nuclear fusion, *Proc. Natl. Acad. Sci. U.S.A.*, 73:3651–3655 (1976).

151. Seki, T., E. H. Choi, and D. Ryu, Construction of killer wine yeast strain, *Appl. Environ. Microbiol.*, 49:1211–1215 (1985).

152. Dutcher, S. K., Internuclear transfer of genetic information in *karl-1*/KARL heterokaryons in *S. cerevisiae, Mol. Cell Biol.*, 1:245–253 (1981).

153. Kielland-Brandt, M. C., T. Nilsson-Tillgren, J. G. L. Petersen, S. Holmberg, and C. Gjermansen, Approaches to the analysis and breeding of brewers yeast, in *Yeast Genetics*, J. F. T. Spencer and D. M. Spencer (Eds.), Springer-Verlag, New York (1983).

154. Carrau, J. L., J. L. deAzevedo, P. Siedbery, and D. Campbell, Methods for recovering fusion products among oenological strains of *S. cerevisiae* and *Sch. pombe, Rev. Brazil Genet.*, 1:221–226 (1982).

155. Russel, I., and G. G. Stewart, Transformation of maltotriose uptake ability into a haploid strain of *Saccharomyces* spp., *J. Inst. Brew.*, 86:55–59 (1980).

156. Williams, S. A., R. A. Hodges, T. L. Strike, R. Snow, and R. E. Kunkee, Cloning the gene for the malo-lactic fermentation of wine from *Lactobacillus debrueckii* in *Escherichia coli* and yeasts, *Appl. Environ. Microbiol.*, 47:288–293 (1984).

157. Lautensach, A., and R. E. Subden, Cloning of malic acid assimilating activity from *Leuconostoc oenos* in *E. coli, Microbios*, 39:29–39 (1984).

6
Yeast Selection for Baking

Tilak W. Nagodawithana and Nayan B. Trivedi
Universal Foods Corporation
Milwaukee, Wisconsin

I. INTRODUCTION

Although cereals have been known to mankind from time immemorial, historians believe that the discovery of leavened bread may have been made accidentally in Egypt around 3000 B.C. Historical evidence in support of this theory stems from numerous scenes of bread making shown among the ruins dating back to the era in Memphis. Prior to this new development, the germinated barley was crushed between stones and the powdered meal was wetted to develop a dough, flattened, and then baked. The first idea of leavening probably originated subsequent to the discovery that the final baked product had better texture and improved digestibility and stability when the dough was left undisturbed for a period prior to baking. With further experience in baking, the beneficial effect of beer sediments on bread leavening was clearly recognized. Even though the bakers in this era were unaware of the causes for such chemical and textural changes in the dough due to the presence of the beer sediments, the experience they acquired was handed down through the centuries to each succeeding generation.

In 1680, A. van Leeuwenhoek, a Dutch lens grinder, with his development of the single-lens microscope, was able to describe for the first time the oval or spherical shape of yeast present in fermenting beer. However, the discovery did not shed any light on the function of yeast in alcoholic fermentations. This important question was finally resolved by Louis Pasteur through his publications in 1866 and 1876. His studies concluded that viable yeast cells cause fermentation under anaerobic conditions during which the sugar in the media is converted to

139

carbon dioxide and ethanol. With these results, the probable role of yeast in baking began to unfold.

During this period, brewers relied on the yeast cultures from a previous fermentation for the inoculation of succeeding brews. These cultures generally contained bacteria and wild yeast in addition to brewer's yeast. Consequently, the beer made from such brews was inferior and was not of uniform quality. This situation was finally resolved around 1881 by the adoption of the pure culture technique developed by Emil Christian Hansen of the Carlsberg Laboratory in Denmark. This technique was later extended to baking, and eventually gave rise to a new industry for the production of baker's yeast to meet the needs of the baking industry.

Despite these advances in science and technology, the brewing and baking industries took little advantage of the newly acquired knowledge for the improvement of industrial strains. It was not until the 1930s that a serious attempt was made to understand the genetics of yeast. This finally led to a burst of activity, especially in the last decade, for strain development in the wine, beer, and baker's yeast industries.

The following brief review on baker's yeast production is intended to provide some basic information on how different baker's yeast products are manufactured to meet the needs of the baker. This will not only enhance the reader's conprehension of the sections that follow it, but will also provide some information on the criteria that a yeast geneticist must take into consideration when defining the objectives of a strain improvement program. The reader can, however, find a more complete treatment of baker's yeast production in Reed and Peppler (1973) and Nagodawithana (1986).

II. COMMERCIAL PRODUCTION OF BAKER'S YEAST

The baker's yeast industry is generally regarded as a mature industry producing approximately 430,000 tons of yeast on a dry solids basis annually for use by the world's bakers. Although the most extensively used baker's yeast is compressed yeast, there are several other types of baker's yeast products used in various applications according to preferences of individual bakers. The performance of a yeast in a dough system is largely determined by its genetic constitution. Hence, it is not surprising that baker's yeast manufacturers rely on different strains to meet the specific needs of bakers.

A. Yeast Propagation

The principal objective of the yeast manufacturer is to propagate a given strain of baker's yeast at minimal cost and make it available to the baker. Strain selection is made by the yeast manufacturer based on the specific needs of a baker. For instance, high-sugar dough formulae require osmotolerant yeast for optimum baking activity, while those dough formulae with no added sugar require yeast with rapid maltose fermenting activity because the predominant fermentable sugar in doughs made from flour, water, yeast, and salt is maltose. Different authenticated starter cultures, made from single-colony isolates, are generally maintained at low temperature on malt agar slants, or slants overlaid with mineral oil. The basic principle of culture maintenance is to restrict growth while maintaining viability.

A yeast propagation, starting from a slant, takes approximately 6–8 days to produce a batch of yeast suitable for distribution to the bakers. In the process, several thousand pounds of crumbled or compressed yeast is produced from a few million cells inoculated at the initial stage of propagation.

The final commercial propagations, carried out in 20,000- to 50,000-gal fermentors, require large quantities of seed yeast for inoculation. At least six different fermentation stages are needed in the building-up process (Pasteur flask, PC1, PC2, F1, F2, and F3) to achieve the required seed to initiate the commercial propagation (Fig. 1) (Reed and Peppler, 1973). The first three or four stages are set batch propagations approaching pure culture conditions. However, the last two seed propagation stages, F2, F3, and commercial propagations (F4) are incrementally fed batch fermentations and are carried out under quasisterile conditions to avoid the high cost of sterilization. The success of a commercial propagation depends on the state of the pitching yeast. For this reason, F3 seed yeast is grown to contain approximately 9% nitrogen, 3% phosphate (P_2O_5) with 17–20% budded cells in the final population to ensure rapid growth during the commercial propagation.

A blend of cane and beet molasses is commonly used in commercial propagations. Yeast respiration is inhibited by high glucose concentrations even under highly aerated conditions (Crabtree effect). Adequate aeration coupled with high sugar levels during a propagation results in an aerobic fermentation with the production of ethanol and carbon dioxide. To prevent the wasteful production of ethanol and maximize biomass production, the wort is added incrementally into the highly aerated medium so that the feed rate of sugar is equivalent to the sugar assimilated by the yeast at any given stage of the propogation. Under these conditions, the residual sugar in the medium is maintained at a minimal level throughout the propagation.

The amounts of ammonia and phosphoric acid that can be used in a given propagation are determined by the desired composition of the yeast. In the case of baker's compressed yeast, it is often necessary to aim at 8–9% N and 2.5–3% P_2O_5 to achieve the desired activity during baking. However, in the case of dry yeast manufacture, the ammonia and phosphoric acid are added in such a manner as to ensure a final yeast nitrogen of 6.5–7% and a P_2O_5 of 2–2.3%. In either case, the

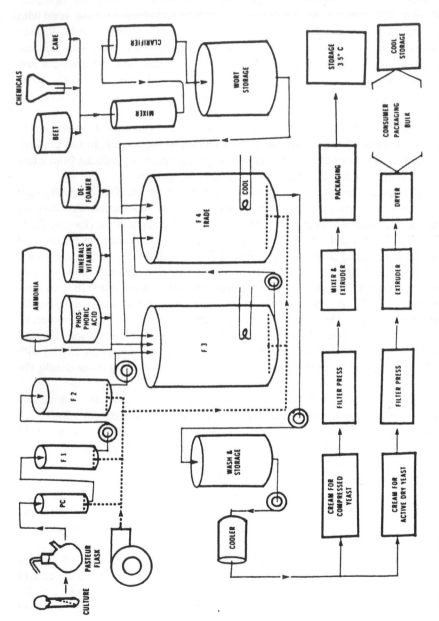

Figure 1. Schematic flow diagram for the production of baker's yeast. PC: Pure Culture; F: Fermentation Series 1–4.

pH of the medium is maintained at 5.2 with a temperature of 30°C throughout the propagation.

The cane molasses used in baker's yeast propagations is an excellent source of mineral and certain vitamins. It cannot, however, meet the Mg, Zn, and S requirements. These minerals along with the vitamin thiamine are also supplied to the growing yeast during the propagation. Beet molasses provides the vitamin mesoinositol for optimal growth of baker's yeast.

The duration of the commercial propagation may vary from 12 to 20 hr. The rate of aeration in the fermentor is governed by the amount of biomass actively propagating in the fermentor. This may vary from 0.5 to 1.5 volumes of air per fermentor volume per minute (VVM), depending on the stage of the propagation. A half hour of aeration without feed is generally provided at the end of propagation. This provides the maturation or ripening of the yeast and imparts greater stability.

The rate of aeration supplied to the fermentor can also be controlled to achieve a specific alcohol pattern. Hydrocarbon analyzers, which can monitor the exit gas, are commonly used for this purpose. More modern equipment, such as mass spectrometer-type gas analyzers coupled to computers, are beginning to provide more advanced control systems to the baker's yeast industry.

The yeast occupies 10–15% of the fermentor volume at the end of a propagation. The yeast is then separated using a continuous nozzle separator and the yeast cream is washed several times with clean water. The cream is cooled to refrigeration temperature and stored in a clean holding tank until it is ready for filtration and subsequent processing.

The final processed yeast cream has 15–20% solids. The removal of extracellular water to achieve crumbled yeast is done with plate and frame presses or in a continuous mode using rotary vacuum filters. Several types of baker's yeasts are made for use in various baking applications. The major types include (a) compressed yeast (or cake yeast), (b) crumbled yeast, (c) active dry yeast, and (d) instant active dry yeast. These four types of yeast products will be discussed further in the following section.

B. Types of Baker's Yeast

1. Compressed Yeast (or Cake Yeast)

Compressed yeast, or cake yeast, has a moisture content around 70% and a protein level in the 48–60% range on a dry solids basis. It also has an emulsifier added to it. Because of the high moisture content, the product is perishable and requires refrigeration (38–45°F) (3–7°C) until used. Compressed yeast is primarily used in the production of bread, buns, rolls, etc.

2. Crumbled Yeast

Crumbled yeast has a moisture content similar to that of compressed yeast but is processed without the emulsifier and is sold in the crumbled form in 50-lb bags. The yeast is used extensively by larger bakeries for the production of bread, buns, rolls, etc.

3. Regular Active Dry Yeast

Special strains of *Saccharomyces cerevisiae* show acceptable baking character-istics, along with good resistance to drying, are generally used for the production of active dry yeast (ADY). Yeast grown to contain high nitrogen is sensitive to heat and is not generally used for the production of ADY. For propagations where the yeast is to be dried, it is highly desirable to achieve a nitrogen content of 6.5–7% on a dry solids basis in the final yeast. It is also preferable to have a bud index of less than 1% in the final yeast population because of the sensitivity of buds to heat. The compressed yeast is extruded to form strands, or noodles, and is generally dried by use of a continuous tunnel dryer or a Rotolouver dryer. These products have a moisture content of approximately 6–8% and a protein content in the 40–45% range. Active dry yeast has the advantage of being relatively stable with a 6–8% activity loss at room temperature when stored for a month in a closed container. Storage in hermetically sealed containers, or under refrigerated conditions, will significantly extend the shelf life of ADY products.

Active dry yeast must be rehydrated by slurrying in water at 40–43°C for 5–10 min. Lower-temperature rehydrations are particularly damaging, resulting in the leaching of yeast solids into the environment. Likewise, dry yeast is adversely affected by rehydration temperatures higher than the specified range.

Active dry yeast can be used in any of the dough systems where compressed or crumbled yeast is used. Bakers substitute ADY for compressed yeast at the rate of 45–50% replacement by weight for achieving equivalent bake results, especially in sponge and dough systems. This conversion ratio, however, becomes lower in sweet dough systems owing to the osmotolerant nature of ADY strains (Thorn and Reed, 1959).

4. Instant Active Dry Yeast

Yeast strains selected for the production of ADY have an acceptable resistance to drying. Nevertheless, there is an appreciable loss in activity during the drying, resulting in a product of lower activity compared to that of compressed yeast. Such shortcomings in ADY products stimulated researchers to search for novel strains and processes for the production of high active dry yeast or instant yeast not significantly different in activity from compressed yeast.

Over the past decade, with the application of new advances in genetics and drying technology, yeast manufacturers have generated a series of high active dry yeast products specifically designed for different dough systems. Such products are referred to as instant or quick-rising yeast. These yeast strains are chosen for their ability to retain maximum activity through the specially designed drying process. The protein content of these yeasts may vary from 38% to 60%, and a moisture content of 4–6% is highly desirable. The drying process produces highly porous particles, which gives them the ability to rehydrate rapidly. One disadvantage of this porosity is that it also provides access to oxygen from the air, resulting in rapid loss of activity during storage. Accordingly, these products must be hermetically packaged in inert gas or under vacuum (Trivedi et al. 1984; Sanderson, 1984).

Unlike ADY, a special advantage of this product is that it can be added directly to the dry ingredients without prior rehydration. Instant active dry yeast can be substituted for compressed yeast at the rate of 33–40% by weight in sponge and dough systems because its activity is very close to that of compressed yeast on an equivalent solids basis.

III. STRAIN IMPROVEMENT FOR BAKER'S YEAST

The history of genetic improvement of industrial strains of yeast has progressed from simple selection of pure cultures through the discovery of the sexual life cycle of yeast and its application for strain improvement. Much of the early work of Winge and Laustsen (1937) has resulted in the development of methods for the isolation of single spores from asci of sporulating yeasts. Subsequent studies on spores led to the discovery of the haploid and diploid phase of the yeast life cycle. In 1938, the same two authors demonstrated the technique of achieving interspecific hybrids in yeast. These studies followed by the discovery of Lindegren and his associates that yeasts are heterozygous for the mating-type allele, marked the beginning of systematic approaches to strain improvement. Their work with cross-breeding culminated in the development of several useful strains that benefited the baking industry (Lindegren, 1949).

Methods of strain improvement through classical genetics include (a) selection, (b) selection following mutagenesis, and (c) cross-breeding or production of new hybrids. Nevertheless, the progress made in the field of yeast genetics during the past several decades has made it possible to set up breeding collections of haploid mating strains with known desirable characteristics. Additionally, yeast gene banks have been constructed from the DNA of appropriate strains, such that strains with some known and desired characters should be able to be constructed with precision and a higher probability of success.

Although the genetics underlying some important yeast characteristics are well understood, most characters of importance to baking are complex. Quantitative characterization has in itself been a major problem. This lack of understanding severely limits the use of the more recent techniques of gene splicing for strain improvement. This does not, however, preclude the use of "shotgun" approaches of recombination to achieve novel strains. Plant and animal breeding provide examples of highly successful recombination programs where selection is for complex characters controlled by large numbers of genes whose biochemical roles have always been obscure and unknown. Similar unpredictable approaches have been successfully applied to strain improvement in baker's yeast.

A.   Techniques Applicable to Yeast Strain Improvement

Types of baker's yeast have been changing in the recent past to accommodate extensive marketing strategies, and it seems possible that the traditionally used strains may no longer adequately satisfy the baker's complex demands. This has provided sound reason for strain improvement through genetic manipulation. The following sections summarize some of the genetic approaches yeast geneticists have used for baker's yeast strain improvement.

1.   Selection

Following Christian Hansen's introduction of the pure culture technique to the brewing industry, there was a significant interest on the part of industrial microbiologists to explore the possibility of extending the concept to the baker's yeast industry. The studies that followed resulted in the identification and use of single-cell clones from naturally occurring yeast populations capable of exhibiting acceptable baking characteristics.

Although Hansen's technique was able to provide cultures derived from a single cell, it failed to provide the expected cultural homogeneity for an extended period, possibly owing to spontaneous mutations and more likely as a result of mitotic recombination during vegetative propagation. Accordingly, there was a need to constantly select for the desired strain or to control the stability of the derived stock to maintain its homogeneity.

Even though the strains that were selected according to the aforementioned procedure were of average baking characteristics, these were of acceptable quality to the bakers at the turn of the century. However, recent developments have placed a greater demand on these traditional strains of baker's yeast. This has ultimately led to extensive strain improvement programs which employ either conventional genetic techniques or more advanced genetic manipulations.

2.   Selection After Mutagenesis

Mutagenesis has not been widely used for the improvement of baker's yeast strains. There are a few scattered reports in the literature, especially from Poland and the Soviet Union, of the use of ultaviolet (UV) and N-methyl-N'-nitro-N-nitrosoguanidone (NTG) mutagenesis for the improvement of baking strains (Johnston and Oberman, 1979). The other common agents used for mutagenizing yeasts have been ethyl methane sulfonate (EMS), nitrous acid, N-nitrosourea, and diethylstilbestrol. The mutations caused by these mutagens are generally recessive and are only expressed after gene conversion or mitotic recombination (Spencer and Spencer, 1983).

There is adequate information about the mechanism by which UV irradiation produces mutations in yeast. When UV light is absorbed by nucleic acids, the absorbed energy can cause alterations in the bond characteristics of the purines and pyrimidines. The bases so altered are called photo products. the pyrimidines are found to be more liable to such changes than purines. One of the consequences of the altered bond characteristics is the formation of covalent bonds between the adjacent pyrimidines of the same DNA strand to form dimers. The thymine dimer is one that forms most readily. Dimerization interferes with the proper base-pairing of thymine with adenine and may result in thymine pairing with guanine. This will produce a T-A to C-G transition. Other photo products, such as cytosine dimers, are formed to a lesser extent and cause mutations by becoming deaminated to uracil dimers. Since uracil acts like thymine, this will eventually produce a G-C to A-T transition. Many of these effects have been found to be followed by a photo reactivation repair mechanism. Such postirradiation conditions that favor the repair process have been reviewed by Jacobson (1981) and Stanier et al. (1970). The mutagenic effects of UV light can thus be enhanced by ensuring that the irradiated culture is not exposed to photo reactivation and by plating immediately after mutagenesis to prevent repair prior to DNA replication.

It is generally recommended that survival curves be established for each industrial strain in order to determine its sensitivity to UV light. Such data can thus help in the determination of the radiation dose required to achieve the desired level of mutagenesis in a particular strain improvement program.

Ultraviolet light has been recommended as the mutagen of first choice (Bridges, 1976). A strain of S. cerevisiae maintained in the stationary phase for 48 hr is thinly streaked on YPD (1% yeast extract, 2% peptone, and 2% dextrose) agar and immediately exposed to UV light of varying dosages (example: 15, 20, 25 sec) at a fluence rate of approximately $0.8.Jm^{-2}sec^{-1}$ as measured by a Lastarject dosimeter. The UV source may be from a mercury vapor lamp emitting radiation in the 253- to- 260-mu range. All experiments need to be carried out under yellow light to avoid photoreactivation. The irradiated culture is then incubated at 30°C in a dark chamber, and the surviving clones can then be picked and streaked onto agar plates for subsequent screening.

Similarly, NTG has been used to produce commercially useful strains of baker's yeast. It has been well established that NTG alkylates base residues in DNA (Lawley and Thatcher, 1970), probably the most important adduct being O^6 methyl guanine (Sklar and Strauss, 1980) and O^4 thymine (Singer, 1975). For this reason, alkylating agents produce a preponderance of G-C to A-T transitions. An interesting property of the alkylating agent NTG is its specific interaction with DNA at the replication fork. These induced base-pair substitution mutations described previously have been known to occur preferentially at the replication forks. NTG is also known to impart multiple mutations in closely linked genes (Guerola et al., 1971). This phenomenon has been referred to as comutation.

Early stationary phase cultures of yeast are centrifuged and resuspended at about 10^8 cells/ml in filter-sterilized TRIS (hydroxymethyl) amino methane-maleate buffer (50 mM, pH 7.8). The nitrosoguanidine is added at at final concentration of 20 μg/ml. The suspension is then incubated at 37°C for 15 min with gentle and intermittent shaking. The yeast is then washed twice by centrifugation, suspended in distilled water, and plated on appropriate media for subsequent screening.

3. Hybridization

Winge and Laustsen (1937) were the first to attempt the production of new strains of yeast by hybridization. However, they encountered difficulties with homothallism, poor mating ability, poor sporulation, and poor spore viability. These authors hybridized yeast by placing a haploid ascospore from one strain in close proximity to an ascospore of the second strain of opposite mating type by use of a micromanipulator. Under favorable conditions, the two spores mated to produce a diploid cell.

Lindegren (1949) improved this procedure based on the observation that ascospores from single mating type heterothallic strains of *S. cerevisiae* persistently produced stable haploid cultures. It was thus possible to hybridize two different haploid cultures of opposite mating types by simply mixing the cells together in an appropriate medium.

Good sporulation depends on yeast cells' being in an active physiological state prior to the onset of sporulation. This implies that the cells must be grown on a medium that ensures they are well nourished. The use of 16% w/v malt extract wort as a presporulation medium for baker's yeast was found to give excellent results (Fowell and Moorse, 1960). The effectiveness of other presporulation media have been reviewed by Fowell (1970).

Ten milliliters of a suitable liquid presporulation medium is inoculated with yeast and incubated at 30°C for 24–48 hr with occasional shaking. Yeast cells are then isolated by centifugation, washed at least twice with distilled water, and then resuspended in 0.5 ml distilled water. A 4-mm inoculating loop, which holds about 0.01 ml of aqueous suspension, is used to inoculate a slant containing 0.5% sodium acetate, 1% KCl, and 1.5% agar. After inoculation, the slant is incubated at 25°C for 3–5 days.

The separation of spores from a vegetative cell suspension can be accomplished by shaking with paraffin oil. The spores enter into the oil phase. The ascus wall may now be ruptured mechanically or by the use of special enzymes, and the individual spores may be isolated by micromanipulation. Selective heat treatment may also be used as a method of isolating haploid spores because the spores are more heat-resistant than the vegetative cells. If the entire population is suspended in sterile water and held at 50°C for 2–4 min, the vegetative cells will be killed and only the spores will survive. These suspensions can then be streaked on agar plates and incubated at 30°C for 2 days. A number of small, well-spaced colonies can then be picked from these plates. The larger colonies are avoided, as they may have arisen from vegetative cells that survived heating.

Lindegren's mass mating technique, which is commonly used in the yeast industry, involves mixing together large numbers of haploid cells of opposite mating type in a nutrient medium. The resultant mixture is left overnight at 16°C and then examined for the presence of zygotes. This is generally determined by transferring the culture directly to gypsum in order to induce sporulation in diploid hybrids. According to Lindegren and Lindegren (1943), such asci are only produced by "legitimate" diploids.

Lindegren's technique of mass mating has been criticized by Winge and Roberts (1948) because of the possibility that some or all of the diploids may arise as a result of self-diploidization. Attention has been directed to modifying Lindegren's mass-mating technique with the intention of minimizing failures and reducing the time needed for the production of hybrid strains. Accordingly, a substantial saving in time, labor, and material has been achieved by preparing a mating mixture and then subculturing it two or three times to encourage the hybrid cells to outgrow the haploid cells. Single large cells are then isolated by use of a micromanipulator. These cells can then be transferred to tubes containing malt extract wort for propagation. Such strains, when sporulated, have always been found to segregate a and α spores, so it is almost certain these have arisen as a result of mating, and not self-diploidization. These strains can then be screened for baking characteristics. Such a modification of Lindegren's mass-mating technique has proved suitable for hybridization of different kinds of yeast by the industry.

4. Protoplast Fusion

As in the case of every organism bred by man, selection of better yeast strains was, for thousands of years, an empirical process that occurred without any knowledge of heredity. With greater understanding of genetics and the yeast life cycle, scientists made use of hybridization techniques to develop suitable traits. However, it was soon found that good industrial yeast strains are often polyploid or aneuploid, and as a consequence, they do not possess a mating type, they have a low degree of sporulation, and they have poor spore viability. While such strains are genetically stable, enabling the manufacture of a consistent product, their superior character could not be genetically analyzed or combined sexually.

Protoplast fusion provides a means of overcoming the genetic barrier of polyploidy in yeast because sexual mating is not required for the formation of fusion products. Using enzymes isolated from the gut of the snail *Helix pomatia*, Eddy and Williamson (1957) reported the first successful preparation of yeast protoplasts. These snail gut enzymes were found to degrade and strip the rigid polysaccharide cell wall material from the cells. This breakthrough spurred the development of the protoplast fusion technique. With further improvements, this basic technique has predominated as a method of choice to improve industrial strains of *S. cerevisiae* during the past decade (Trivedi et al., 1986; Stewart et al., 1983; Van Solingen and Van Der Plaat, 1977).

Recently, the baker's yeast industry has become active in the construction of new yeast strains. The industry is interested in improving a number of characteristics of yeast strains, including greater carbon dioxide for baked goods, osmotolerance, and high maltase activity. However, most of these characteristics either are polygenic or are not well understood at the molecular level. Therefore, the improvement of many of these characteristics cannot be achieved by cloning or DNA transformation techniques, because these techniques are generally used to transfer only a few well-characterized genes at a time. Protoplast fusion circumvents the problem of specific gene identification and isolation, and does not require the use of molecular probes. Furthermore, fusions between different genera of yeast are possible. Fusions between different genera may further expand the genetic diversity available in strains over that which can be obtained by interspecies hybridization or fusion.

Protoplast fusion has successfully been used to construct a number of novel baker's yeast strains. For instance, fusion techniques have been used to construct strains that exhibit improved leavening characteristics, as well as higher levels of osmotolerance. The higher osmotolerance allows these strains to perform better in sweet dough (up to 20% sugar) baked products. The procedure and basic techniques used in protoplast fusion are described here in detail.

First, both parental strains are used to isolate spontaneous petite (respiratory deficient) mutants. The strains to be used in the fusion experiment are characterized by heterozygosity at the mating type locus, so neither the parental strains nor their petites could be mated in nature. The two parental strains are cultivated on a medium containing dextrose or sucrose as the sole source of carbon. The spontaneously occurring petite colonies are distinguished from the *grande* colonies by size, shape, and color. One technique that is useful in identification is to use an agar overlay containing the dye 3–5-triphenyl tetrazolium chloride. This dye stains the normal *grande* colonies pink, while petites remain white. The petites are then isolated and subcultured on agar media containing 1% yeast extract, 2% peptone, and 2% dextrose (YPD). Following subculturing, colonies are transferred to an agar media containing 1% yeast extract, 2% peptone, and 3% glycerol. Therefore, this step offers a final confirmation of the respiratory-deficient nature of these mutants to be used in the fusion experiment.

The selected petite mutants from each parental strain are grown in YPD broth for 48 hr, harvested by centrifugation, and washed twice with sterile distilled water. Both washed cultures are then resuspended in a hypertonic buffer and individually treated with the enzyme zymolyase (β- glucuronidase). This step removes the cell walls and thus creates the protoplasts. The protoplasts are then washed with hypertonic buffer, gently resuspended in a fusion buffer, and combined in a 1:1 radio of protoplasts. The fusion buffer contains polyethylene glycol (molecular weight 4,000 to 6,000), sorbitol, and calcium salt. Following a 30-min incubation at 30°C, the protoplast mixture is gently centrifuged to remove the fusion buffer. The resulting pellet is resuspended in a recovery broth and incubated overnight. The recovery broth cotains 1% yeast extract, 0.8M sorbitol, and 0.2% dextrose. Following incubation in the recovery broth, the regenerating protoplasts are plated on hypertonic glycerol agar containing 1% yeast extract, 2% peptone, 3% glycerol, 0.8M sorbitol, and 3% agar. Only fusion products, which have recombined to complement each of the parental types' respiratory deficiency, will proliferate on this agar. Colonies of individual fusion products are then picked and employed in a subsequent screening program.

Recently, several new methods have been developed in the field of fusion technology. One such technique that deserves mention is electrofusion. Polyethylene glycol-mediated electrofusion first employs a short treatment of the protoplasts with alternating current. This aids in "lining up" or orienting respective protoplasts in a side-by-side fashion. This is followed by a very short (microseconds) burst of direct current which facilitates membrane fusion. Electrofusion has been shown to greatly enhance the yield of intrageneric fusion products (Schnettler et al., 1984).

5. Transformation and Recombinant DNA

Transformation is widely used for achieving genetic recombinations while offering a means of overcoming the nonspecificity inherent in protoplast fusion. Beginning in 1978, advances in the application of recombinant DNA methodology to investigations of yeast provided important new tools for examining the structure and the mechanism of gene expression of several important genes within the yeast cell. Development of appropriate plasmid cloning vectors has made it possible to clone virtually any yeast gene for which a mutation can be identified. This has generally been accomplished by introducing clone banks through the use of plasmid cloning vectors into mutant strains of interest and selecting and screening for complementation of the mutant phenotype. Transformations made in this manner have opened the way for genetic improvement of industrial yeasts by direct introduction of one or two genes from any desired source.

Hinnen et al. (1978) were the first to describe a successful yeast transformation by treating the yeast protoplasts with foreign DNA in the presence of polyethylene glycol. They used a chimeric ColE1 plasmid carrying the yeast LEU 2 gene (pYeleu10) to transform spheroplasts of a yeast strain carrying a leu 2- double mutation. thus, by suitable selection, these investigators were able to identify the first yeast transformants by screening for strains that were leucine prototrophs.

Four different types of vectors are available that facilitate the introduction of genetic information into the yeast cell (Botstein and Davis, 1982). These have been named for the way in which they are maintained in yeast after transformation. The vectors that can be maintained only by integration into homologous chromosomal DNA are designated YIp (yeast integrating plasmid); those that use a fragment of the 2 μM plasmid for maintenance are termed YEp (yeast episomal plasmid); those that are maintained because they contain an autonomously replicating sequence are called YRp (yeast replicating plasmid); and the plasmids that contain a functional centromere are called YCp (yeast centromere plasmid). Essentially, the YCp vectors are YRp vectors that contain functional centromeric DNA which provides a site of attachment for microtubules and regulates the proper chromosomal movement during mitosis and meiosis. Hence, YCp vectors behave in yeast as stable minichromosomes without being affected by cell division. These vectors generally contain both yeast and bacterial selection and maintenance elements for ease of handling and manipulation. A complete description of the different systems available for transformation is beyond the scope of this review. However, the reader can find a more complete treatment of these systems in Stiles et al. (1983).

The aforementioned procedures are useful for mutating and altering small segments of the *S. cerevisiae* genome that do not exceed one or two genes. Nevertheless, most yeast characteristics important to the baker are known to be complex, and a precise understanding of these genes and their control mechanisms is a prerequisite for successful application of rDNA technology. Hence, an immense data base has to be developed, particularly with regard to genetic systems associated with baking characteristics, before some of the more imaginative gene manipulations can be attempted.

Strain selection and strain improvement are ongoing research programs in the baker's yeast industry. These programs are aimed at supporting two principal objectives: achieving bake activities superior to those of competitive products, and manufacturing a low-cost baker's yeast. Under the first category the baker's yeast industry has shown considerable interest in developing new strains with (a) rapid maltose fermenting ability, (b) better osmotolerance, (c) faster fermentation rates, (d) improved brew dough activity, and (e) better freeze tolerance.

In support of process economics, the production of baker's yeast can be made more cost effective through the use of novel strains that have (a) rapid melibiose utilizing ability, (b) rapid lactose utilizing ability, and (c) increased temperature tolerance.

B. Strain Improvement for Superior Baking Quality

1. Rapid Maltose Adaptation

Dough systems are generally characterized by the level of sugar included in the respective formulations. Some baked goods are produced primarily from flour, water, salt, and yeast with no added sugar. These are lean dough formulations where the main source of fermentable sugar is produced when the amylolytic or diastatic enzymes (α and β amylases), which are naturally present in the flour, catalyze the hydrolysis of the starch of damaged starch granules. The maltose derived by such enzymatic reactions is fermented during the final proofing period, and the gas thus formed prior to baking is responsible for the final loaf volume of the bread. French and Vienna breads are types of baked goods made from lean dough formulations. They are characterized by their coarse, tough crumb and hard crust. The adaptability of yeast to maltose fermentation assumes importance not only in lean dough formulation, which contains little or no added sugar, but also in the sponge stage of sponge and dough systems and preferments containing flour to which little or no sugar has been added.

The total fermentable sugar naturally present in the flour amounts to approximately 0.25% and consists of glucose, fructose, sucrose, and maltose. Maltose is the predominant fermentable sugar in the dough, primarily due to the enzymatic degradation of starch subsequent to hydration of the flour. Its utilization by an unadapted yeast cell requires an induction period after the glucose level in the medium has been reduced to a negligible proportion (<0.1%). The fermentation of maltose by *Saccharomyces* strains requires at least two enzymes: α-D-glucosidase (maltase), and maltose permease. When cells are grown in the absence of the inducer maltose, or under glucose repressing conditions, only very low basal levels of the enzymes are detected. Addition of maltose leads to a coordinate increase in both maltase and maltose permease levels. Generally, there is a lag in the appearance of the two enzymes after the initiation of the induction. This lag period is entirely strain-dependent.

The development of yeast strains that utilize maltose under repressing conditions would be highly beneficial in a lean dough system. The genetic approach has been to develop strains that either exhibit constitutive maltase and maltose permease synthesis or have an increased copy number of MAL genes to take advantage of their additive effect. Khan and Eaton (1971) have, in fact, been able to isolate mutants that have constitutive maltase and maltose permease synthesis.

The presence of any one of the family of five unlinked MAL genes (MAL1, MAL2, MAL3, MAL4, and MAL6) confers on yeast the ability to utilize maltose as a carbon source. However, most haploid strains that have the ability to utilize maltose contain only a single functional copy of one of the MAL genes. This is often referred to as the dominant gene. Naumov (1971, 1976) concluded that each MAL locus was composed of two linked genes termed MALp and MALg. Naumov (1976) and ten Berge et al. (1973) suggested that the MALp gene is associated with a regulatory function. Subsequent studies of Federoff et al. (1982) and Needleman and Michels (1983) showed that the MALg gene encodes the maltase structural gene. Needleman et al. (1984) subsequently reported that the MALg segment contained not one but two maltose-inductible transcripts, one for maltase (MALs) and the other for maltose permease (MALt), both being regulated at the transcriptional level by the MALp gene product.

Khan and Eaton (1971) reported that yeast strains containing the MAL4 gene synthesize the maltase enzyme constitutively even under repressing conditions. On the contrary, MAL1, MAL2, MAL3, and MAL6 appear to be maltose-inductible and glucose-repressible, and are directly under the control of their respective regulatory genes.

Selection, or selection after mutagenesis, has been used by many investigators to develop strains that appear to constitutively synthesize maltase and maltose permease. Although many of these strains were shown to be unsuitable for commercial baking when crossed with baking strains of poor lean dough quality, the hybrids were found to ferment maltose rapidly and without a lag period. These strains were sometimes referred to as "quick" strains of baker's yeast because of their rapid adaptability to maltose. These yeasts are described in British patents 868, 621 (Burrows and Fowell, 1961a), 868, 633 (Burrows and Fowell, 1961b), and 989, 247 (Koninklijke Nederlandsche, Gist-en Spiritusfabriek, 1965). Although these quick strains are highly adapted to maltose, they are generally osmosensitive and are significantly slowed by mold inhibitors such as acetic, sorbic, or propionic acid or their salts. Recent work done in the United States, France, and the Netherlands has shown that it is possible to improve the usefulness of such quick strains by the incorporation of other desirable baking characteristics (example: osmotolerance, acid tolerance, etc.) either by classical hybridization or by protoplast fusion. Micromanipulation has been the preferred method when crosses are made in which one of the haploids in the crossing is a member of a *Saccharomyces* sp. other than *S. cerevisiae*.

Geneticists have often considered constitutive synthesis of maltase and maltose permease as a character vitally important in the construction of strains for instant active dry yeast (IADY). Rapid adaptability to maltose by yeast not only improves lean dough characteristics but has also been found to impart greater stability to the yeast during drying. A high level of trehalose in yeast is also known to improve the tolerance of yeast to drying. According to the data reported by de Oliveira et al. (1981) and Operti et al. (1982), trehalose accumulation in yeast was evident in glucose-mediated propagations, provided the strain used carries constitutive MAL alleles. Strains with such a genetic makeup have accordingly shown good dryability together with acceptable leavening (Giesenschlag and Nagodawithana, 1982). Although the constitutive synthesis of the MAL system enzymes greatly influences the leavening in a lean dough system, it has, obviously, very little effect on sweet dough systems supplemented with sucrose or corn syrup.

2. Strain Improvement for Osmotolerance

Most baking strains of yeast show a decline in leavening activity when sugar concentration in the dough exceeds 4% based on the weight of the flour. Likewise, salt concentrations in the 2–2.5% range behave similarly, causing considerable inhibition in yeast activity. This is a common osmotic phenomenon observed with yeast in high-sugar or high-salt dough systems. With sugar concentrations sometimes as high as 25%, products such as Danish pastries, coffee cakes, doughnuts, etc. belong in this category. The poor yeast activity in these high-sugar systems is compensated for by the use of higher levels of yeast (7–10% on a flour basis) to achieve satisfactory leavening. Since the baker prefers a more cost-effective way of leavening sweet dough systems, there exists a need to develop yeast strains with greater osmotolerance.

The basic understanding about the response and adaptation of osmotolerant yeasts to higher sugar concentrations is based on physical observations and, to some degree, on the biochemical changes taking place within the cell. Although the underlying genetics associated with osmotolerance is not fully explored, this does not preclude the use of certain recombination techniques to bring together the prerequisite genes. Most investigators have found the following features in baker's yeast highly desirable for achieving osmotolerance and, consequently, good sweet dough activity. Accordingly, the yeast strain must have (a) the ability to accumulate high intracellular levels of glycerol to rapidly equilibrate with the high osmotic environment, (b) a high trehalose level in the yeast, probably to maintain membrane stability, and (c) a low invertase activity for use in those doughs containing sucrose as the sweetener.

The exact manner in which these gene complexes interact in an orderly fashion to provide the necessary osmotolerance in yeast is not precisely understood. However, a review of the literature on the progress made to date should broaden our understanding of the subject.

Glycerol Production and Osmotolerance. Osmotolerance in yeast is determined by the genetic constitution of the organism and the conditions under which the organism is grown. At the molecular level, there is a chain of events that occur upon exposure to high osmotic environments. This leads to osmotic adaptation involving osmotically modulated enzymes coupled with appropriate adjustments in the membrane.

Much of the information available on osmoregulation of yeast is the result of work done on the osmotolerant yeast *Saccharomyces rouxii*. This is one of the few species of yeast that can withstand high osmolality in its environment, but it has very poor baking characteristics. The more osmosensitive *S. cerevisiae* strains of yeast are able to grow in sugars at water activity (Aw) levels down to about 0.9. *S. rouxii*, on the other hand, will grow in the range of approximately 0.6–1.00 Aw. *S. rouxii* responds to diminished water activity by the production and subsequent retention of glycerol within the cell. At low water activity, the *S. rouxii* plasma membrane is known to become remarkably impermeable to glycerol. *S. cerevisiae*, on the contrary, synthesizes a greater amount of glycerol under similar conditions but can retain only a low level because of the permeability of its plasma membrane to glycerol (Brown and Edgley, 1980). There is evidence to show that glycerol transport is active in *S. rouxii*, but not in *S. cerevisiae* (Brown, 1974). It is not certain whether the active transport of glycerol is in any way associated with the higher level of linoleic acid (C 18:2) present in the membranes of *S. rouxii*. In a previous study with erythrocytes, Walker and Kummerow (1964) reported that an increase of linoleic acid in the membranes resulted in a loss of permeability to glycerol. These findings may be helpful in elucidating the role of linoleic acid in the membranes of *S. rouxii*.

Although osmotolerant and nonosmotolerant yeast strains produce glycerol, their regulatory controls and modes of glycerol synthesis are known to be different (Brown and Edgley, 1980). *S. cerevisiae* depends almost exclusively on glycolysis to generate NADH for glycerol synthesis whereas *S. rouxii* appears to depend primarily on NADPH generated by the pentose phosphate pathway. It would seem probable that the two systems are regulated by two different complex mechanisms. Hence, the number of genes taking part in each system, besides being different, may be too great for genetic manipulation through cloning.

Despite *S. rouxii* being a highly osmotolerant yeast, the strain was originally found unsuitable for hybridization because of its poor mating and sporulating characteristics. Kosikov and Miedviedieva (1976) have isolated stable mutants of *S. cerevisiae* with improved osmotolerance by mutagenesis and enrichment techniques. However, these investigators were unable to use these mutants for hybridization with commercial baker's strains because of their inability to sporulate. Despite such problems, hybridization has been used in Japan and Europe to obtain osmotolerant yeasts for baking (Gunge, 1966; Windisch et al., 1976). Some investigators have relied on other osmotolerant baker's or distiller's yeasts to achieve osmotolerance by hybridization.

In the mid 1970s, researchers recognized the great potential of the newly developed protoplast fusion technique. This brought about an upsurge of interest among those who were seeking a means of hybridizing yeasts such as *S. rouxii* that were difficult to mate or sporulate. The feasibility of constructing a stable hybrid between highly fermentative *S. cerevisiae* and osmotolerant *Saccharomyces mellis* strains by protoplast fusion has been demonstrated by Legmann and Margalith (1983). The hybrids showed a significant improvement in their fermentative ability at high sugar concentrations. Although this strain has been tested in the distilling industry, there are no reports to indicate its applicability to the commercial baking industry. More recently, Spencer et al. (1985) were able to construct a hybrid between *Saccharomyces diastaticus* and *S. rouxii* by protoplast fusion. The hybrid was found to be similar to the *S. diastaticus* parent. These hybrids have been found to perform exceptionally well in sweet dough systems. In fact, they perform better than commercial baker's yeast under identical conditions.

A recent development in the yeast manufacturing industry is the introduction of instant active dry yeast (IADY) or quick-rising yeast which have baking characteristics remarkably superior to other active dried yeast (ADY) products in the market. For example, Clement and Hennette (1982) patented a process for obtaining new strains of yeast for bread making by mutation and/or hybridization. According to this invention, the hybridization consisted of systematic crossings of haploids derived from "quick" strains of *S. cerevisiae* adapted to maltose with haploids derived from very "slow" strains not adapted to maltose, but well adapted to high-sugar doughs. In some crossings, quick strains adapted to maltose have been used as parents. To improve the osmotolerance of the yeast, the parents were first mutagenized to obtain isolates that had lower invertase levels.

Similarly, Jacobson and Trivedi (1986) reported the successful development of a novel strain for use in instant active dry yeast production by use of the protoplast fusion technique. Following the construction of new hybrid yeast strains by protoplast fusion, initial tests were performed on each of the isolates to select those showing rapid growth potential. The next step of selection involved measurement of carbon dioxide production by individual fusion products in three test dough systems. The systems included tests for lean and sweet dough activity. Approximately 50 mg of dry yeast solids was added to the model mixes listed below. The yeasts were suspended in 15 ml of water before addition to the systems.

Test A	Test B	Test C
Flour 20 g	Flour 20 g	Flour 20 g
Water 15 ml	Water 15 ml	Water 15 ml
	NaCl 0.4 g	NaCl 0.4 g
		Sucrose 4 g

The doughs were mixed for 45–60 sec and incubated at 30°C for 4 hr. Gas volume was measured, using a fermentometer, at half-hour intervals. Strains that produced more than 300 ml of carbon dioxide per 100 mg calculated dry yeast solids in test A, and at least 200 ml of gas in tests B and C, were selected for further study. Following a subsequent series of screenings, using shake flasks, bench-top fermentors, and pilot scale propagations, two strains were selected for industrial-scale production. These two strains were designated NRRL Y-15338 and NRRL Y-15339 (Jacobson and Trivedi, 1983). Table 1 shows the baking characteristics of these two strains obtained after commercial production runs.

The minimum standards of performance, based on commercial averages, are: lean dough—116 min; regular dough (white pan bread type)—140 min; and sweet dough—155 min.

Trehalose Accumulation and Membrane Stability. Oura et al. (1974) demonstrated that stopping the sugar feed toward the final phase of a commercial baker's

Table 1. Gassing Characteristics in Regular, Lean, and Sweet Dough Fermentations of Two Commercially Acceptable Fusants Constructed by Protoplast Fusion

	Rise Time (min)		
Yeast strain	Regular dough	Lean dough	Sweet dough
NRRL Y-15338	122	117	124
	121	116	124
NRRL Y-15339	124	112	121
	116	115	101

yeast propagation, without change of aeration, resulted in a yeast product of greater stability and improved leavening characteristics. Numerous studies in our laboratory and others have provided conclusive evidence that higher levels of reserve carbohydrate in yeast make them most resistant to adverse conditions. Great importance is attributed to trehalose whose actual concentration is, currently, an important criterion for evaluating the hardiness of dry baker's yeast. In this connection, the ability of anhydrobiotic organisms, such as macrocysts of the slime mold *Dictyostelium,* cysts of *Artemia salina,* and some nematode worms and larvae, to survive after complete dehydration, has also been attributed to a high concentration of trehalose (Crowe et al. 1984).

In baker's yeast, trehalose (α-D glucopyranosyl(1-1)D-glucopyranoside), a nonreducing disaccharide, can account for more than 20% or, in some cases, less than 1% of the dry weight. The level of trehalose is entirely governed by the growth conditions and the stage at which the sample was taken during the growth cycle. Early studies suggested an intracellular compartmentalization of trehalase and its substrate trehalose (Avigrad and Neufeld, 1965). Keller et al. (1982) have shown that trehalose is located in the cytosol and trehalase in vacuoles. When cells accumulate trehalose, the intracellular concentration can easily exceed 0.1 M based on total cell volume. It has been observed that trehalose often accumulates in a transitory response to adverse conditions. Lillie and Pringle (1980) and Keller et al. (1982) have suggested the possibility that trehalose plays a role not only in osmoregulation but also as a protective agent in maintaining the delicate structure of the macromolecules and the cytoplasm under conditions of stress. The good leavening activity observed in high sugar doughs by yeast containing high trehalose levels may thus be due to the high degree of osmotolerance acquired by the yeast.

When biological membranes are dehydrated in the absence of stabilizers, highly destructive changes take place during the phase transition, causing irreversible damage to the membrane. Such changes are generally avoided by dehydration in the presence of trehalose (Crowe et al. 1984). These investigations suggest that hydrogen bonding between hydroxyl groups of the trehalose molecule and the phosphate groups of the phospholipids may play a role in this protective effect. These hydrogen bonds may replace the same or similar hydrogen bonds between the lipids and the water that would be expected in a naturally occurring hydrated membrane. The trehalose molecules would thus perform a function similar to that of the water molecule in maintaining the integrity of the plasma membranes. The protective effect of trehalose to yeast membranes would then become additionally important under conditions of dehydration or rehydration.

In recent studies, it has been shown that trehalose accumulation, or TAC (+) phenotype, is closely associated with maltose utilization (Panek et al., 1979; De Oliveira et al., 1981). An essential requirement for expression of the TAC (+) phenotype is that a given MAL gene be in the constitutive mode, MAL^c. These investigators demonstrated that mutation of a constitutive MAL allele to a maltose-inductible or maltose-nonfermenting (mal) state results in a decline, or a complete loss, of trehalose accumulation during growth in a glucose medium. Results of these studies confirm that the TAC (+) trait is a direct function of MAL^c genes and not associated with different, but closely linked genes.

In yeast strains containing an inducible MAL gene, the degree of trehalose accumulation varies with the type of carbon source. In glucose media, trehalose accumulation is inhibited in the presence of nitrogen sources, whereas yeast growing on maltose under similar conditions accumulates trehalose readily during growth (Panek et al., 1979). It is conceivable that maltose induces the formation of a "nitrogen-insensitive" component (NIC) of the trehalose-accumulating system which replaces the "nitrogen-sensitive" component found in cells grown on glucose. Accordingly, NIC formation would be regulated by MAL genes similar to the regulation of maltase, maltose permease, and $\propto$-methyl glucosidase, as shown in Fig. 2. Hence, MAL gene products behave like a positive regulator coordinating the expression of four MAL gene products, including the gene which codes for a nitrogen insensitive component of the trehalose accumulation system.

Khan and Eaton (1971) were able to identify a strain labeled 1403-7A which carries a MAL4 allele that was found to be dominant and constitutive. Maltose constitutive hybrids have also been constructed (Koninklijke Nederlandsche Gist-en Spiritusfabriek N.V., 1965). Although these strains were found unsuitable for baking, haploid crosses with other baking strains permitted the development of improved strains of baker's yeasts with the ability to adapt rapidly to maltose. Similar strains have been developed and patented by Clement and Hennette (1982), with one of the important criteria for selection being satisfactory trehalose accumulation. More recent patents by Jacobson and Trivedi (1983) have made use of protoplast fusion to develop strains capable of accumulating high levels of trehalose for use in the instant active dry yeast market. Under repressed or

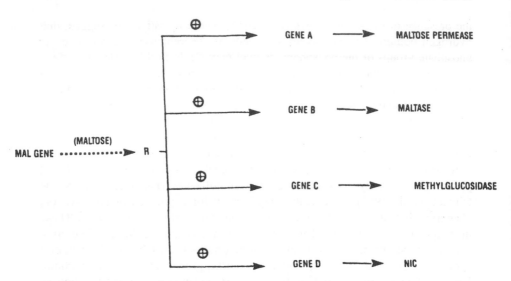

Figure 2. A proposed mechanism for coordinate regulation of maltose utilization, alpha-methylglucosidase, and a component of a trehalose accumulation system. R: Gene product of MAL gene, a positive regulator. NIC: A "nitrogen-insensitive" component (NIC) of trehalose accumulation system.

nonrepressed conditions, these strains have been found to contain higher levels of maltase activity than that normally associated with glucose-repressible strains.

Low Invertase Activity and Osmotolerance. Most major baker's yeast manufacturers use cane and beet molasses as their principal sugar substrate in their manufacturing practices. The relative proportions of the two substrates used are basically dictated by availability and economics. Both substrates contain sucrose as the predominant fermentable sugar. Sucrose is also commonly used at varying levels in a variety of dough systems for improvement of flavor and as a source of sugar for leavening. Hence, the ability of baker's yeast strains to utilize sucrose rapidly, while being unaffected by its osmotic pressure, is of immeasurable importance to those bakers whose primary interest is making sweet bakery products.

During the assimilation process, sucrose is first rapidly hydrolyzed outside the cytoplasm by the enzyme invertase, located in the cell wall or the periplasmic space or bound to the outer surface of the cytoplasmic membrane. The products formed as a result of invertase activity, glucose and fructose, are transported into the cell by facilitated diffusion. Six unlinked polymeric genes (SUC1 to SUC5 and SUC7) have been identified with sucrose utilization. Most strains do not carry all six SUC genes, but any one of the SUC genes is known to confer sucrose fermenting ability to the yeast cell (Carlson et al., 1981). Hence, a given industrial strain may contain one or more SUC genes, with the number of copies dictated by the ploidy of the cell. Grossmann and Zimmermann (1979) have demonstrated the effect of gene dosage on invertase activity.

Each SUC gene can encode an internal, nonglycosylated invertase enzyme which becomes glycosylated during its passage to the cell surface. The latter enzyme is largely under glucose regulation and confers on the yeast strain the ability to utilize sucrose. At low glucose concentrations, the external invertase is known to increase by a factor of 1,000 (Gascon and Lampen, 1968).

Excessive invertase levels in baker's yeast have been found to adversely affect the yeast activity and consequently their leavening power in sweet dough systems. This is primarily due to the rapid doubling of osmotic pressure that results from the rapid hydrolysis of sucrose to glucose and fructose. However, this phenomenon is of no significance when glucose or high-fructose corn syrup is used instead of sucrose in sweet-dough formulations. Some yeast manufacturers, however, reduce the level of invertase in their yeast by the application of an acid wash at the cream stage prior to pressing. Clement and Hennette (1982), in their patented process for obtaining improved baking strains, describe a technique for obtaining haploid strains with low levels of invertase for use in hybridization. The maltose adapted haploid or diploid strains used in this investigation were treated with mutagenic agents such as EMS or NTG, to achieve low invertase- producing mutants. In order to select isolates with low invertase activity, these mutants were subjected to a preliminary screening using O-dianisidine followed by a more elaborate colorimetric assay. These selected mutants were then used for hybridization. The hybrids obtained contained low invertase activity and were highly osmotolerant.

3. Improvement of Fermentation Activity in Yeast

In most genetic studies, screening of yeast strains for use in baking has primarily been based on two important criteria: (a) the ability to ferment fast, and (b) the ability to achieve high growth rates which result in high biomass yields by the end of a regular propagation. Hence, the strain must be produced economically to benefit the yeast manufacturer, but it should also satisfy the baker with respect to its leavening ability. The precise mechanism by which these genes are controlled, and why they impart different fermentation rates on different yeast strains, is still not well understood. Nevertheless, an absolute understanding of the genetic control of a given character is not critically important for strain improvement. Under these circumstances, genetic recombination can be an enormously efficient procedure for generating desired genetic variation. However, recombinants must still be sorted by suitable and efficient selection procedures.

Although cross-breeding or hybridization is considered purely empirical, because the outcome of such an event is highly unpredictable (Harrison, 1971), the procedure has been used extensively by early geneticists for improving fermentation activity of baker's yeasts. One such breeding scheme between a commercial strain of baker's yeast (strain A) and a brewer's strain (strain B) has been described in detail by Burrows (1979) (Fig. 3). Although strain A was a weak fermenter with slow adaptation to maltose, the hybrid (H1) produced by crossing with strain B (brewer's strain) exhibited a faster adaptation to maltose and a 10% increase in fermentation activity over that of parental strain A. The H1 hybrid was again crossed with a different baker's strain (strain C), exhibiting high fermentation activity but poor adaptability to maltose. Such a cross resulted in a hybrid (H2) with excellent fermentation activity at both early and late stages of dough fermentation. Subsequent to further testings, both the H1 and H2 hybrids were introduced into commercial production in the 1950s (Burrows, 1979).

The protoplast fusion technique, first demonstrated by Van Solingen and Van der Plaat (1977), shows the greatest potential for genetic recombination in yeasts. This procedure shows complete disregard for either ploidy or for mating type. Although the general outcome of a given fusion cannot be precisely predicted, it no doubt offers a recipient cell a means of receiving an entire pool of genes, associated with a character such as high fermentation activity, without any knowledge of the location or the nature of the genes in question. In one such study, Jacobson and Trivedi (1986) constructed fusion products exhibiting superior leavening characteristics in lean, regular, and sweet dough systems. This was accomplished by the fusion of two baking strains, one with good lean and the other with good sweet activity. The overall superior fusion products, subsequent to appropriate testing, have now been introduced into commercial production.

While a considerable effort is being made at present to develop improved baking strains of yeast by use of conventional methods, there is a substantial level of research in progress, specifically in the yeast industry, to identify genes or gene products associated with several baking characteristics. In particular, studies on the rate-limiting enzymes of the glycolytic pathway in different strains of baker's yeast are currently under way in many laboratories, including the present authors' laboratory. These studies may provide information on the extent of genetic manipulation necessary to achieve faster fermentation activity. It is possible, however, that the elimination of one limiting step in a pathway may lead to the appearance of a second limiting step at a different location, perhaps nullifying the effect of any previous improvement. Hence, the ability to identify, clone, and amplify genes associated with rate-limiting steps in the glycolytic pathway may not necessarily increase carbon flow and simultaneously improve leavening characteristics of the yeast in a given dough system.

The way in which carbon metabolism is regulated in yeast cells has also come under increasingly close scrutiny in the past several decades. Of particular interest to many yeast biochemists is the interplay between glycolysis and gluco-neogenesis, the principal pathways of carbon metabolism. Glycolysis is essential for growth of microorganisms utilizing carbohydrates as a carbon source, whereas

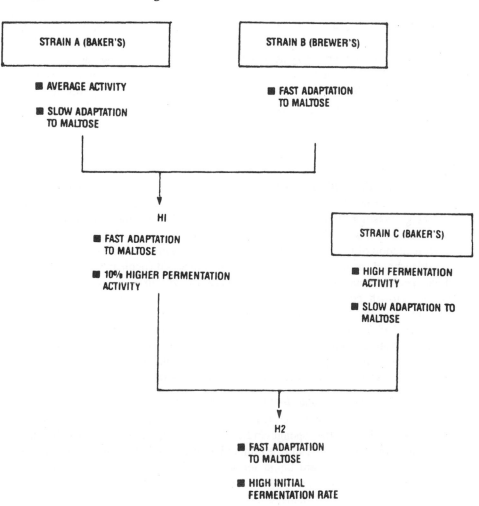

Figure 3. Breeding scheme for baker's yeast (as suggested by Burrows, 1979).

gluconeogenesis is essential for growth in the presence of nonsugar carbon sources, such as ethanol, glycerol, or peptone. Although some of the reactions are reversible and catalyzed by the same enzyme in both glycolytic and gluconeogenic directions, three of the steps—(a) glucose $\rightleftarrows$ glucose-6-phosphate, (b) fructose-6-phosphate $\rightleftarrows$ fructose-1,6-biphospate, and (c) phosphoenolpyruvate $\rightleftarrows$ pyruvate—are catalyzed by metabolically antagonistic obligate glycolytic and obligate gluconeogenic enzymes. Of these three steps, the fructose-6-phosphate $\rightleftarrows$ fructose-1,6-biphosphate step is unique. Phosphofructokinase (E.C.2.7.1.11) catalyzes this reaction in the glycolytic direction with the concomitant hydrolysis of ATP while fructose-1,6-bisphosphatase (E.C.3.1.3.11) catalyzes this reaction in the gluconeogenic direction with the concomitant production of ATP. The simultaneous presence and catalytic activity of these two antagonistic enzymes might result in an energy wasting "futile cycle" (Holzer, 1976). The operation of such a cycle would cause a rapid and repetitive hydrolysis and production of ATP, resulting in a net reduction of the energy level of the cell. The manner in which the yeast cell would respond during such an energy-wasting mode is difficult to assess, but some have speculated that it might result in increased gasing rates. Although under normal growth conditions the yeasts have complex regulatory systems that preclude wasteful "futile cycling" (Foy and Bhattacharjee, 1981; Banuelos and Fraenkel, 1982), a controlled deregulation of the appropriate enzymes may bring about the restoration of this phenomenon. This may be accomplished through the use of molecular biology and genetic engineering techniques. However, whether such strains would perform better in dough systems remains to be seen.

4. Strain Selection for High Brew Dough Activity

According to recent statistics, nearly half of the white bread produced in the United States is made by the sponge-and-dough system, somewhat less by the continuous-mixing process, and the remainder by the liquid ferment process. There is, however, a trend toward liquid ferments, which are becoming increasingly popular in the baking industry. Because of the rapid growth and commercial incentives in this area, new demands are beginning to be placed on yeast geneticists for strain improvement to meet the needs of this important market segment.

Generally, a preferment contains 0–70% flour, yeast, sugar, salt, yeast nutrients, and water. A pumpable brew may contain up to 50% of the flour in the brew. When the flour is completely eliminated in the aforementioned formulation, it is referred to as a water brew. As the level of the flour in the preferment is decreased, a gradual decline in the buffering capacity becomes apparent which, in most instances, has an adverse effect on the activity of the yeast. Furthermore, there is a corresponding reduction of fermentable sugar with gradual reduction of the flour which, at one stage, becomes evident by poor gasing activity. The inclusion of sugar in the preferment then becomes mandatory to alleviate this condition. A finished product with too low a pH will taste sour, bitter, sharp, etc., and could very well result in "cripples" due to the excessive mellowing or weakening action of the acid on the flour gluten. Accordingly, the bakers prefer to maintain the end pH of their preferments in the neighborhood of 4.5.

Brew buffers are commonly added to the ferments, especially to water brew systems, to keep the pH within the acceptable range for good yeast activity. Bakers often refer to calcium carbonate as a "buffer," which is inaccurate from a chemical point of view. However, the calcium carbonate can react with acids in the brew to generate an efficient buffer system. Most often, monocalcium phosphate (MCP) nonfat dry milk or flour is used to maintain brew pH in the desired range.

The changes in pH value found during the course of bread making are readily explainable on the basis of the effects of yeast fermentation. The acids formed by the yeast during fermentation, principally lactic, succinic, and acetic acids, may be expected to lower the pH of the dough because these acids have a substantial degree of ionization. The pH of the dough is also strongly affected by the presence of ammonium salts (ammonium sulfate, ammonium chloride, etc.). As shown by the following equations, yeasts readily assimilate ammonia as a source of nitrogen thereby liberating acids.

$$(NH_4)_2SO_4 \xrightarrow{\text{Yeast activity}} 2NH_3 + H_2SO_4$$

$$NH_4Cl \xrightarrow{\text{Yeast activity}} NH_3 + HCl$$

Sulfuric and hydrochloric acids are both strong acids which ionize almost completely and hence yield practically all their H^+ ions in the dissociated form. To avoid the harmful effects of a highly acidic environment, a large usage of yeast nutrients should be avoided.

High acidity in the brew may also be due to the presence of lactic and/or acetic acid bacteria that may have entered the dough system through contaminated flour, yeast, or other ingredients used in the formulation. Any carbonic acid formed in the dough during fermentation is weak and highly unstable. Since it is poorly ionized, it contributes minimally to any fluctuations in pH.

A freshly mixed preferment has a pH of approximately 5.3. Yeast strains that are known to perform satisfactorily in a brew system generally lower the pH to approximately 4.5, as a result of the various reactions described previously, while maintaining their fermentative activity at a desirable level. Strains that perform poorly in a brew system have been found to make excessive levels of lactic, acetic, and succinic acids at a faster rate, causing the pH to quickly dip well below 4.5. These conditions can have an adverse effect on yeast activity and dough rheology.

Baker's yeast shows marked impairment of respiratory capacity under anaerobic conditions, or even under aerobic conditions when abundant fermentable sugar is present in the medium. This results in the channeling of much of the pyruvate, generated by glycolysis, toward the production of ethanol and carbon dioxide. However, there is some low-level conversion of pyruvate either directly to lactic acid or indirectly to succinic acid by the TCA cycle under anaerobiosis. Rossi et al. (1964) have also suggested an alternate pathway for the generation of succinic acid under oxygen limitation via adaptive fumarate reductase activity on the TCA cycle intermediate oxaloacetate.

The relative proportions of succinic, lactic, and acetic acids formed by yeasts are also determined by such conditions as pH of the environment and the source of nitrogen in the medium (Harrison and Graham, 1971). Accordingly, the ability of yeasts to produce succinic acid in the presence of ammonium salts was found to be higher than with amino acids under identical conditions. This could be the indirect result of the channeling of more pyruvic acid into the TCA cycle in an attempt to first synthesize keto acids and subsequently amino acids for the ultimate production of proteins. Unfortunately, the regulatory controls associated with pyruvate channeling are poorly understood, and the outcome, as in the case of a preferment, cannot be predicted with precision.

In addition to good fermentation characteristics, either the resistance to low pH or the ability to maintain conducive pH environments in unbuffered or a weekly buffered brew system have been the principal criteria taken into consideration in strain selection for commercial production. Although acid production in yeast is associated with fermentative ability, the precise regulatory mechanism involved for one yeast to produce more organic acids than the other within a given time is complex and not well understood. Given these constraints, hybridization and protoplast fusion are the only viable techniques available for use in a strain improvement program. Until the genetic elements controlling acid production or pH resistance are fully understood, no attempt can be made to take full advantage of rDNA technology for strain improvement.

5. Strain Improvement for Freeze Tolerance

Frozen dough, which has been produced and marketed for many years, has provided several advantages to both the consumer and the frozen-dough manufacturer. This industry has shown steady growth as a result of significant quality improvements made in meeting the diverse needs of the baker. A large number of in-store bakery operations, which were opened to take the advantage of the newly developed technology, now represent the principal marketing channel. This industry is currently enjoying a significant market share of the bakery business with continued yearly increases anticipated. A small number of in-store bakeries freeze their own dough, but the remainder have bake-off facilities to process the dough purchased from other wholesale frozen-dough manufacturers. Such a shift from complete bakery to less cumbersome bake-off operations has provided the baker significant savings in time, labor, and maintenance costs.

In the past, a considerable research effort has been directed to improve all phases of frozen-dough manufacture. Accordingly, the critical requirements for frozen-dough manufacture as summarized by Lorenz (1974) are (a) to use increased yeast and shortening levels, (b) to achieve reduced absorption, (c) to use winter or spring flour of good quality, (d) to use properly developed cool dough made by straight dough process, (e) to have short fermentation prior to freezing, (f) to freeze immediately after molding, (g) to select suitable packaging material, and (h) to conduct frozen storage at −10°C. Although these criteria have largely helped to guide the frozen-dough industry, there is significant research activity in progress to further improve product quality to meet the challenges that exist in the competitive market. A significant emphasis is now being paid to strain improvement and to optimizing rates of freezing and thawing to further improve product quality.

Although freezing is lethal to most living systems, it can also preserve cells and their constituents. The manner in which yeasts respond to subzero temperatures and to the freezing of liquids has been extensively reviewed (Mazur and Schmidt, 1968; Mazur, 1970; Lorenz, 1974). These findings are based on the performance of yeast cell suspensions rather than actual yeast in a frozen dough. However, it is very possible that these results will also apply to frozen-dough products.

The freezing point of cytoplasm is usually above −1°C based on osmolal concentrations below 0.5. However, cells generally remain unfrozen and therefore supercooled, to −10 or −15°C even when ice is present in the external medium (Mazur, 1970). This implies that cell membranes can help prevent the growth of external ice into the supercooled interior, probably because of the lack of nucleators within the cell. Under these conditions, the supercooled water within the cell tends to have a higher vapor pressure than the ice present outside the cell. Such conditions favor the flow of water out of the cell in response to the vapor pressure gradient. The resulting dehydration concentrates the cell solutes, lowering the intracellular liquid vapor pressure. Such events during the freezing process depend primarily on the cooling velocity and on membrane permeability to water (Mazur, 1963).

Since cooling velocity affects the physical and chemical events in cells during freezing, it is not surprising that it also affects their survival. In slowly cooled cells, where the cooling rate is less than optimal, the freezing is entirely extracellular and the major cause of cell damage is prolonged exposure to concentrating solutes in the surrounding medium. At optimal cooling rates, the period of exposure to surrounding concentrated solutes becomes progressively shorter, and survival increases to a maximum. According to Mazur (1970), a cooling rate of approximately 1°C/min is considered optimal. At higher than optimal cooling rates, intracellular ice begins to form with a rapid decline in yeast viability. Thus, the avoidance of internal ice seems to be one prerequisite for higher survival rate in yeast under frozen conditions.

The relationship between freezing rate and thawing rate is well documented.

Mazur and Schmidt (1968) observed that cells cooled to −10°C had higher survival rates regardless of how the thawing was conducted. However, when the cells were cooled to −30 to −75°C the final viable count was largely governed by cooling and warming rates. Optimal low cooling (1°C/min) resulted in many more viable cells than did rapid cooling. The rate of thawing has not been shown to be important except after very rapid freezing, where rapid thawing resulted in greater survival than slow thawing. One explanation for this is that, during slow thawing, the very small intracellular ice crystals formed as a result of rapid freezing to −30 to −75°C will grow in size, and in doing so cause greater damage to the cell. Rapid thawing prevents crystal growth and also minimizes the time spent in contact with concentrated solutes. Accordingly, the best results can be achieved when doughs are frozen to −10°C. If the product is cooled to a temperature lower than −10°C, a cooling rate of approximately 1°C/min and a rapid thaw rate are recommended. However, this is contrary to the procedures generally practiced in the frozen-dough industry.

The objective of a frozen-dough strain improvement program is to obtain novel strains of yeast having the same excellent gasing power as a conventional baker's yeast that also have the capability of long-term storage under frozen conditions without a decrease in gasing power after thawing. At present, the genetics associated with freeze tolerance is not understood, and any new knowledge that would promote our understanding on the subject should facilitate efforts on strain improvement. It seems likely that the genetic elements responsible for imparting osmotolerance in yeast may also impart, at least to some degree, the ability to withstand freeze-thaw conditions. This is demonstrated by the high viability shown by some osmotolerant yeasts such as *Saccharomyces rosei* and *S. rouxii* under freeze-thaw conditions. However, such strains do not generally ferment maltose rapidly and show inadequate gasing power in dough systems containing low concentrations of sugar. Such strains show a poor performance when used as a leavener in the lean dough systems. Nevertheless, these osmotolerant strains have been used in hybridization studies to improve osmotolerance in regular baking strains. Because of the apparent close relationship that exists between osmotolerance and freeze tolerance, any attempt to improve osmotolerance in a baking strain should simultaneously enhance the strain's ability to withstand freezing conditions. Under somewhat difficult conditions, inter- and intraspecific hybrids have been constructed by crossing baking strains of *S. cerevisiae* with such strains as *Saccharomyces logos, S. uvarum, S. rosei*, and *S. rouxii* (Windisch et al., 1976). These studies could perhaps be of value to the frozen-dough industry.

Some special strains with freeze tolerance, developed by hybridization, have also been the subject of some recent patent claims. In this regard, a patented strain of a *Saccharomyces* species, which has been developed by Nakatomi et al.(1985), has apparently shown most of the characteristics desired in a frozen-dough yeast. These authors have described a long and laborious breeding scheme for obtaining yeast hybrids with freeze tolerance characteristics, for use in the commercial production of frozen dough (Fig. 4). The new strain, which they have designated FD 612, was essentially bred by selective hybridization between a baking strain and a hybrid obtained by crossing a brewing strain (*S. uvarum*) and an unspecified *Saccharomyces* species (IFO 1426) presumably having the genetic makeup for imparting the character for freeze tolerance. The resulting hybrids were back-crossed with the parental baker's strain to improve the stability of the hybrid. These matings were carried out by the mass-mating technique or the rare-mating method as described by Tubb (1979).

Some investigators have, in addition, found that freeze tolerance in yeast is closely associated with alcohol tolerance. Consequently, ethanol tolerant strains for use in frozen dough have also become the subject of a patent claim (Kawai and Kazuo, 1983). According to this patent, certain alcohol-tolerant yeast strains like *S. cerevisiae* and *Saccharomyces chevalieri* have been shown to possess freeze-tolerance characteristics in a dough system.

Although there are a few examples of the use of hybridization for the improvement of freeze tolerance in yeast, it seems likely that similar hybrids of industrial importance can also be constructed by using protoplast fusion techniques. Nevertheless, the literature indicates that such research efforts have been either scarce or nonexistent. Besides, there is no information available pertaining to the structural and regulatory genes associated with freeze tolerance in yeast. As previously described this appears to be the case for most of the complex characteristics, such as osmotolerance or brew dough activity, associated with a desirable yeast product. The lack of knowledge of the genetic elements associated with freeze tolerance makes it impossible to immediately apply recently developed genetic engineering techniques for strain improvement. Hence, there is a definite need to elucidate and characterize the important structural and regulatory genes associated with freeze tolerance in order to take full advantage of the latest technology for strain improvement.

C. Improvement of Process Economics

1. Strain Development for Melibiose-Utilizing Ability

A mixture of beet and cane molasses is commonly used by yeast manufacturers for the production of baker's yeast. The proportions of each of the components are dictated by economics and general availability of the two products. Although the predominant sugar in both types of molasses is sucrose, beet molasses contains raffinose in the 0.5–5.2% range. Only one third of this trisaccharide molecule is utilized by baker's yeast (Burrows, 1979). Baker's yeast contain the β- fructosidase (invertase) enzyme which allows the cell to utilize the fructose portion of the raffinose molecule.

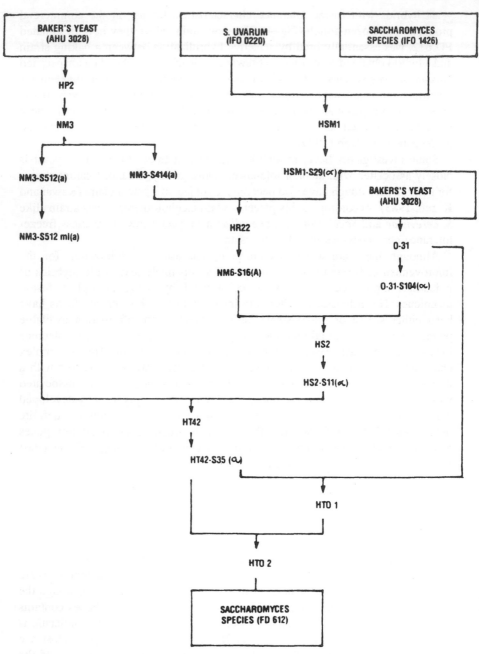

Figure 4. A scheme for breeding of freeze-tolerant yeast (as suggested by Nakatomi et al., 1985).

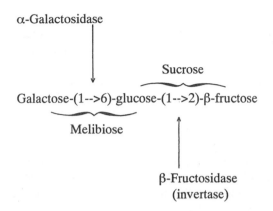

The presence of both the β-fructosidase and α-galactosidase (melibiase) enzymes in brewer's yeast (*S. uvarum*) permits utilization of the entire raffinose molecule.

It is known that at least six unlinked polymeric genes (SUC genes) provide the ability to produce the invertase enzyme. Accordingly, all baker's yeasts carry one or more of these SUC genes. In addition to the SUC genes, *S. uvarum* is known to carry a MEL gene (Gilliland, 1956), which confers the ability to utilize melibiose. Thus, the use of a melibiose fermenting baker's yeast strain should result in improved sugar yields depending on the level of beet molasses used for baker's yeast production. It would not only increase the sugar yield, and consequently the productivity, but result in a significant saving through reduction of biological oxygen demand (BOD) in the effluent stream.

Hybrid baking strains of yeast capable of completely utilizing raffinose have been developed in the past by interspecies mating with strains of *S. uvarum* which could readily ferment melibiose (Lodder et al., 1969; Kew and Douglas, 1976; Winge and Roberts, 1956).

Matings are most easily made between haploid cultures that have been derived from single ascospore cultures. Accordingly, strains of *S. cerevisiae* and *S. uvarum* are separately subjected to a preliminary anaerobic growth phase on a rich medium followed by washing and sporulation on KI sodium acetate agar of defined composition. Since the spores are more heat-resistant than the vegetative cells, the cell suspensions are held at 58°C for 2–4 min to completely eliminate the vegetative cells. The suspensions are then cooled rapidly and plated on solid media such as malt agar.

Each haploid culture, if not sterile, is either of the *a* or α mating type. The spores produced from each parent strain are grown separately so that the stocks can be held and used for subsequent tests if found to be of value. According to the procedure developed by Lindegren and Lindegren (1943), the spore cultures of opposite mating types are mixed together, shaken aerobically, and centrifuged to a pellet form to provide intimate contact between cells to promote mating. All hybrids are then selected for their ability to ferment melibiose rapidly. To preserve and stabilize baking characteristics, several backcrossings with the parental baking strain are often necessary. Strains meeting the necessary criteria in the preliminary screenings are then propagated at the bench level following standard baker's yeast propagation procedures. Strains that exhibit improved sugar yields are then evaluated for baking characteristics. Yeast hybrids meeting the requirements are then tested at the pilot (500-gal) level. The strain that is judged superior is then recommended for commercial (50,000-gal) propagation. There are few detailed studies in the literature on the use of hybridization for the development of baking strains with complete raffinose-utilizing ability. Nevertheless, there are scattered reports of hybrids with raffinose-utilizing ability of potential commercial value in the baking and distilling industries.

Baker's yeasts, like most other galactose-utilizing organisms, require three important enzymes to convert galactose to glucose-1-phosphate. These are referred to as the Leloir pathway enzymes. They include galactokinase (E.C.2.7.1.6), galactose-1-phosphate uridyltransferase (E.C.2.7.7.10), and uridine diphosphogal-actose-4-epimerase (E.C.5.1.3.2), which are encoded by the GAL1, GAL7, and GAL10 genes, respectively. The enzyme phosphoglucomutase (E.C.2.7.5.1), produced constitutively in yeast, converts the glucose-1-phosphate to glucose-6-phosphate, thus permitting its entry into the glycolytic pathway. The induction of the three galactose-metabolizing enzymes in the Leloir pathway is regulated by the highly coordinated interactions of galactose or metabolites of galactose with the products of two major regulatory genes, GAL4 and GAL80 (Kew and Douglas, 1976). The GAL4 gene synthesizes a protein that is required for the transcriptional induction of Leloir pathway genes. The action of the GAL4 protein is inhibited by the GAL80 protein. The GAL80 protein is thought to bind to the GAL4 protein or at its site of action on the DNA in the absence of galactose. In the presence of galactose, this binding is prevented and all the associated Leloir pathway genes are induced (Fig. 5). A recessive gal80 or a dominant GAL81 mutation gives rise to constitutive expression of the three Leloir pathway enzymes (Douglas and Hawthorne, 1966). The same three structural genes are also regulated by the GAL3 gene, as indicated by the long-term adaptation to galactose by a recessive gal3 mutation (Winge and Roberts, 1948).

Recent studies have shown that the GAL4 and GAL80 proteins also regulate transcriptional expression of the α-galactosidase gene (Kew and Douglas, 1976). Unlike the genes associated with galactose metabolism, the MEL1 gene responsible for the α-galactosidase enzyme is expressed at detectable basal levels in the absence of galactose (Post-Beittenmiller et al., 1948). Thus, it seems likely that although the induced and basal modes of MEL1 expression are GAL4-dependent, the unique behavior of the MEL1 gene reflects a different mode of interaction between the DNA sites and the protein that binds to them.

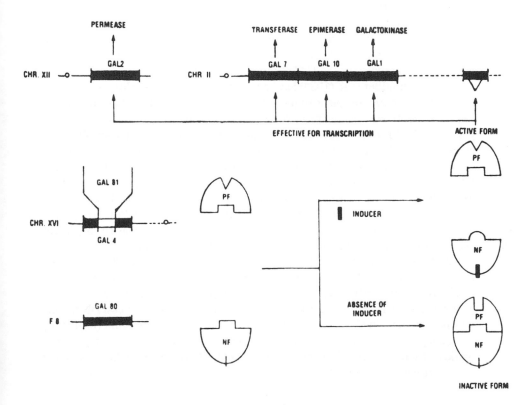

Figure 5. Model for the role of the GAL4 (positive factor [PF]) and GAL80 (negative factor [NF]) gene products in the regulation of the Leloir pathway enzymes (Oshima, 1983).

Sumner-Smith et al. (1985), in a recent study, have sequenced the entire MEL1 gene and its flanking regions. They were able to identify an 18 amino acid N-terminal signal sequence for secretion and an upstream activation sequence having certain areas of sequence homology with sites found upstream of the structural genes GAL1, GAL7, and GAL10. These investigations also resulted in the identification of eight potential glycosylating sites. Six of these sites are located in the C-terminal half of the polypeptide chain. These studies have provided a greater understanding of the expression of the MEL1 gene as regulated by the action of the products of the GAL4 and GAL80 genes.

Post-Beittenmiller et al. (1984) have been able to successfully clone and transfer the MEL1 gene to a recipient strain of baker's yeast that did not have the ability to utilize melibiose. The recipient yeast was a wild type for all known regulatory genes required for the production of α- galactosidase and galactose-catabolizing enzymes except that it did not contain the gene for α-galactosidase activity.

According to the procedure, a DNA pool was first prepared using the genomic DNA from a diploid strain of *S. uvarum* which was homozygous at the MEL1 locus. Subsequent to a Sau 3A partial digest, the DNA fragments of 10–15 kb in length were ligated into Bam H1 sites of a YEp 24 shuttle vector. Prior to ligation, the DNA fragments were treated with bacterial alkaline phosphatase to prevent self-ligation. Yeast transformations were performed with the plasmid DNA using either the spheroplasting method (Hinnen et al., 1978) or the lithium acetate method (Ito et al., 1983) (See Chapter 11). A clonal isolate was recovered that conferred on the transformed yeast the ability to produce α-galactosidase. The α-galactosidase gene was found to be contained on a 12-kb insert of yeast DNA. This was further subcloned to obtain an α- galactosidase positive plasmid (pMP 550) containing a 3.4-kb insert of yeast DNA (Post-Beittenmiller et al., 1984). These workers have also described mutations at GAL80 which result in constitutive (gal80) α-galactosidase activity.

2. Lactose-Fermenting Baker's Yeast

At the turn of the century, most baker's yeast manufacturers used cereal-based sugar substrates for the propagation of baker's yeast. Although this practice continued for some time, toward the end of World War I there was an acute shortage of grains for human consumption. This made the yeast manufacturers search for cheaper sugar substrates for the manufacture of baker's yeast. Blackstrap molasses, which was abundantly available at the time, was an obvious choice for use as a test substrate. The extensive research that followed ultimately led to the successful replacement of cereal mashes by molasses mashes for the propagation of baker's yeast.

Although molasses was in wide use for a considerable period of time as the principal substrate for yeast manufacture, the quality of molasses declined during the last decade as a result of newer technologies in the sugar industry which have improved sugar extraction efficiency. The decline in molasses quality coupled with correspondingly higher prices has prompted yeast manufacturers to search again for cheaper and viable alternate substrates. Whey, a by-product of the cheese industry, is now produced in large quantities in the United States. On a sugar basis, whey is a cheaper substrate than molasses. Lactose is the major sugar component in whey, but baker's yeast, *S. cerevisiae*, does not have the enzymes necessary to utilize lactose sugar as a carbohydrate source. There are a few species of yeasts like *Kluyveromyces fragilis* or *K. lactis* capable of utilizing lactose, but these strains have shown very poor baking characteristics (Bruinsma and Nagodawithana, 1986). To take advantage of the cheaper whey available from the cheese industry, it is apparent that there is a definite need to develop new strains of baker's yeast with the extra ability to metabolize lactose. Such intrageneric genetic transfers, which were previously difficult if not impossible by conventional means, are becoming more and more common by the use of genetic engineering.

Lactose utilization within the genus *Kluyveromyces* is accomplished by induction of both a lactose transport protein and an intracellular β-galactosidase (E.C.3.2.1.23) more than 100-fold above a moderate basal level. The lactose is transported across the cell membrane by means of the induced permease, and it is subsequently hydrolyzed to glucose and galactose by the enzyme β-galactosidase. Galactose is catabolized by inducing enzymes of the Leloir pathway as described previously. The structural gene for β-galactosidase has been identified and designated LAC4 (Sheetz and Dickson, 1981).

The induction of β-galactosidase in *K. lactis* involves a complex regulatory system whose mechanism has been partially elucidated. Using cloned genes as probes, it has been shown that induction of the LAC4 gene by lactose is regulated at the transcriptional level (Lacy and Dickson, 1981). Dickson et al. (1981) hypothesized that regulation is partly due to the action of a LAC10 gene product which regulates the transcription of the LAC4 gene in a negative manner, preventing rather than facilitating transcription. The function of the LAC10 gene product may thus be analogous to that of the GAL80 gene of *S. cerevisiae* as described previously. These studies also demonstrated that the LAC10 gene is not closely linked to the LAC4 gene, although both loci have been mapped to chromosome II. Subsequent deletion mapping of the 5' regulatory region of the LAC4 gene has revealed a DNA element required for its induction, presumably for the binding of a positive regulator. The search for this positive regulatory gene product is currently under way (Das et al., 1985).

Although protoplast fusion technology has been applied to the development of baker's yeast strains with lactose-utilizing ability during the past decade, no confirmed reports of successful hybrids are available. In almost all instances, there was the usual problem of genetic instability of the fused strains. This may be due to the fact that genetic recombination is still dependent on the degree of DNA homology between the strains involved.

The early reports using protoplast fusion for the development of yeast strains with lactose-fermenting ability were published by Stewart (1981). Despite their success in achieving fusion, the fusion products were unstable and subsequently reverted to original parental types. Results of similar studies conducted in our laboratory confirmed the above findings. Nevertheless, Farahnak et al. (1986) have reported successful use of the protoplast fusion technique to improve ethanol tolerance and lactose assimilation of yeast strains. Such hybrids were obtained by the fusion of auxotrophic strains of lactose-fermenting *K. fragilis* isolated after EMS mutagenesis. These investigators have not, however, reported long-term stability of this fusant.

The genes responsible for lactose utilization in *K. lactis* are not clustered and are characterized by the lack of an operon (Sheetz and Dickson, 1980). These investigators have also estimated that there are at least seven unlinked genes involved in lactose utilization. Although the manner in which the lactose-utilizing genes of *K. lactis* are regulated is not fully established, recombinant DNA techniques may have some advantage over protoplast fusion techniques in constructing hybrids of baker's yeast with lactose-fermenting capability.

Several investigators have recently developed a convenient system for cloning DNA segments containing the β-galactosidase gene into the yeast *S. cerevisiae*. One such study has been reported by Dickson (1980) who has demonstrated the expression of the *K. lactis* β-galactosidase gene LAC4 in *S. cerevisiae*. The level of β-galactosidase expression in the transformants was found dependent on whether the recombinant DNA vector carrying the LAC4 gene integrated into the chromosome or was maintained as an episome. The cells transformed with episomal vectors had significantly higher levels of β-galactosidase activity then the transformants containing vectors, which subsequently integrated into the chromosome.

There is probably only one copy of the LAC4 gene per haploid genome in strains transformed with an integrated vector in contrast to the episomal vector which may be present in several copies per cell. Hence, a gene dosage effect can be regarded as the most likely explanation for elevated activity. There was, however, a serious stability problem with the transformants containing the episomal vectors, even though the gene as measured by the β-galactosidase activity was very stable when integrated into the genome. It is assumed that the instability is a result of the loss of episome rather than the inactivation of the specific gene. Although the genetically engineered *S. cerevisiae* strain produced β-galactosidase enzyme intracellularly, it was not able to grow on lactose, probably because of the lack of a functional lactose transport system. The location of the gene(s) coding for the lactose permease was not known at the time of these studies.

The interest in this subject that existed at the time led to a search for more effective vectors to achieve more efficient transformations. Das and Hollenberg (1982) observed that certain yeast vectors like YRp 7, which contained *S. cerevisiae* autonomously replicating sequences (ARS), behaved differently in *K.*

lactis and led to an integrative transformation. Likewise, the 2-μm circle-derived DNA vectors, which gave high-frequency transformation in *S. cerevisiae,* were only able to transform *K. lactis* at very low frequency. Despite these drawbacks, these investigators were able to show high-frequency transformation in *K. lactis* with recombinant plasmids containing a DNA fragment from *K. lactis* presumably carrying an autonomously replicating sequence called KARS.

Without a full understanding of the regulatory controls associated with the LAC4 structural gene, the chances of constructing hybrid strains using genetic engineering techniques is rather low. However, a recent communication by Sreekrishna and Dickson (1985) has described the construction of a strain of *S. cerevisiae* that could utilize lactose. In this study, an *S. cerevisiae* laboratory strain was transformed with a plasmid (pKRIB LAC4-1) carrying the *K. lactis* LAC4 gene and its flanking sequences. The plasmid also contained the kanamycin-resistant gene from *E. coli* which confers resistance to the antibiotic G 418 in yeast. Transformants were then selected on the basis of growth on lactose and resistance to the antibiotic G 418. Southern hybridization experiments have shown that the LAC⁺ transformants contained 15–25 tandem copies of the vector integrated into the host chromosome. Good growth characteristics shown by the transformants in lactose media imply that the transformants received both the β-galactosidase and lactose permease genes from the pKRIB LAC4-1 plasmid. Furthermore, the authors have suggested that the location of the permease gene is 2–8.6 kb upstream from LAC4 gene (Fig. 6). The permease gene has been designated LAC12. Although the *S. cerevisiae* used in this study was a laboratory strain, the same procedure might be applied to industrial strains to achieve the ability to utilize lactose while retaining good baking characteristics.

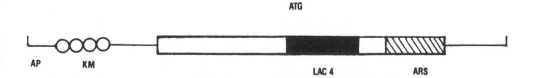

Figure 6. Structure of yeast shuttle vector PKRIB-LAC4-1 Ap: Plasmid carries genes conferring resistance to ampicillin (AP) and kanamycin (KM). Open bars: *K. lactis.* Hatched bars: autonomous replication sequence (ARS) of *K. lactis* that allows replication of *L. lactis* and *S. cerevisiae.*

IV. SUMMARY

During recent decades, the baking industry has undergone substantial development away from small-scale traditional baking to highly mechanized, large-scale technological systems. At the early stages, as the scientific know-how related to leavening began to unfold, bakers began to recognize the impact of good-quality yeast on their process economics. The traditional strains of yeast that were in use at the time presented sizable problems because they did not meet the high standards expected of them in a highly mechanized baking process. These inadequacies prompted yeast geneticists to embark upon ambitious strain improvement programs to meet the growing needs of the baking industry.

Although much of the early work was accomplished through the use of hybridization techniques, this approach had several limitations in its application to industrial baking strains. The protoplast fusion technique developed during the past decade provided a means to overcome some of the barriers, thus opening the way to the creation of new genomes. This new approach has also offered a wide range of applications for strain improvement owing to its complete disregard for ploidy. This technique also made it possible to obtain interspecific and intergeneric hybrids. Although this technique has the disadvantage of nonspecificity and unpredictability, it has been shown in this chapter that hybrids of commercial value have been developed using this technique.

Although a few yeast characteristics are mediated by a single gene, many characteristics important for baking are polygenic and are not well characterized genetically. Hence, the most obvious applications for recombination in genetic improvement are in the transfer of single genes. The least obvious are in the construction of desirable genotypes containing characters inadequately understood and controlled by a large number of scattered genes. However, the genetics and biochemistry of a few traits—for example, maltose-utilizing ability or invertase production—are less complex and adequately understood for transformations to be executed using currently available genetic techniques. Likewise, new hybrids of *S. cerevisiae* with melibiose-fermenting ability have emerged, providing a cost advantage to certain manufacturers through improved molasses yields and reduced BOD load. Although lactose-fermenting genes have been successfully cloned into baking strains of yeast, the production economics and the acceptability of leavening characteristics of the corresponding yeast in different dough systems remain to be established.

Despite these developments, the knowledge of the genetics responsible for a number of baking characteristics is extremely limited. This has severely limited the use of recently developed genetic engineering techniques for strain improvement. When one considers the importance of the brew dough segment of the baking industry, it is almost unbelievable how little geneticists know about the key elements controlling good brew dough characteristics. For instance, little is known about the location of the structural genes involved, not to mention possible regulatory genes or the manner in which these genes interact to impart the desired effect. Likewise, what little is known about the control mechanisms that allow osmotolerant yeast to perform well in sweet-dough systems has been described elsewhere in this chapter. Although the precise genetic mechanisms responsible for these phenomena are fragmentary, in such complex situations a random or "shotgun" approach may prove successful. However, to achieve the desired objectives in a completely predictable way rather than depending on chance, a knowledge of the precise genes and their locations will be highly valuable. This knowledge would be extremely helpful to plan strategies for the construction of chimeric plasmid vectors carrying combined gene clusters specific for a given characteristic. The plasmid DNA could then be used to transform spheroplasts of baking strains needing improvement.

Attempts in the past to construct baking strains of yeast with improved thermotolerance have been unsuccessful. Nevertheless, it has generated a great deal of interest within the industry because of the impact it would have on process economics and stability of products in storage. Although this property is strain-dependent, there is very little information available regarding the genetics of this trait. This may be largely due to the scarcity of thermotolerant yeasts in the family Saccharomycetaceae for comparative studies. Although there are several types of bacteria that are thermotolerant, there is very little understanding of the nature of the gene cluster associated with this trait. Obviously, a knowledge of the role of these bacterial genes, their location, and how they are linked will be invaluable for the direct introduction of such gene complexes into baker's yeast through the use of cloning vectors.

Although protoplast fusion techniques, coupled with laborious screening procedures, have resulted in some desired objectives, much more emphasis is being placed on research designed to help identify genes and the regulatory mechanisms associated with superior baking characteristics. A greater understanding of glycolytic enzymes and the regulatory functions of glycolytic intermediates on the appropriate enzymes, as in the control of futile cycling, may in fact, be necessary to increase gasing rates or the flow of carbon through glycolysis. The extent of genetic manipulation needed to achieve improvement in the fermentation characteristic cannot be predicted with certainty because of the limited knowledge available on the subject. Eventually, these studies should help to reveal the underlying genetic and biochemical basis of high fermentation activity, which could, in turn, facilitate strain improvement through the use of more predictable techniques such as cloning.

The potential of genetic engineering is difficult to overestimate, particularly since it offers an efficient means of achieving specific genetic changes formerly achieved by chance, if attained at all. The tremendous advances that have been made in the short time that the rDNA technologies have been in use indicate that the future holds profound possibilities. As it stands today, it is clear that industries with leadership in this technology will be able to retain their product position as they move new and genetically improved strains into the marketplace. Needless to say, the possibilities that exist for the application of rDNA technology to baker's yeast improvement are manyfold and exciting. However, the future will tell us how many of these expectations were realistic.

ACKNOWLEDGMENTS

The authors wish to express their profound appreciation to Dr. Gerald Reed and Dr. James Foy for their suggestions and critical review of the manuscript. The expert clerical assistance provided by Mrs. Betty Blue in the preparation of the manuscript is also gratefully acknowledged.

REFERENCES

Avigrad, G., and Neufeld, E. (1965). Intracellular trehalose of a hybrid yeast. *Biochem. J.* 97:715–722.

Banuelos, M., and Fraenkel, D. G. (1982). *Saccharomyces cerevisiae fdp* mutant and futile cycling of fructose 6-phosphate. *Mol. Cell. Biol.* 2:921–929.

Botstein, D., and Davis, R. W. (1982). Principles and practice of recombinant DNA research with yeast. In *The Molecular Biology of the Yeast Saccharomyces—Metabolism and Gene Expression*, J. N. Strathern, E. W. Jones, and J. R. Broach (Eds.). Cold Spring Harbor Laboratory, Cold Spring Harbor, NY, pp 607–636.

Bridges, B. A. (1976). In *Second International Symposium on Genetics of Industrial Microorganisms*, K. D. MacDonald (Ed.). Academic Press, New York, pp. 7–14.

Brown, A. D. (1974). Microbial water relations: Features of the intracellular composition of sugar-tolerant yeasts. *J. Bacteriol.* 118:769–777.

Brown, A. D., and Edgley, M. (1980). Osmoregulation in yeast. In *Genetic Engineering of Osmoregulation*, D. W. Rains, R. C. Valentine, and A. Hollaender (Eds.). Plenum Press, New York, pp. 75–90.

Bruinsma, B., and Nagodawithana, T. W. (1987). Comparison of *Saccharomyces cerevisiae*and *Kluyveromyces fragilis* in gas production, dough rheology and bread making. *Cereal Food World* (unpublished data).

Burrows, S. (1979). Baker's yeast. In *Economic Microbiology,* Vol. 4. *Microbial Mass*, A. H. Rose (Ed.). Academic Press, New York, pp. 31–64.

Burrows, S., and Fowell, R. R. (1961a). Improvements in yeast. British patent no. 868,621.

Burrows, S., and Fowell, R. R. (1961b). Improvements in yeast. British patent no. 868,633.

Carlson, M., Osmond, B. C., and Botstein, D. (1981). *SUC* genes of yeast: A dispersed gene family. *Cold Spring Harbor Symp. Quant. Biol.* 45:799–812.

Clement, P., and Hennette, A. L. (1982). Strains of yeast for bread making and novel strains of yeast thus prepared. U.S. patent no. 4,318,930.

Crowe, J. H., Crowe, L. M., and Chapman, D. (1984). Preservation of membranes in anhydrobiotic organisms: The role of trehalose. *Science* 223:701–703.

Das, S., and Hollenberg, C. P. (1982). A high frequency transformation system for the yeast *K. lactis. Curr. Genet.* 6:123–128.

Das, S., Breunig, K. D., and Hollenberg, C. P. (1985). A positive regulatory element is involved in the induction of the β-galactosidase gene for *K. lactis. EMBO J.* 4:793–798.

DeOliveira, D. E., Rodrigues, E. G. C., Mattoon, J. R., and Panek, A. D. (1981). Relationship between trehalose metabolism and maltose utilization in *S. cerevisiae.* II. Effect of constitutive MAL genes. *Genetics* 3:235–242.

Dickson, R. C. (1980). Expression of a foreign eucaryotic gene in *Saccharomyces cerevisiae*: β-galactosidase from *K. lactis. Gene* 10:347–356.

Dickson, R. C., Sheetz, R. M., and Lacy, L. R. (1981). Genetic regulation of yeast mutants constitutive for β-galactosidase mRNA. *Mol. Cell. Biol.* 1:1048–1056.

Douglas, H. C., and Hawthorne, D. C. (1966). Regulation of genes controlling synthesis of the galactose pathway enzymes in yeast. *Genetics* 54:911–916.

Eddy, A. A., and Williamson, D. H. (1957). A method of isolating protoplasts for yeast. *Nature* 179:1252–1253.

Farahnak, F., Seki, T., Ryu, D. D. Y., and Ogrydziak, D. (1986). Construction of lactose-assimilating and high ethanol producing yeast by protoplast fusion. *Appl. Environ. Microbiol.* 51:362–367.

Federoff, H. J., Cohen, J. D., Eccleshall, T. R., et al. (1982). Isolation of a maltase structural gene from *S. carlsbergensis. J. Bacteriol.* 149:1064–1070.

Fowell, R. R., and Moorse, M. E. (1960). Factors controlling the sporulation of yeasts. I. The presporulation phase. *J. Appl. Bacteriol.* 23:53–68.

Fowell, R. R. (1970). Sporulation and hybridization of yeast. In *The Yeasts*, Vol. 1, A. H. Rose and J. S. Harrison (Eds.). Academic Press. New York, pp. 303–383.

Foy, J. J., and Bhattacharjee, J. K. (1981). Concentration of metabolites and the regulation of phosphofructokinase and fructose-1,6-bisphosphatase in *Saccharomyces cerevisiae. Arch. Microbiol.* 129:216–220.

Gascon, S., and Lampen, J. O. (1968). Purification of the internal invertase of yeast. *J. Biol. Chem.* 243:1567–1572.

Giesenschlag, J., and Nagodawithana, T. W. (1982). Effect of reserve carbohydrates on the stability of yeast. (Unpublished data.)

Gilliland, R. B. (1956). Maltotriose fermentation in the species differentiation of *Saccharomyces. C. R. Trav. Lab. Carlsberg Ser. Physiol.* 26:139–148.

Grossman, M. K., and Zimmermann, F. K. (1979). The structural genes of internal invertases in *S. cerevisiae. Mol. Gen. Genet.* 175:223–229.

Guerola, N., Ingraham, J. L., and Cerda-Olmedo, E. (1971). Introduction of mutations by nitrosoguanadine. *Nature (New Biol.)* 230:122–125.

Gunge, N. (1966). Breeding of baker's yeast: Determination of the ploidy and an attempt to improve practical properties. *Jpn. J. Genet.* 41:203–214.

Harrison, J. S., and Graham, J. C. J. (1970). Yeast in distillery practices. In *The Yeasts*, Vol. 3, A. H. Rose and J. S. Harrison (Eds.). Academic Press, New York. pp. 283–344.

Harrison, J. S. (1971). Yeasts in baking: Factors affecting changes in behavior. *J. Appl. Bacteriol.* 34:173–179.

Hinnen, A., Hicks, J. B., and Fink, G. R. (1978). Transformation of yeast. *Proc. Natl. Acad. Sci. U.S.A.* 75:1929–1933.

Holzer, H. (1976). Catabolite inactivation in yeast. *Trends Biochem. Sci.* 1:178–180.

Ito, H., Fukuda, Y., Murata, K., and Kimura, A. (1983). Transformation of intact yeast cells treated with alkali cations. *J. Bacteriol.* 153:163–168.

Jacobson, G. K. (1981). Mutations. In *Biotechnology*, Vol. 1, H. J. Rehm and G. Reed (Eds.). Verlag Chemie, Weinheim, West Germany, pp. 280–304.

Jacobson, G. K., and Trivedi, N. (1983). U.S. patent appl. 503,323.

Johnston, J. P., and Oberman, H. (1979). Yeast genetics in industry. In *Progress in Industrial Microbiology*, Vol. 15, M. J. Bull (Ed.). Elsevier, New York. pp 151–191.

Kawai, M., and Kazuo, U. (1983). Doughs comprising alcohol resistant yeasts. European patent no. 78182.

Keller, F., Schellenberg, M., and Wiemken, A. (1982). Localization of trehalase in vacuoles and of trehalose in the cytosol of yeast. *Arch. Microbiol.* 131:298–301.

Kew, O. M., and Douglas, H. C. (1976). Genetic co-regulation of galactose and melibiose utilization in *Saccharomyces*. *J. Bacteriol.* 125:33–41.

Khan, N. A., and Eaton, N. R. (1971). Genetic control of maltase formation in yeast. I. Strains producing high and low basal levels of enzymes. *Mol. Gen. Genet.* 112:317 –322.

Koninklijke Nederlandsche Gist-en spiritusfabriek N. V. (1965). British patent no. 989,247.

Kosikov, K. V., and Miedviedieva, A. A. (1976). Experimental increase in the osmophilic properties of yeast. *Mikrobiologiya* 45:327–328.

Lacy, L. R., and Dickson, R. C. (1981). Transcriptional regulation of *K. lactics* β-galactosidase gene. *Mol. Cell. Biol.* 1:629–634.

Lawley, P. D., and Thatcher, C. J. (1970). Methylation of deoxyribonucleic acid in cultured mammalian cells by N-methyl-N'-nitro-N-nitrosoguanidine. *Biochem. J.* 116:693–707.

Legmann, R., and Margalith, P. (1983). Interspecific protoplast fusion of *S. cerevisiae* and *S. mellis*. *Eur. J. Appl. Microbiol. Biotechnol.* 18:320–322.

Lillie, S. H., and Pringle, J. R. (1980). Reserve carbohydrate metabolism in *S. cerevisiae*. Response to nutrient limitation. *J. Bacteriol.* 143:1384–1394.

Lindegren, C. C., and Lindegren, G. (1943). Selecting, inbreeding, recombining and hybridizing commercial yeasts. *J. Bacteriol.* 46:405–419.

Lindegren, C. C. (1949). *The Yeast Cell. Its Genetics and Cytology*. Education Publishing, St. Louis, pp. 10 (2-8).

Lorenz, K. (1974). Frozen dough. *Bakers Digest* 4:14–22.

Lodder, J., Khoudokormoff, B., and Langejan, A. (1969). Melibiose fermenting baker's yeast hybrids. *Antonio Van Leeuwenhoek Yeast Symp.* 35:F9.

Mazur, P. (1963). Kinetics of water loss from cells at sub-zero temperatures and the likelihood of intracellular freezing. *J. Gen. Physiol.* 47:347–3369.

Mazur, P., and Schmidt, J. J. (1968). Interaction of cooling velocity, temperature and warming velocity on the survival of frozen and thawed yeat. *Cryobiology* 5:1–17.

Mazur, P. (1970). Cryobiology: The freezing of biological systems. *Science* 168:939–949.

Nagodawithana, T. W. (1986). Yeasts: Their role in modified cereal fermentations. In *Advances In Cereal Science and Technology*, Vol. VIII, Y. Pomeranz (Ed.). American Association of Cereal Chemistry, St. Paul, MN, pp. 15–104.

Nakatomi, Y., Saito, H., Nagashima, A., and Umeda, F. (1985). *Saccharomyces* species FD 612 and the utilization thereof in bread production. U.S. patent no. 4,547,374.

Naumov, G. I. (1971). Comparative genetics in yeast. V. Complementation in the MAL gene in *S. cerevisiae* which do not utilize maltose. *Genetika* 7:141–148.

Naumov, G. I. (1976). Comparative genetics of yeast. XVI. Genes for maltose fermentation in *Saccharomyces carlsbergensis*. *Genetika* 12:87–100.

Needleman, R. B., and Michels, C. A. (1983). A repeated family of genes controlling maltose fermentation in *Saccharomyces carlsbergensis*. *Mol. Cell. Biol.* 3:796–802.

Needleman, R. B., Kaback, D. B., Dubin, R. A., et al. (1984). MAL6 of *Saccharomyces*: A complex genetic locus containing three genes required for maltose fermentation.*Proc. Natl. Acad. Sci. U.S.A.* 81:2811–2815.

Operti, M. S., DeOliveira, D. E., Freitas-Valle, A. B., Oestreicher, E. G., Mattoon, J. R., and Panek, A. D. (1982). Relationships between trehalose metabolism and maltose utilization. III. Evidence for alternative pathways of trehalose synthesis. *Curr. Genet.*5:69–76.

Oshima, Y. (1983). Regulatory circuits for gene expression: The metabolism of galactose and phosphate. In *Yeast Genetics*, J. F. T. Spencer, D. M. Spencer, and R. R. W. Smith (Eds.). Springer-Verlag, New York, pp. 159–169.

Oura, E., Suomalainen, H., and Parkkinen, E. (1974). *Proc. Fourth Int. Symp. Yeast* B25:125–126.

Panek, A. D., Sampaio, A. L., Braz, G. C., Baker, S. V., and Mattoon, J. R. (1979). Genetic and metabolic control of trehalose and glycogen synthesis. New relationship between energy reserves, catabolite repression and maltose utilization. *Cell. Mol. Biol.*2 5:345–354.

Post-Beittenmiller, M. A., Hamilton, R. W., and Hopper, J. E. (1984). Regulation of basal and induced levels of the MEL1 transcript in *Saccharomyces cerevisiae*. *Mol. Cell. Biol.*4:1238–1245.

Reed, G., and Peppler, H. J. (1973). Baker's yeast production. In *Yeast Technology*. Avi, Westport, CT, pp. 53–102.

Rossi, C., Hauber, J., and Singer, T. P. (1964). Mitochondrial and cytoplasmic enzymes for the reduction of fumarate to succinate in yeast. *Nature* 204:167–170.

Sanderson, G. W. (1984). Yeast products for baking industry of today and tomorrow. Paper presented at the International Symposium on Advances in Baking Science and Technology, Kansas City, KS.

Schnettler, R., Zimmermann, U., and Emeris, C. C. (1984). Large-scale production of yeast hybrids by electrofusion. *FEMS Microbiol. Lett.* 24:81–85.

Sheetz, R. M., and Dickson, R. C. (1980). Mutations affecting synthesis of β-galactosidase activity in yeast, *K. lactis*. *Genetics* 95:877–890.

Sheetz, R. M., and Dickson, R. C. (1981). LAC4 is the structural gene for β-galactosidase in *Kluyveromyces lactis*. *Genetics* 98:729–745.

Singer, B. (1975). The chemical effects of nucleic acid alkylation and their relation to mutagenesis and carcinogenesis. *Prog. Nucl. Acid Res. Mol. Biol.* 15:219–284.

Sklar, R., and Strauss, B. (1980). Role of uvrE gene product and of inducible O^6-methylguanine removal in the induction of mutation by N-methyl-N'-nitro-N-nitrosoguanidine in *E. coli*. *J. Mol. Biol.* 143:343–362.

Spencer, J. F. T., and Spencer, D. M. (1983). Genetic improvement of industrial yeasts. *Annu. Rev. Microbiol.* 37:121–142.

Spencer, J. F. T., Spencer, D. M., Bizeau, C. Martini, A. V., and Martini, A. (1985). The use of Mitochondrial mutants in hybridization of industrial yeast strains. *Curr. Genet.*9:623–625.

Sreekrishna, K., and Dickson, R. C. (1985). Construction of strains of *S. cerevisiae* that grow in lactose. *Proc. Natl. Acad. Sci. U.S.A.* 82:7909–7913.

Stanier, R. Y., Doudoroff, M., and Adelberg, E. A. (1970). *The Microbial World.* Prentice-Hall, Englewood Cliffs, NJ, pp. 417–447.

Stewart, G. G. (1981). The genetic manipulation of industrial yeast strains. *Can. J. Microbiol.* 27:973–990.

Stewart, G. G., Russell, I., and Panchal, C. (1983). Current developments in the genetic manipulation in brewing yeast strains. A review. *J. Inst. Brew.* 89:170–188.

Stiles, J. I., Clarke, L., Hsiao, C., Carbon, J., and Broach, J. R. (1983). Cloning of genes into yeast cell. In *Methods in Enzymology—Recombinant DNA*, Vol. 101. R. Wu, L. Grossman, and K. Moldave (Eds.). Academic Press, New York, pp. 290–325.

Sumner-Smith, M., Bozzato, R. P., Skipper, N., Davies, R. W., and Hopper, J. E. (1985). Analysis of inducible MEL1 gene of *S. carlsbergensis* and its secreted products, alpha-galactosidase (melibiase). *Gene* 36:333–340.

tenBerge, A. M. A., Zoutewelle, G., and Van de Poll, K. W. (1973). Regulation of maltose fermentation in *S. carlsbergensis*. I. The function of the gene MAL6 is recognized by MAL6 mutants. *Mol. Gen. Genet.* 123:233–246.

Thorn, J. A., and Reed, G. (1959). Active dry yeast. *Cereal Sci. Today* 4:198–201.

Trivedi, N. B., Cooper, E. J., and Bruinsma, B. L. (1984). Development and application of quick rising yeast. *Food Technol.* 51–57.

Trivedi, N. B., Jacobson, G. K., and Tesch, W. (1986). Baker's Yeast. *CRC Crit. Rev. Biotechnol.* 24:75–109.

Tubb, R. S. (1979). Applying yeast genetics in brewing. A current assessment. *J. Inst. Brew.* 85:286–289.

Van Solingen, P., and Van der Plaat, K. W. (1973). Fusion of yeast spheroplasts. *J. Bacteriol.* 130:946–947.

Walker, D. L., and Kummerow, F. A. (1964). Erythrocytes fatty acid composition and apparent permeability to non-electrolytes. *Proc. Soc. Ex. Biol. Med.* 115:1099–1103.

Windisch, S. Kowalski, S., and Zander, I. (1976). Dough-raising tests with hybrid yeast. *Eur. J. Appl. Microbiol.* 3:213–221.

Winge, O., and Laustsen, O. (1937). On two types of spore germination and on genetic segretation of *Saccharomyces* demonstrated through single spore cultures. *C. R. Trav. Lab. Carlsberg Ser. Physiol.* 22:99–117.

Winge, O., and Roberts, C. (1948). Inheritance of enzymatic character in yeast and the phenomenon of long term adaptation. *C. R. Trav. Lab. Carlsberg Ser. Physiol.* 24:263–315.

Winge, O., and Roberts, C. (1956). Complementary action of melibiase and galactozymase on raffinose fermentation. *Nature* 177:383–384.

7

Yeast Strain Development for Extracellular Enzyme Production

Malcolm A. J. Finkelman
Genex Corporation
Gaithersburg, Maryland

I. INTRODUCTION

The utilization of microbially derived enzymes in a wide spectrum of industrial processes is well established (Halpern, 1981; Fogarty, 1983; Terry, 1983). Many excellent reviews, detailing the sources, characteristics, and applications of these enzymes, are available (Birch et al., 1981; Fogarty, 1983; Godfrey and Reichert, 1983; Laskin, 1985). The advent of breakthroughs in the molecular biology of nucleic acids, as well as steady progress in the fields of genetics, physiology, and biochemical engineering, has provided the basis for a rapid expansion of processes based on microbial enzymes.

Inspection of the sources of industrially important microbial enzymes reveals that the bulk of production is derived from a relatively small group of species, principally aspergilli and bacilli (Fogarty, 1983). Although yeast is the leading industrial microorganism by a number of measures including product value, volume, and diversity (Peppler, 1979; Stewart, 1984), it is not prominent in the ranks of enzyme producers (Johnson and Oberman, 1981; Fogarty, 1983). This is perfectly understandable, given the overwhelming capability of the aspergilli, bacilli, and other species, for secreting very high levels of proteins. There are, however, certain incentives for developing yeasts capable of high-titer secretion of enzymes. A primary incentive has been the need to develop strains of greatly increased efficiency in the assimilation or fermentation of various polymeric substrates. This category encompasses the production of yeast biomass or ethanol

185

as the end product. Of significance in this aspect are the enzymes amylase, inulinase, cellulase, xylanase, and protease. A second category of interest is the production of the enzymes as the end products. This may be due to the particular characteristics of the enzymes or to their origin in a GRAS (generally regarded as safe) microorganism. This category includes lactase, invertase, and lipase. These and other applications give every sign of assuming greater industrial significance. It is therefore of some interest to review recent progress in the development of yeast strains for the production of extracellular enzymes.

II. YEAST EXTRACELLULAR ENZYMES OF INDUSTRIAL INTEREST

A. Amylase

1. Introduction

The industrial importance of the amylolytic enzymes is well established (Luenser, 1982) and has been extensively described (Aunstrup et al., 1979; Norman, 1979; Tubb, 1986). The family of starch-degrading enzymes includes α-amylase (α-1,4-D-glucan glucanohydrolase, E.C. 3.2.1.1); β-amylase (α-1,4-D-glucan maltohydrolase, E.C. 3.1.2.2); amyloglucosidase (α-1,4-D-glucan glucohydrolase, E.C. 3.2.1.3); pullulanase (α-dextrin-6-glucanohydrolase, E.C. 3.2.1.4); cyclodextrin glycosyltransferase (1,4-α-D-glucan 4-α-D-transferase, E.C. 2.4.1.19); and α-glucosidase (α-D-glucosideglucohydrolase, E.C. 3.2.1.20) (Kelly and Fogarty, 1983). The modes of action of these enzymes have been studied very intensively over the past decade, and a number of reviews are available (Fogarty and Kelly, 1979, 1980; Norman, 1979).

Starch-degrading enzymes are produced by an extremely broad spectrum of microorganisms. Microbial sources and the characteristics of the enzymes involved are the subject of an excellent and extensive review by Fogarty (1983). Although many genera have been shown to produce these enzymes, industrial production has been largely dominated by the bacilli and aspergilli (Ingle and Boyer, 1976; Aunstrup et al., 1979; Luenser, 1982; Fogarty, 1983). Interest in developing yeast strains with the capability of producing amylolytic enzymes stems from a variety of applications including single-cell protein (Spencer-Martins and Van Uden, 1977; Touzi et al., 1982), alcohol production (Calleja et al., 1982; Wilson et al., 1982), low-carbohydrate beer (Sills et al., 1984), and starchy waste conversion (Skogman, 1976; Moresi et al., 1983). These and other applications have stimulated a number of groups to search large collections of yeasts and yeastlike organisms for promising strains (Augustin et al., 1978; DeMot et al., 1984a–c).

While these studies have demonstrated that although many yeasts can grow on a variety of starchy carbon sources, a rather smaller number have the capability to produce potentially useful types and amounts of amylolytic enzymes (Augustin et al., 1978). Among these, the more intensively investigated have included

Schwanniomyces sp. for α-amylase and glucoamylase (Clementi et al., 1980; Wilson and Ingledew, 1982; Sills et al., 1984), *Endomycopsis fibuligera* for glucoamylase (Clementi et al., 1980; Steverson et al., 1984), *Lipomyces* sp. for isoamylase and glucoamylase (Spencer-Martins, 1982; Kelly et al., 1985), *Saccharomyces diastaticus* for glucoamylase (Modena et al., 1986; Sakai et al., 1986), *Candida tsukubaensis* for glucoamylase (DeMot et al., 1985), and Filobasidium capsuligenum for α-amylase and glucoamylase (DeMot and Verachtert, 1985).

2. Enzyme Characterization

The purification and general characteristics of the amylolytic enzymes from a number of yeasts have been reported, allowing a number of similarities and differences to emerge (Table 1). The pH and temperature optima for all of the enzymes are relatively similar, being mainly moderately acidic and between 40 and 55°C, respectively. Notable exceptions include the relatively thermotolerant *L. starkeyi* α-amylase and the more acidopilic *C. tsukabaensis* glucoamylase.

Inspection of the various values reveals a fairly broad molecular weight range for both the α-amylases and the glucoamylases. The determination of molecular weight for these enzymes is an area of uncertainty owing to the heterogeneity of glycosylation. Careful analysis of deglycosylated proteins is required to assess molecular weight values unequivocally (Meaden et al., 1985). The molecular weights of the glucoamylases of two strains of *S. diastaticus* and one of *S. pombe* were compared before and after treatment with Endo-H (endo-β-N-acetylglucosaminidase) for removal of N-linked oligosaccharides (Yamashita et al., 1985b). While the *S. pombe* enzyme was not reduced in molecular weight, inferring the absence of N-linked oligosaccharides, one of the *S. diastaticus*enzymes showed extensive glycosylation in both subunits. No subunit structure was found in the glucoamylase of the second *S. diastaticus* strain. Its glucoamylase had a molecular weight of 250,000 daltons of which 70,000 daltons were lost after Endo-H treatment (Yamashita et al., 1985a).

The STA gene product in *S. diastaticus* was found to be a two-domain protein, consisting of a catalytic domain and a noncatalytic domain that was rich in threonine and serine residues (Yamashita et al., 1986). The model proposed by the authors held that the two-domain protein precursor was, under some culture conditions, proteolytically processed to yield a functional domain consisting of two unequal subunits of molecular weight 41,000 and 3,400 daltons. Modena et al. (1986) described a glycoamylase II enzyme of *S. diastaticus* composed of two equal subunits of 300,000 daltons molecular weight. These were reduced, upon Endo-H treatment, to 56,000 and 170,000 daltons, respectively.

Tanaka et al. (1986) examined the amino acid sequences of the glucoamylases of *Rhizopus, Aspergillus,* and *S. diastaticus* and found homology averaging 25–36%. Within the homologous regions, the homology rose to 30–44%

Table 1. Properties of Yeast Amylases

Strain	Enzyme	Molecular weight	pH optimum	Temperature optimum	Reference
E. capsularis	α-amylase	—	4.5	40–50°C	Ebertova, 1966
E. capsularis	amyloglucosidase	—	4.5	40–50°C	Ebertova, 1966
F. capsuligenum	α-amylase	64,000	5.6	55°C	DeMot and Verachtert, 1985
F. capsuligenum	glucoamylase I	60,000	5.0–5.6	55°C	DeMot and Verachtert, 1985
F. capsuligenum	glucoamylase II	60,000	4.8–5.3	50°C	DeMot and Verachtert, 1985
L. starkeyi	α-amylase	76,000	4.0	70°C	Kelly et al., 1985
L. kononenkoae	α-amylase	38,000	5.5	40°C	Spencer-Martins and Van Uden, 1979
L. kononenkoae	glucoamylase	81,500	5.5	40°C	Spencer-Martins and Van Uden, 1979
P. polymorpha	α-amylase	—	4.0	40°C	Moulin et al., 1982
S. diastaicus	glucoamylase I	250,000	—	—	Yamashita et al., 1986
S. diastaicus	glucoamylase II	600,000	5.1	—	Modena et al., 1986
Sch. alluvius	α-amylase	62,000	6.3	40°C	Wison and Ingledew, 1982
Sch. alluvius	glucoamylase	155,000	5.0	50°C	Wison and Ingledew, 1982
Sch. castelli	α-amylase	40,000	6.0	60°C	Oteng-Gyang et al., 1981
Sch. castelli	glucoamylase I	90,000	6.0	60°C	Oteng-Gyang et al., 1981
Sch. castelli	glucoamylase II	45,000	6.0	60°C	Oteng-Gyang et al., 1981

(*Rhizopus/Aspergillus*, 44%; *Rhizopus/Saccharomyces*, 36%; *Saccharomyces/Aspergillus*, 39%).

3. Mutation, Selection, and Screening

The capacity of a large variety of yeasts to assimilate carbon from starch has been amply demonstrated using simple growth tests (Spencer-Martins and Van Uden, 1977; Belin, 1981). Using the criterion of soluble starch assimilation, Augustin et al. (1978) showed that of 177 strains tested, only 20 failed to grow. Because the simple observation of growth does not provide adequate information with respect to the type and utility of starch-degrading enzymes produced by yeasts, additional techniques have been used to differentiate these characteristics. Spencer-Martins and Van Uden (1977) measured the growth yield upon starch (gram of biomass/per gram of starch) for 56 strains in 15 genera. A yield value range of 0.043 (*Torulopsis colliculosa*, IGC 2916) to 0.482 (*Lipomyces starkeyi*, IGC 3944) was observed. That the differences in growth were due to variable starch-hydrolyzing ability was shown by adding commercial amylases to the broths in stationary phase. With this treatment, all strains tested produced yields around 0.532. Further tests with 25 *Lipomyces* strains showed that two, *L. kononenkoae* CBS 2514 and CBS 5608, used nearly 100% of the available starch and gave yield values of 0.590 and 0.580, respectively. Identical specific growth rates on glucose and starch ($\mu = 0.12$) showed that the rate of starch hydrolysis was not limiting, indicating an amylolytic capability well adapted for the complete hydrolysis of starch, and the rapid growth of the species.

To directly screen yeasts for amylolytic activity, Augustin et al. (1978) used the technique of growing strains on a medium partially consisting of epichlorhydrin-cross-linked amylose. Strains possessing α-amylase activity were differentiated by their ability to liquefy the gel. Of the 177 strains examined in the study, only eight strains from five species showed this capability (*Endomycopsis capsularis, Schwanniomyces occidentalis, S. alluvius, Leucosporidium capsuligenum*, and *Cryptococcus luteus*). Thus, by this analysis, endoamylase activity is a property of a minor fraction of the yeasts tested. In similar screening studies for starch degradation and amylase production by 73 ascomycetous (DeMot et al., 1984a) and 79 nonascomycetous yeasts (DeMot et al., 1984b), differences in growth and amylase activity were partially attributed to variability in the quality of the starches used. DIFCO-soluble starch was analyzed and found to contain glucose, maltose, and maltotriose, while Merck-soluble starch contained no reducing sugars. In one extreme case, *Pichia nakazawae* produced high levels of amylase activity when grown on DIFCO-soluble starch and none on the Merck-soluble starch (DeMot et al., 1984b). The species determined to produce the highest levels of extracellular amylase activity as judged by reducing sugar release from Merck starch included *Endomycopsis capsularis, E. fibuligera, Lipomyces kononenkoae, L. starkeyi, L. tetraporus, Pichia burtonii, Schwanniomyces alluvius, S. castelli, and S. occiden-*

talis, among the ascomycetes. *Candida ishiwadae, C. naeodendra, C. tropicalis, C. tsukabaensis, C. viswanathi, Cryptococcus gastricus, Filobasidium capsuligenum, L. gelidum, Phaffia rhodozyma, Sporobolomyces roseus, Torulopsis milischiana, Trichosporon cutaneum,* and *T. pullulans* were the highest producers among the nonascomycetes tested.

One of the best studied of the amylolytic yeasts is *S. diastaticus,* a strain closely related to *S. cerevisiae.* The ability of *S. diastaticus* to ferment starch to ethanol was observed over three decades ago (Andrews and Gilliland, 1952; Lindgren and Lindgren, 1956). The production of starch-degrading glucoamylase by this strain is controlled by the presence of one of three unlinked genes; STA1 (DEX2), STA2 (DEX1), and STA3 (DEX3) (Erratt and Stewart, 1978; Pretorius et al., 1986; Tamaki, 1978). These genes code a glucoamylase with activity against the α-1,4 linkages of starch, and their gene products have been extensively characterized (Yamashita et al., 1985a–d; Pretorius et al., 1986).

Utilization of *S. diataticus* strains for the direct fermentation of starch to ethanol and for single-cell protein production with the avoidance of costly enzyme treatment has been the object of a number of studies (Azzoulay et al., 1980; Russel et al., 1986; Sakai et al., 1986). The Symba process, a mixed culture fermentation of *Candida utilis* and *Endomycopsis fibuliger,* was developed to reduce, on a commercial scale, starchy waste release while producing a valuable product (Skogman, 1981). Amylolytic enzyme production by the *Endomycopsis* strain permits abundant growth by the more desirable *C. utilis.*

A key aspect of complete starch assimilation by any organism is the ability to hydrolyze α-1,6 branch points. The occurrence of debranching activity in 39 starch-degrading yeasts was analyzed using an assay for reducing sugar release from pullulan (DeMot et al., 1984c). By this assay all the strains tested caused reducing sugar release from pullulan, with the highest activities recorded for three *Lipomyces konoenkoae* strains (IGC 4052, 4051-I, and 4052-II) and a *Schwanniomyces castelli* mutant, R-91, of ATCC 26077. These strains had pullulan-degrading activities that were 10- to 15-fold above those of the positive-testing but less active strains from the various genera tested; *Candida, Cryptococcus, Pichia, Schwanniomyces,* and *Torulopsis.* The detection of low levels of debranching activity has been reported in a wide range of yeasts including *Sch. alluvius* (Wilson and Ingledew, 1982), *Candida tsukabaensis* (DeMot et al., 1985), *Filobasidium capsuligenum* (DeMot and Verachtert, 1985), and *Trichosporon pullulans* (DeMot and Verachtert, 1986b). A novel debranching activity was observed in *Lipomyces kononenkoae* CBS 2514 by Spencer-Martins (1982). This strain produced an extracellular isoamylase (glycogen 6-glucanohydrolase, E.C. 3.2.1.68). The demonstration of widely held ability to produce low levels of debranching activity should encourage studies aimed at obtaining hyperproducing mutant strains of industrial significance.

The development of yeast strains hyperproducing amylase has been undertaken for several species: *Lipomyces kononenkoae* (Van Uden et al., 1980), *Schwanniomyces castelli* (Dhawale and Ingledew, 1983; Sills et al., 1984), and *Saccharomyces diastaticus* (Sakai et al., 1986). Using a combination of ultraviolet light mutagenesis and selection for resistance to 2-deoxyglucose (2DG)-induced catabolite repression (Zimmerman and Scheel, 1977), Van Uden et al. (1980) were able to obtain a number of mutant strains of *Lipomyces kononenkoae* that were derepressed for α-amylase production in the presence of glucose. The mutants obtained in the simple two-step process gave up to nearly fourfold more activity in liquid culture than the parent strain. The growth of the mutants on soluble starch was characterized by the loss of a diauxic growth pattern and the rapid loss of iodine-staining capacity by the supernatant compared to the parent strain.

Using similar techniques, Dhawale and Ingledew (1983) and Sills et al. (1984) generated mutants of *Schwanniomyces castelli* that showed improvements in both α-amylase and glucoamylase production of the same order as obtained by Van Uden et al. (1980). In addition to ultraviolet mutagenesis, ethylmethyl sulfonate (Sills et al., 1984) and nitrosoguanidine (Dhawale and Ingledew, 1983) were also employed to produce mutants. As with the *Lipomyces* strain, the use of 2DG as a glucose analog was shown to produce strong repression of growth and production of α-amylase and glucoamylase. This effect is not universal among yeasts, however, as *Endomycopsis fibuligera* and *Pichia burtonii* were shown to be resistant to 2DG (Sills et al., 1984).

Sakai et al. (1986) have employed protoplast fusion in an effort to improve the starch-fermenting abilities of *Saccharomyces diastaticus*. This approach was adopted over mating because of repression of amylase STA1/STA3 expression in diploids heterozygous for mating type (Yamashita and Fukui, 1983; Yamashita et al., 1985d). Measurement of the amylase activities of the fusants showed an increase that approximated the increase in ploidy, indicating a gene dosage effect. Strain BC-3, a fusant of the expected genotype STA2/STA3, fermented starch three times faster than the parent strains and much faster than strains with the expected genotype STA1/STA3, which fermented starch more slowly than parental strains. These results were unexpected, since the STA1/STA3 expected genotypes also produced higher levels of amylase. These results indicate a higher level of complexity in the development of starch fermentation improvements and mean that simple amylase activity measurements are insufficient for predictive purposes.

4. Methods of Detecting Extracellular Amylolytic Activity

Two different chromogen-altering systems were used to quantitate and characterize the amylolytic activities in the studies by DeMot et al. (1984a,b). Both were specific for endoamylolytic activity. The first was loss of iodine-staining capacity and was used as a rapid, semiquantitative measure of substrate degradation. With the exception of species whose amylolytic activity is cell-bound (*Lipomyces starkeyi, L. tetrasporus,* and *Pichia burtonii*), the rapid decline in iodine staining correlated positively with high extracellular amylase activity (DeMot et al., 1984a). The second technique, used for differentiating α-amylase activity, was that described by Ceska (1971a,b). The formation of haloes around colonies on media containing dye-linked starch allowed the detection of strains secreting α-amylase, with halo size reflecting enzyme activity. This procedure allowed the direct identification of α-amylase-producing strains. Three species of Candida (*C. tsukabaensis, C. homilenta, C. silvanorum*) were thus newly identified as α-amylase producers, the last two at high level. A non-chromogen-based zone clearing assay for α-amylase and glucoamylase was developed by Dhawale et al. (1982), which used washed wheat starch as the indicator material. A quantitative semilogarithmic relationship between zone size and enzyme activity was demonstrated. Searle and Tubb (1981) have described a screening test for amyloglucosidase producers involving the incubation of "toothpick transferred" single colonies in microtiter plate wells containing maltotriose solutions. Strains positive for amyloglucosidase released glucose from the maltotriose, and the wells turned green on additon of glucose detection reagent.

5. Other Factors Affecting Amylase Titers

A major concern in a screening study for highly active amylolytic yeasts is the effect of medium and environmental factors on enzyme production. These aspects were investigated for 10 candidate, high-producing strains of *Candida homilenta, C. silvanorum, C. tsukabaensis, Cryptococcus flavus, Leucosporidium capsuligenum, Filobasidium capsuligenum,* and *Trichosporon pullulans* (DeMot et al., 1984d). Although the optimal pH was the same for a given strain with all of the carbon sources tested (maltose, Maldex 15 [Amylum, Aalst, Belgium], soluble starch [Merck], and glucose), large differences in activity were seen at the high and low ends of the pH range studied, pH 3.5–7.5. In all cases, very little activity was seen at pH 7.5. In the majority of cases, at pH 3.5, activity was less than 50% of the maximum. This was particularly true with glucose as the substrate. Glucose was quite repressive of activity, and in all cases, no activity was detected outside the middle range of pH 5.6–6.5. For all strains tested, maximum activities were observed at pH 4.5–6.5 with Maldex 15 or soluble starch. With *L. capsuligenum*CCY 64-2-4, maltose was as good as inducer as Maldex 15 or soluble starch.

In a separate study of 18 amylolytic yeast strains capable of assimilating β-cyclodextrin (BCD, Janssen Chimica, Belgium) as a carbon source, DeMot et al. (1986) observed that BCD could stimulate glucoamylase production. With one strain of *Candida antarctica,* CBS 6678, α-amylase activity was stimulated fivefold (0.48 U/ml to 2.51 u/ml) while glucoamylase rose eightfold (0.89 u/ml to 7.29 U/ml). Other yeasts showed a potentially more dramatic stimulation with BCD, but results may be skewed by the extremely low control values (e.g., *Schwanniomyces occidentalis* ATCC 26074 α-amylase — 0.01 u/ml to 0.49 U/ml). About half the strains tested showed barely detectable α-amylase activity on soluble starch. These activities rose substantially with BCD as the carbon source, indicating the potential utility of BCD in determining the presence of α-amylase in the enzyme repertoire of starch-utilizing yeasts.

Another factor investigated and shown to be of great importance was the effect of the nitrogen source (DeMot et al., 1984d). Using soluble starch as the carbon source and the previously determined optimum pH, DeMot et al. (1984d) tested YNB (DIFCO, 0.67%), yeast extract (YE) (0.3%), and corn steep liquor (CSL) (3.0%) as nitrogen sources. Significant differences were observed with the YNB, which gave better results with the *Candida* strains than YE or CSL. CSL gave surprisingly good results with the *Leucosporidium* strains, allowing fourfold higher activity than with YNB or YE.

6. Cloning and Expression of Amylase Genes

The development of amylolytic yeasts is currently being revolutionized by the successes achieved in cloning and expressing amylase genes from diverse sources in yeast strains. A recent review lists 10 different *S. cerevisiae* strain constructions using genes obtained from various sources including wheat, mouse, *Aspergillus*, *Rhizopus*, and *S. diataticus* (Tubb, 1986). Varying levels of extracellular activity have been observed. Erratt and Nasim (1986) reported 100-fold lower glucoamylase activity in *Schiz. pombe* transformants receiving an *S. diastaticus* gene, while Meaden et al. (1985) produced fivefold more activity from a *S. cerevisiae* strain transformed with an *S. diastaticus* DEX1 gene. In the latter study, it was also observed that excess glucose did not repress production of extracellular glucoamylase by the transformant as it did in the parent. Yamashita and Fukui (1984) cloned the STA1 gene of *S. diastaticus* into *Schiz. pombe* and observed a threefold increase in activity.

The practical applications of high-level glucoamylase secretors, resistant to glucose inhibition, are many. Additional benefits will accrue from the expression of enzymes providing the complete hydrolysis of starch. Progress toward such strains is evident in the introduction and expression of glucoamylase from *Aspergillus awamori* (Innis et al., 1985) and mouse pancreatic-amylase (Filho et al., 1986). The first area where the benefits of amylolytic *S. cerevisiae* strains will be felt is in the production of ethanol (Tubb, 1986). Since this will require the secretion of cloned amylases, it is encouraging to note that 90% of the activity produced by recombinant *S. cerevisiae* strains expressing *Aspergillus awamori* glucoamylase (Innis et al., 1985) and mouse α-amylase (Filho et al., 1986) was secreted.

B. Invertase

1. Introduction

Invertase (β-D-fructofuranoside fructohydrolase, E.C. 3.2.1.26) is one of the best studied of the yeast enzymes. The biochemical and catalytic properties of invertase have been thoroughly reviewed by a number of authors (Newberg and Mandl,

1950; Myeback, 1960; Lampen, 1971; Kulp, 1975). Invertase catalyzes the hydrolysis of the terminal fructosyl moiety of sucrose and various oligosaccharides, such as raffinose and stachyose. Additionally, invertase catalyzes the transfer of fructose to free sugars and primary alcohols (Straathof et al., 1986; Kulp, 1975). Invertase is commercially significant in the confection industry for the production of cream-centered products and in the sweetener industry for the production of invert sugar (Bucke, 1981; Wiseman and Woodward, 1975; Kulp, 1975). Commercial sources of invertase are *Saccharomyces cerevisiae, S. carlsbergensis,* and *S. diastaticus* (Godfrey, 1983; Frommer and Rauenbusch, 1975; Kulp, 1975).

2. Enzyme Characterization and Regulation of Expression

In yeast, invertase occurs as a hexose repressible enzyme (Gascon and Lampen, 1968; Ottolenghi, 1971; Mormeneo and Sentandreu, 1982). It is coded for by a group of six genes labeled *SUC* (Mortimer and Hawthorne, 1969; Carlson et al., 1980). The inclusion of any single intact *SUC* gene in the yeast genome provides for the expression of invertase (Carlson et al., 1980). Invertase occurs in both a cytoplasmic, nonglycosylated form and as a secreted, heavily glycosylated form (Gascon and Lampen, 1968). The external form has been characterized by Gascon et al. (1968) as a glycoprotein of approximately 270,000 daltons, of which one-half is glycan, principally mannan. It is typically tightly associated with the yeast cell wall and periplasmic space (Ballou, 1982). Trimble and Maley (1977) further analyzed the properties of external invertase, after removing most of the carbohydrate by treatment with endo-β-N-acetylglucosaminodase, and established that invertase consisted of two 60,000-dalton subunits.

A single gene has been hypothesized to encode both the internal and external species (Perlman and Halvorson, 1981). This gene was shown to be transcribed into two mRNA species differing in their 5` regions (Carlson and Botstein, 1982). These authors proposed a model for the regulation of invertase secretion in which staggered transcriptional initiation sites result in transcripts either with a signal sequence (secreted form) or without (cytoplasmic form). Subsequent sequencing of the invertase gene, *SUC2*, by Taussig and Carlson (1983) inferred amino acid sequences for the presumed external and internal coding regions which were quite similar to those previously described. The external sequence provided for a polypeptide of 58,567 daltons, and the internal sequence provided for a polypeptide of 58,480 daltons. This compares favorably with the value of Trimble and Maley (1977) of 60,000 daltons. Additionally, the amino-terminal and carboxy-terminal sequence data corresponded to those obtained earlier (Carlson et al., 1983; Trimble and Maley, 1977). The hypothesis of single gene origin has been confirmed by analysis of a cloned *SUC2* gene product and analogs (Williams et al., 1985). The expression of the external form of invertase is regulated by its repression in the presence of hexose (Gascon and Lampen, 1968; Gascon et al.,

1968). The internal form appears to be expressed constituitively and at low levels (Sarokin and Carlson, 1984). Relief from carbon catabolite repression by hexoses has been observed to be mediated by the *SNF1* (sucrose nonfermenting) gene product (Carlson et al., 1981; Celenza and Carlson, 1984a,b; Carlson et al., 1984; Neigeborn and Carlson, 1984). The *SNF1* gene product has been shown to be a serine-threonine protein kinase, suggesting that derepression of glucose-repressible functions is mediated by protein phosphorylation (Celenza and Carlson, 1986). Mutations in the *SNF1* locus prevent derepressions of invertase synthesis. This defect is relieved by suppressor mutations in another locus, *SSN6*, which results in constitutive, high-level invertase systhesis. Thus, the regulation of external invertase systhesis is the result of a highly complex set of interactions between regulatory proteins and glucose or its catabolites.

3. Mutation, Selection, and Screening Studies

The development of yeast strains capable of secreting high levels of invertase has been the subject of a variety of research efforts over the past two decades. These efforts have stemmed both from a basic research interest in the regulation of invertase secretion and from practical concerns dealing with the commercial applications of hyperproducing strains.

In a prototypical study, Montenecourt et al. (1973) investigated invertase production in strains mutagenized to resistance against hexose repression. A *Saccharomyces* strain (303-67) carrying a single invertase gene, *SUC2*, and susceptible to hexose repression, was mutagenized to 99.99% lethality by ultraviolet light. Surviving cells were plated on an invertase succinate-fructose medium. Colonies producing invertase were detected by a filter paper overlay onto which a reducing sugar-detecting solution containing sucrose was sprayed. Colonies positive for invertase turned red. These were picked and incubated overnight in the repressing medium and screened for invertase levels. The resulting single-step colonies were moderately repression resistant, giving three- to fourfold increases in invertase production. A second round of mutagenesis and screening with these strains yielded mutants with much higher invertase activity under repressing conditions. Of these second-stage mutants, one, FH4C, produced 35–40 times as much as the original parent strain, 303-67. FH4C produced 1–2% of the cell protein as invertase, and was subsequently used in studies requiring high-level production (Trimble and Maley, 1977; Perlman and Halvorson, 1981). A similar method for screening mutagenized *S. cerevisiae* was employed by Trumbly (1984). Increases of up to 45-fold in log-phase invertase activity were observed in single-step mutants grown on YEP medium supplemented with 2% glucose (30°C).

Multiple rounds of mutagenesis and screening were also used by Frommer and Rauensbusch (1973) to develop a higher-yielding strain of *S. carlbergensis*. Of 129 isolates from the first round of ultraviolet mutagenesis, three were substantially better, the best of which gave a 66% increase in activity. A second and third round of mutagenesis and screening with the best isolates resulted in only marginally more activity, with the final strain yielding about 193% of the original invertase titer. In addition to the technique of mutagenesis and screening, a number of groups have utlized the more powerful tool of selection pressure to obtain derepressed mutants of yeast (Toda, 1976; Zimmerman and Scheel, 1977; Hackel and Khan, 1978). These groups used 2-deoxyglucose (2DG), a nonmetabolizable analog of glucose, and raffinose in a selection medium. The 2DG mimics hexose in eliciting repression (Witt et al., 1966). The utility of raffinose-based coselection stems from its poor availability to the cell in the absence of high invertase activity. Zimmerman and Scheel (1977) used various 2DG/raffinose (2%) media containing 150–300 μ/ml 2DG to select spontaneous mutants of a *SUC3* carrying strain of *S. cerevisiae*. This procedure resulted in a high proportion of the mutants obtained demonstrating elevated invertase and maltase titers in a repressing medium (yeast extract, peptone, glucose 8%). Mutants producing up to several thousandfold higher activity than the highly repressed parental strains were obtained. Substitution of galactose for raffinose in the selection medium dramatically reduced the proportion of mutants showing the derepressed phenotype. Hackel and Khan (1978) employed the same technique for selecting mutants of *S. cerevisiae* carrying the *SUC3* locus. Invertase activity up to 50- to 60-fold higher than that of parental strains in YEPD broth were obtained. In both these studies, and that of Zimmerman and Scheel (1977), the levels of α-glucosidase activity rose in parallel to the invertase activity in most strains tested.

Toda (1976) examined the production of invertase in continuous culture by a spontaneous mutant of *S. carlsbergensis* LAM1068, isolated on 2DG/raffinose-containing media. The wild-type strain showed a bell-shaped peak of activity at a dilution rate of 0.15 hr^{-1}. The specific invertase activity at this point, the activity per mg cells, was 12-fold higher than that obtained in batch culture. Analysis of the response of the mutant to varying dilution rates showed that its activity also reached a peak at a dilution rate of 0.15 hr^{-1}, but at twice the level obtained with the wild type. More importantly, there was no subsequent decline in activity as the dilution rate increased.

4.   Cloning and Expression of Yeast Invertase

The cloning and expression of both the internal and external *SUC2* gene products in *S. cerevisiae* S288C was the subject of a study by Williams et al. (1985). In a wild-type host background, the expression of external invertase was increased up to 10-fold. This was further increased by the use of a derepressed host, strain RTY110. In this strain, plasmid-mediated expression resulted in invertase with an activity level three- to fourfold higher than strain FH4C, obtained by Montenecourt et al. (1973). The increase in activity produced with the derepressed strain is indicative of a strong regulatory activity able to influence expression from multiple copies of the *SUC2* gene in the wild-type host.

Hohmann and Zimmerman (1986) cloned five of the six invertase structural genes of *S. cerevisiae, SUC1–SUC5*. Large differences in the titers of invertase activity were observed when these genes were individually expressed in an invertase-deficient host under broth repressing and nonrepressing conditions. Maximum activity was observed with the *SUC1* transformant. *SUC2* was the least active. Up to 50-fold more activity was observed with the transformants under derepressed conditions. The observation that glucose repression remained effective with the multicopy plasmid is consistent with the data of Williams et al. (1985). This is an important observation from the standpoint of strain development for maximum productivity, and it is very likely that the development of new, high-yielding, recombinant strains will occur with some rapidity. The introduction of recombinant invertase into foods and food processing may pioneer the application of recombinant products to this area.

C. Inulinase

1. Introduction

The enzyme inulinase (β-D-fructofuranoside fructohydrolase, E.C. 3.2.1.26) hydrolyzes terminal β-(2,1)- and β-(2,6)-fructofuranosidic bonds (Snyder and Phaff, 1960, 1962). Typical substrates include the polyfructan, inulin, and oligosaccharides such as raffinose (Phaff and Snyder, 1962; Demeulle et al., 1981) and sucrose. Yeast inulinase sequentially cleaves terminal fructosyl units from the polymer, leaving, finally, a mixture of monomeric fructose and the penultimate glucose moiety of the fructan (Snyder and Phaff, 1962; Workman and Day, 1983). Inulinase activities have been found associated with a variety of fungi (Kim, 1975; Nakamura et al., 1978) and yeasts (Snyder and Phaff, 1960, 1962; Negoro and Kito, 1973; GrootWassink and Fleming, 1980; Chautard et al., 1981). *Aspergillus* inulinase differs from those of the yeasts in that it is able to cleve internal linkages producing oligofructans, as does the inulinase from *Arthrobacter ureafaciens*(Nakamura et al., 1978; Tanaka et al., 1972).

Inulinase has received increasing attention in recent years due to rising industrial interest. This interest includes the potential of using inulin, the storage polysaccharide of plants of the Compositae family (e.g., Jerusalem artichoke) as a source of fermentable sugar for the production of alcohol (Fleming and GrootWassink, 1979) and as a sweetener after conversion to a high-fructose syrup (Fleming and GrootWassink, 1979; Bajpai and Margaritis, 1985). Additionally, inulinase from *K. fragilis* has been demonstrated to compare favorably with yeast invertase in terms of properties useful in the manufacture of confections (Fleming and GrootWassink, 1979).

2. Characterization of Inulinase

The yeasts demonstrated to produce inulinase include *K. fragilis* (Snyder and Phaff, 1960, 1962), *K. lactis* (GrootWassink and Fleming, 1980), *Candida kefyr*

(Negoro and Kito, 1973), *C. salmenticensis* (Guiraud et al., 1980), *Pichia polymorpha* (Chautard et al., 1981), and *Debaromyces* sp. (Beluche et al., 1980; Demeulle et al., 1981). In virtually all cases, the inulinase activity is found both cell-bound and extracellular. Lam and GrootWassink (1985) have reported that up to 75% of the *K. fragilis* inulinase is cell-bound in batch culture. Disruption of the integrity of the cell wall by mechanical or chemical means has resulted in the release of all of the inulinase activity. In some cases, there is an increase in observed activity with respect to inulin as a substrate (Beluche et al., 1980). This is thought to reflect diminished access of the substrate to the cell-wall-entrapped enzyme (Beluche et al., 1980; Guiraud et al., 1982). Protoplasts of inulinase-producing strains have been demonstrated to be completely without activity, indicating the lack of both an intracellular form and any attachment of the secreted enzyme to the plasma membrane (Beluche et al., 1980). Yeast inulinases are generally thought of as glycoproteins; the degree of glycosylation is observed to be variable. The inulinase of *K. fragilis* was reported to contain 66% carbohydrate by weight (Workman and Day, 1983). Two inulinases obtained from *Pichia*, E1 and E2, which differed only in elution properties from DEAE-cellulose and Sephadex G-150 columns, contained 0.016% and 0.0% carbohydrate, respectively (Chautard et al., 1981).

Inulinase is an inducible enzyme in yeasts and is subject to carbon catabolite repression (GrootWassink and Fleming, 1980; Beluche et al., 1981; GrootWassink and Hewitt, 1983). Tsang and GrootWassink (1985) observed that high levels of glucose were required to completely eliminate inulinase production by colonies grown on an indicator medium. At 4% glucose, clearing zones were produced, but these were suppressed in medium containing 8% glucose. The glucose analog 2-deoxyglucose (2DG) has been used extensively to provide catabolite repression selection pressure in strain development protocols for inulinase overproducers (Bajon et al., 1983; Tsang and GrootWassink, 1985).

The inulinase enzymes of a variety of yeast species are quite similar in terms of pH and temperature optima. These are, respectively, for the following species: *Candida salmenticensis*, pH 4 and 50°C; *Debaromyces phaffia*, pH 4 and 45°C; *Debaromyces cantarelli*, pH 4 and 45°C; *Kluyveromyces fragilis*, pH 5 and 55°C; and *Pichia polymorpha* (two subtypes), pH 2 and pH 5 and 45°C (Guiraud and Galzy, 1981). Representative k_m values for inulin are: *C. salmenticensis*, 17 mM (Guiraud et al., 1981); *D. phaffia*, 12 mM (Demeulle et al., 1981); *D. cantarelli*, 15 mM (Guiraud et al., 1982). In these studies, a molecular weight of 5,000 daltons for inulin was assumed.

There is a good deal of interest in production methods for inulinase, and a variety of techniques for its production have emerged including batch culture (Guiraud and Galzy, 1981), continuous culture (GrootWassink and Hewitt, 1983), and live-cell extraction with recycle potential (Lam and GrootWassink, 1985).

3. Mutagenesis, Selection, and Screening Studies

There are relatively few published reports describing the mass screening of yeasts for high extracellular inulinase titers. The existing reports have focused upon relatively few of the yeast genera.

In a study examining the hydrolysis of both purified inulin and inulin from extracts of chickory and Jerusalem artichoke, Guiraud and Galzy (1981) examined five yeast strains: *C. salmenticensis* CBS 5121, *D. cantarelli* CBS 4349, *D. phaffia* CBS 4346, *Pichia polymorpha* CBS 186, and *K. fragilis* CBS 1555. Inulinase was prepared from cell-free media by dialysis and concentration. Inulin served as both carbon and energy source in the broth. All treatments used equivalent levels of enzyme (1 unit defined as the amount of enzyme releasing 1 μmole of fructose per minute from 0.25% inulin). pH and temperatures were adjusted to the published optima of the source species. With pure inulin at a concentration of 25 g/liter and inulinase at 0.36 U/ml, inulinase from all strains, with the exception of the pH 5 enzyme of *Pichia polymorpha*, hydrolyzed 80–100% of the inulin by 10 h. At 150 g/liter, hydrolysis was slower and less complete, with the exception of the *D. cantarelli* enzyme, which achieved 100% hydrolysis at 24 h. The *K. fragilis* enzyme gave 95% hydrolysis under those conditions. In all cases, the initial rate of hydrolysis was very rapid but slowed dramatically as the reaction proceeded past 50% completion. Presumably, inhibition by reaction products was the reason for the decline in activity.

A similar pattern of rapid hydrolysis followed by a decline in reaction rate was observed with Jerusalem artichoke extract containing 150 g/liter carbohydrate. Only the enzymes of *P. polymorpha* (optima at pH 2 and pH 5, respectively) gave 100% hydrolysis. The others gave 80–90% hydrolysis by that point but did not reach 100%. Very similar results were obtained with the chickory extract at 150 g/liter carbohydrate. In all cases tested, the use of 0.36 u/ml gave faster and more complete hydrolysis than 0.012 U/ml.

In addition to demonstrating the utility of inulinase from a variety of yeast sources in providing extensive hydrolysis of inulin and other fructosides in plant extracts, this study illustrates the differences between enzymes in their response to both pure and complex substrates. While enzyme titer is important, the response to potential complex substrates is also very significant in choosing a source of enzyme.

Negoro (1981) compared inulinase production by four yeasts: *K. fragilis, C. kefyr, C. utilis,* and *S. cerevisiae.* The latter two strains produced negligible amounts of inulinase compared to the former two. *K. fragilis* produced about 46% more extracellular inulinase than *C. kefyr.*

In an effort to improve the production of the *P. polymorpha* CBS 186 enzymes, Bajon et al. (1983) used the techniques of mutagenesis, enrichment, and screening for derepression in the presence of 2DG to find a hyperproducting mutant for use in an immobilized live-cell reactor. After mutagenesis with ethylmethylsulfonate, mutants were enriched by sequential growth in YNB-glucose (0.5%) and YNB-inulin (0.5%) to take advantage of the theoretically reduced growth lag of derepressed mutants. Following this procedure, mutants derepressed for inulinase were positively selected on YNB-inulin (0.5%) plates containing 0.01% 2DG, which inhibited growth of the wild type. A mutant strain, F-5, was obtained in this manner and selected for further study. The mutant produced almost eightfold more inulinase with inulin as the sole carbon source and produced it by the end of log phase of growth rather than in late stationary phase as for the wild type. Interestingly, the use of 2DG to select derepressed hyperproducers of *K. fragilis* did not result in stable strains (Tsang and GrootWassink, 1985).

F-5 inulinase activity was 0.087 µmole fructose released/min/mg protein at 48 hr with inulin as the carbon source. With glucose as the carbon source, 0.035 µmole/min/mg was released after 24 hr. The mutant produced sixfold more activity on a glucose/inulin mixture and nearly fourfold more activity on a fructose/inulin mixture as a carbon source. Importantly, under non-growth-supporting conditions produced by resuspending the cells in phosphate-buffered inulin, inulin/glucose, or inulin/fructose, similar levels of inulinase were produced, nearly sevenfold more than wild-type cells. This indicates little or no repression by glucose or fructose under non-growth-supporting conditions.

The F-5 mutant was further analyzed to determine the distribution of the inulinase activity. Chautard et al. (1981) demonstrated two forms of inulinase from *P. polymorpha* CBS 186, an early-eluting E1 (DEAE cellulose and Sephadex G-150) and a late-eluting E2. Analysis of the wild-type and the F-5 mutant revealed that while both forms were present in the parent strain, the E2 form was absent in the mutant. A similar protocal involving ethylmethylsulfonate mutagenesis and 2DG selection was used to obtain a higher-yielding, partially derepressed mutant of *K. (fragilis) marxianus* (Bourgi et al., 1986). Ethanol production rates on inulin substrate were about one third better under aerobic and anaerobic conditions.

Constitutive producers of inulinase were isolated from mixed carbon source continuous cultures of *K. fragilis* (GrootWassink and Hewitt, 1983). Analysis of colonies isolated from cultures grown on fructose and galactose revealed that 80% of the isolates had higher activity than the original wild-type strain (*K. fragilis*, ATCC 12424). One isolate, PRL Y53A, demonstrated higher productivity, which was stable to 12 serial transfers. The mutants were generated in a mixed carbon source continuous-culture system in which the fructose was gradually replaced with galactose. Although it is largely unclear why such a process should select for high inulinase producers, the authors speculate that coregulation of the fructose metabolism apparatus and that of inulinase may exist. Thus, positive selection for mutants better able to cope with lower levels of fructose may be better inulinase producers. The mutant obtained produced higher inulinase activity on all of a series of 11 carbon sources including ethanol and malate. That the mutant was constitutive was demonstrated by its ability to produce high levels of inulinase on substrates that allowed the wild type to produce only basal levels.

One of the standard methods of screening for hyperproducers of extracellular enzymes is the zone-clearing assay on indicator plates. Tsang and GrootWassink (1985) developed such an assay and have used it to find second-generation mutants of the *K. fragilis* strain PRL Y53A described above. The characteristic sought was resistance to carbon catabolite repression. Strains isolated from continous culture of the original hyperproducing mutant were mutagenized with N-methyl-N-nitro-nitrosoguanidine and plated on an inulin-containing medium (0.5% yeast extract, 0.5% peptone, 4% or 8% glucose, 4% inulin, and 2% agar). The inulin was in an undissolved form and produced an opaque field on the plate. Colonies were grown for 4–6 days at 30°C and then shifted to 40°C for 1 day to allow clearing zones to develop. Of 6,000 colonies screened, 56 were selected for testing in culture broth containing 8% glucose. Three high-yielding isolates producing greater than 120 U/ml (parental titer: 12 U/ml) were obtained. Of these, only one demonstrated high yield in carbon limited culture (1% glucose). Further analysis revealed that this mutant (PRL Y53E) was not carbon catabolite repression resistant. Grown on 8% glucose, the mutant used glucose much faster than the parent strain with no difference in growth rate or yield. The authors speculated that the mutant was likely converting the glucose to nonrepressing by-products which supported growth and high inulinase production.

4. Cloning of Yeast Inulinase

There have not as yet been, to this reviewer's knowledge, any reports published describing the cloning of inulinase genes of yeast. It is likely that such studies will emerge quite quickly, given the level of interest in this enzyme.

D. Beta-Galactosidase from Yeast

1. Introduction

Lactase (β-D-galactoside galactohydrolase, E.C. 3.2.1.23) catalyzes the hydrolysis of the milk disaccharide, lactose, to its constituents, glucose and galactose. The investigation of the production, properties, and methods of application of this enzyme has been greatly intensified over the past two decades. This is due to a variety of factors evolving from the huge economic significance of dairy products in the food industry. The applications of lactase stem from widespread lactose intolerance in humans, lactose crystallization problems in processed foods, cheese manufacture, cheese whey waste reduction and protein upgrading, and ethanol production (Nijpels, 1981; Kierstan and Corcoran, 1984; Maiorella and Castillo, 1984; Hahn-Hagerdal, 1985; Swaisgood, 1985; Gianetto et al., 1986).

A number of reviews have been published describing the commercial applications of β-galactosidase (Richardson, 1975; Kilara and Shahani, 1979; Nijpels, 1981; Gekas and Lopez-Leiva, 1985). Although most milk treatment is performed by batch hydrolysis, current research is heavily focused on the use of immobilized lactase or lactase-rich whole cells (Burgess and Shaw, 1983; Swaisgood, 1985; Gianetto et al., 1986; Linko and Linko, 1986; Sarto et al., 1986). A major impediment to the implementation of immobilized cell or enzyme technology for milk processing is the problem of microbial contamination of process equipment and streams (Burgess and Shaw, 1983; Gekas and Lopez-Leiva, 1986).

The main commercial sources of yeast lactase are *Kluyveromyces (Saccharomyces) lactis* and *Kluyveromyces (Saccharomyces) fragilis* (Nijpels, 1981; Burgess and Shaw, 1983). The synthesis and properties of β-galactosidase from these microorganisms are described below.

2. Characterization of Yeast β-Galactosidase

β-Galactosidase is an intracellular enzyme in yeast (Caputto et al., 1984; Davies, 1964; Dickson et al., 1979; Dickson and Barr, 1983). The utilization of lactose depends on both the enzyme and an active transport system for the uptake of the disaccharide (Dickson and Barr, 1983). The production of β-galactosidase in yeasts is inducible. Using *K. lactis*, NRRL Y1140, Dickson and Martin (1980) have shown that the enzyme activity began to increase 10–15 min after the addition of a suitable inducer (lactose, galactose, or lactobionic acid). Levels up to 150-fold higher than basal could be obtained. The enzyme level was growth-phase-dependent, and dropped when stationary phase was entered. The continuous presence of inducer (lactose), at concentrations of 1–2 mM, was required for maximum activity. Glucose addition produced only transient repression of enzyme production, and lactose could be utilized in its presence.

The β-galactosidase from *K. lactis* Y1140 was purified 80-fold on nondenaturing gels (Dickson et al., 1979). On SDS gel electrophoresis, a single band was observed and a subunit size of 135,000 daltons was inferred. A different pattern was observed for the enzyme from *S. fragilis* by Mahoney and Whitaker (1978). The *S. fragilis* enzyme appeared to consist of two unequal subunits of 90,000 and 120,000 daltons. No carbohydrate was found in the *S. fragilis* enzyme. Itoh et al. (1982) characterized the molecular weight of lactase from four yeast species by gel electrophoresis. All were in the range of 200,000–233,000 daltons. This is roughly consistent with the holoenzyme values of Dickson et al. (1979) and Mahoney and Whitaker (1978). The k_m values for lactose and ortho-nitrophenyl-galactopyranoside (ONPG) occurred at 12–17 mM and 1.6 mM, respectively. Maximal activity with the *K. lactis* enzyme was found at pH 7.25. This pH optimum is higher than those described by others, pH 6.0–6.7, for the hydrolysis of lactose in natural substrate solutions (Kulp, 1975; Nijpels, 1981; Burgess and Shaw, 1983). Guy and Bingham (1978) reported a pH optimum of 6.5 for *S. lactis* lactase on sweet whey.

The regulation of β-galactosidase activity is complex, showing sensitivity to the concentration of a variety of ions (Davies, 1974; Uwajima et al., 1972; Mahoney and Adamchuk, 1980). The *S. lactis* enzyme has maximum activity in the presence of 40–100 mM sodium or potassium and 0.1–1.0 mM manganese (Dickson et al., 1979). The effects of sodium and potassium ions on β-galactosidase activity in whole cells were demonstrated with *K. bulgaricus* (Van Huynh and Declaire, 1985). Preincubation of whole cells with 500 mM NaCl or KC1 resulted in up to 10-fold more activity than in controls.

Analysis of the reaction products from lactose hydrolysis with the *S. lactis* enzyme revealed, in addition to glucose and galactose, several transglycolytic products, allolactose, galactobiose, trisaccharides, and tetrasaccharides (Dickson et al., 1979). This pattern was identical to that produced by the *E. coli* enzyme under the same conditions.

3. Mutagenesis, Selection, and Screening Studies

β-Galactosidase is widely distributed in nature (Wierzbicki and Kosikowski, 1973; Kulp, 1975; Rao and Dutta, 1978; Nijpels, 1981). Commercial exploitation of lactase sources is restricted to relatively few microorganisms. Among the yeasts, these include *Kluyveromyces (Saccharomyces) fragilis, K. lactis,* and *Candida pseudotropicalis.* The best-characterized production strains are *K. fragilis* and *K. lactis.* Several groups have conducted studies examining a broad range of strains in efforts to identify those with superior characteristics with regard to lactase production or lactose utilization (Mahoney et al., 1975; Rao and Dutton, 1978; Itoh et al., 1982; Bothast et al., 1986). The range of lactase production varies greatly between strains of the same species. Mahoney et al. (1975) reported a 60-fold difference of activity in 41 strains of *K. fragilis* grown on a 10% lactose medium. Lactase activity was growth-associated and was maximal at the beginning of stationary phase, the point at which activity was measured for comparison purposes. The best strains produced slightly more than 5.1 lactase units per milligram dry weight (1 μmol of glucose liberated per minute per milligram cells) when grown on supplemented whey media. Wierzbicki and Kosikowsky (1973) grew four yeast strains (*S. lactis* ATCC 10689, *S. fragilis* ATCC 8582, *S. fragilis* C-17, and *Candida pseudotropicalis* ATCC 2540) on supplemented deproteinized acid whey and examined the lactase activity of the freeze-dried biomass. The yields of dry matter varied greatly, from 0.5 g/liter for *S. lactis* ATCC 10689 to 6.9 g/liter for *S. fragilis* C-16. No differences were observed in the enzyme characteristics or the extent of hydrolysis observed when whey solutions were treated with 80 mg of dried yeast per 100 g of whey. No specific activity data were presented for the various strains.

One potentially excellent source of lactase-producing yeast strains is raw milk. Twelve strains from this source were examined by Itoh et al. (1982). These consisted of three strains of *Torulopsis versatilis*, two of *K. lactis*, two of *T. sphaerica*, and five of *C. pseudotropicalis*. The specific activities of these strains, cultivated on an 8% lactose medium, varied over eightfold, ranging from 0.2 to 1.67 units/mg protein for *C. pseudotropicalis* C-54 and *K. lactis* M13, respectively. Maximum enzyme yield was obtained from *K. lactis* M11 owing to higher biomass yield on the lactose medium.

Bothast et al. (1986) examined 107 yeast strains including *K. marxianus (K. fragilis)*, *K. lactis*, *C. pseudotropicalis*, *C. versatilis*, *B. ettanomyces anomalus*, *B. classenii*, and *Trichosporon melibiosum* for their ability to convert lactose to ethanol. Wide variations in production rates and final yields of ethanol were observed. Of the yeasts examined, several *K. marxianus* strains were shown to be superior, producing ethanol at levels approaching theoretical yield and at rapid rates. It would be of interest to know whether there was any correlation between lactase activity and the relative ability to produce ethanol from lactose.

Barbosa et al. (1985) examined two strains of *K. fragilis* (145 and 276) and one strain of *K. lactis* in a study of lactase production on cheese whey. The three strains varied greatly in their production of lactase during the growth cycle on whey. Of particular significance in the choice of strain for further development work was the relative drop in the specific activity at the end of exponential phase. Although *K. fragilis* 276 showed better specific and volumetric activity through the exponential phase of growth on whey, the drop in specific activity at the end of log phase was extreme compared to that seen for *K. fragilis* 145. These observations underscore the importance of characterizing activity retention in the development of strains for industrial applications.

Sheetz and Dickson (1980) isolated several strains of *K. lactis* with mutations in the *LAC5* locus, which conferred constitutive synthesis of β-galactosidase in the absence of inducer. At a growth temperature of 23°C, activity levels were approximately eight- to 12-fold higher than the wild-type background. The mutations had complex effects as these strains were unable to grow on glucose in the presence of lactose and galactose.

Analysis of additional mutants, constitutive for β-galactosidase production, revealed eight- to 10-fold higher levels of messenger RNA for that enzyme (Dickson et al., 1981). These constitutive mutants occurred at a frequency of 1 in 4,000. The increase in mRNA levels was attributed to mutations in the *LAC10* gene, a locus unlinked to the β-galactosidase locus, *LAC4*. The mutation in *LAC10* was found to be recessive to wild type. Analysis of other genes in the galactose catabolic pathway showed that galactokinase was elevated severalfold while galactose-1-phosphate uridyltransferase was only slightly elevated. The authors proposed that the *LAC10* product might be a negative regulator involved in differential coregulation of the galactose catabolic pathway.

In addition to maximizing lactase yields by careful strain selection, optimization of growth conditions offers potential gains. Bales and Castillo (1979) increased the volumetric production of lactase from *C. pseudotropicalis* from approximately 5 U/ml (μmol o-nitrophenol released from ONPG per minute) to nearly 68 U/ml by optimizing medium composition. Interestingly, the effects of temperature and pH on lactase production by *S. fragilis* are pronounced (Wendorff et al., 1970). On a mineral-supplemented deproteinized 7.5% whey medium, the specific activity at pH 4.5 was nearly double that at pH 6.0 and about 50% higher than at pH 3.5. The activity also exhibited a sharp peak at 28°C, with a specific activity nearly five times that observed at 20°C. The specific activity leveled off between 32°C and 37°C at about 65% of the maximum.

Complex nitrogen supplements have been demonstrated to have a beneficial effect on volumetric lactase activity, indicating a general nitrogen deficiency in many deproteinized whey media. Both Wendorff et al. (1970) and Sonawat et al. (1981) reported increases of severalfold by supplementation of whey media with corn steep liquor. This emphasizes the need to carefully evaluate media effects and minor nutrient limitations in strain screening.

Continuous culture has been used to obtain yeast mutants derepressed for inulinase production (GrootWassink and Hewitt, 1983). Pedrique and Castillo (1982) observed no changes in specific lactase activity (per unit biomass) in cultures that had been maintained in continuous culture for up to 15 days at a dilution rate of 0.1 hr. Increases in the dilution rate did not result in increases in the specific activity.

4. Cloning of Yeast β-Galactosidase

Yeast β-galactosidase has been cloned and expressed in *E. coli* (Dickson and Markin, 1978) and *K. lactis* (Das et al., 1985) and found to be active. The *K. lactis* β-galactosidase gene (*LAC4*) was cloned into *S. cerevisiae* under the control of a novel hybrid promoter (Velati-Bellini et al., 1986). The enzyme was produced in *S. cerevisiae* at up to 15% of total cellular protein, placing it among the highest production levels for heterologous gene expression in yeast. Introduction of the β-galactosidase gene alone does not confer the ability to utilize lactose since *S. cerevisiae* lacks the ability to transport lactose (Dickson, 1980). Sreekrishna and Dickson (1985) have reported the construction of *S. cerevisiae* strains transformed with both the β-galactosidase (*LAC4* gene; Sheetz and Dickson, 1981) and lactose permease (*LAC12*) of *K. lactis*. Development of the LAC+ phenotype was shown to be insertion-site-dependent, and an unintegrated vector or integration at sites different from those found for the LAC+ phenotype did not produce the ability to grow on lactose. LAC+ strains transported lactose at rates 20–40% of those of *K. lactis* controls (Sreekrishna and Dickson, 1985). Further evidence supporting the utilization of the *K. lactis* permease by the *S. cerevisiae* transformants was demonstrated by the similar inhibition response to dinitrophenol-mediated energy uncoupling. Under nonselective conditions, the LAC$^+$ phenotype was reduced by 82%. DNA analysis showed *K. lactis* sequence deletion in the LAC$^-$ revertants. Das et al. (1985) cloned the *K. lactis* β-galactosidase gene and retransformed *K. lactis* with a vector that was present at an estimated copy number of 10. Induced activity levels in the transformant were two- to threefold higher than in the induced wild type. The ability to clone and express these genes in *S. cerevisiae* and in *K. lactis* should have profound effects on both the economics of lactase production and potential applications.

E. Lipase from Yeasts

1. Introduction

Lipase (triacylglycerol acylhydrolase, E.C. 3.1.1.3) catalyzes the hydrolysis of glycerides to free fatty acids and glycerol. Great variability in specificity for fatty-acid type and position exists among lipases. The characteristics of lipases from a wide variety of plant, animal, and microbial sources have been excellently reviewed by Brockerhoff and Jensen (1974) and Macrae (1983).

The investigation of microbial lipases has been stimulated in recent years by both academic and applied interests (Shahani, 1975; Macrae, 1983). The latter interest is due to the potential uses of lipases in, among others, digestive aids, hydrolysis of oils, interesterification of oils, flavor modification, and esterification of fatty acids to glycerol, alcohols, and carbohydrates (Shahani, 1975; Macrae, 1983; Hoq et al., 1985; Kilara, 1985; Parviainen et al., 1986; Uchiboro, 1986). Although the ability to produce lipases is widely distributed among microorganisms (Macrae, 1983), the lipases of relatively few yeast species have been extensively studied. These include *Saccharomycopsis lipolytica* and synonymous species (*Candida lipolytica C. paralipolytica, C. deformans,* and *Yarrowia lipolytica* [Ota et al., 1968, 1972; Muderhwa et al., 1985] and *Candida cylindracea* [Tomizuka et al., 1966a,b]).

2. Characterization of Yeast Lipase

One of the best-studied yeast lipases is that of *Saccharomycopsis lipolytica*. The activity occurs as a cell-associated and as an extracellular enzyme. The relative proportions are highly dependent on culture conditions and age. Grown in the absence of an inducer, the culture produces very little lipase. In the presence of olive oil or oleic acid, a large increase in the cell-associated form is observed (Ota et al., 1968; Sugiura et al., 1975). The increase in extracellular activity is coincident with a decline in the cell-associated activity, suggesting that the former is not produced de novo, but is the result of enzyme release from the cell (Ruschen and Winkler, 1972). The appearance of extracellular lipase coincides with the onset of stationary phase (Ruschen and Winkle, 1972). The stimulation, by medium additives, of lipase release from the cell has been demonstrated by Ruschen and Winkler (1972) and Ota et al. (1968a,b, 1977, 1978). Ruschen and Winkler (1972) showed a time- and concentration-dependent release of cell-bound lipase by the addition of potassium hyaluronate at 1–5 mg/ml. Similarly, Ota et al. (1978) reported that an alkaline extract of defatted soybean meal stimulated the release of lipase in a medium that normally supported the production of only the cell-bound form. The increase in extracellular activity was approximately 130-fold. Because no significant increase in the total lipase activity occurred, the activity of the soybean fraction appeared to be specific for stimulating the release of the cell-bound form.

Muderhwa et al. (1985) studied the production and characteristics of the extracellular lipase from *Candida deformans (S. lipolytica)* grown on rapeseed oil. Growth on rapeseed oil and rapeseed oil/glucose media showed that the lipase activity rose sharply at the onset of stationary phase in a manner similar to that observed by Ruschen and Winkler (1972). The repression of lipase synthesis by glucose was inferred from the delay of lipase appearance until the glucose was depleted from the medium. Further experiments with both growing and resting cells confirmed the repressive effect of glucose. Two extracellular lipases (lipases I and II) were reported for *Rhodotorula pilmanae* CBS 5804 grown on rapeseed oil medium (Muderhwa et al., 1986). In contrast to the production profile seen with *S. lipolytica*, both were produced in a growth-associated manner in either rapeseed oil or glucose media. On the glucose medium, the production time course for both enzymes was very similar, except for the level of activity. On the rapeseed oil medium, lipase I peaked somewhat earlier.

Ota et al. (1982) purified two forms of cell-bound lipase from *S. lipolytica* to homogeneity. The molecular weights were 39,000 daltons (lipase I) and 44,000 daltons (lipase II). Both had an absolute requirement for an activator (oleic acid) for activity on tributyrin or triolein, a requirement that was absent in the cell-bound state (Ota et al., 1972). The two forms had very similar pH optima (pH 8.2), activity versus substrate profiles, pH and temperature stability profiles, and specific activities. Because of their extreme similarity, it was concluded that they represented a modified and an unmodified version of the same protein. The modification resulted in a difference in molecular weight and elution profile from CM-Sepharose CL-6B. These enzymes appear to be quite different from the extracellular lipase purified from *C. deformans (S. lipolytica)* purified and characterized by Muderhwa et al. (1985). In that study, the extracellular enzyme had a molecular weight of 207,000 daltons (by gel filtration) and a pH optimum of 7.0. A temperature optimum between 40°C and 50°C was reported.

The lipase produced by *Candida cylindracea* was purified to homogeneity and shown to be a glycoprotein with a molecular weight of 110,000–120,000 daltons (Tomizuka et al., 1966a,b). Optimum activity occurred at pH 7.2 and 45°C. The sugars associated with the enzyme were identified as xylose and mannose. This enzyme shows a complete lack of positional specificity with respect to natural oils (Benzonana and Esposito, 1971). Inter-esterification of triglycerides in butter fat using *C. cylindracea* lipase also showed no positional specificity (Parviainen et al., 1986). The optimum temperature and pH for this enzyme were 45°C and pH 7.2 (polyvinyl alcohol emulsifier, olive oil substrate) (Benzonana and Esposito, 1971).

In contrast to the two highly similar cell-bound forms of lipase reported for *S. lipolytica* (Ota et al., 1982), the two forms of extracellular lipase from *R. pilmanae*are quite distinct (Muderhwa et al., 1986). Lipases I and II were of molecular weight of 172,800 and 21,400 daltons, respectively. Lipase I had a pH optimum at 7.0 while lipase II had an optimum at 4.0. Temperature optima were similar however, falling between 45°C and 55°C. Analysis of the rapeseed oil substrate during growth indicated that triglycerides disappeared first, followed by diglycerides, eventually leading to an increase in the monoglyceride content of the remaining lipid. The authors concluded that the lipases were specific for the primary position of glycerides. For both lipases, activity was higher with decreasing carbon chain length with saturated triglycerides. The reverse was true with monounsaturated fatty acids. Activity also rose with increased sites of unsaturation in unsaturated C18 fatty-acid triglycerides. Much attention has been focused on the activation and substrate specificity of the *S. lipolytica* lipase. A variety of substances have been shown to positively influence the activity of this enzyme. These include activation by anionic surfactants such as bile salts (Ota and Yamada, 1966a), calcium, strontium, barium (Ota and Yamada, 1966b), and oleic acid (Ota et al., 1972). The substrate specificity of *S. lipolytica* lipase was investigated by Ota et al. (1978). Using triglycerides composed of identical fatty acids, they observed that maximum activity occurred with a carbon chain length of 8. At a chain length of 12, activity was reduced by almost 50%. Further increase diminished activity to negligible levels. With methyl esters as test substrates, a different pattern emerged. Maximum activity remained at 8 carbons, declined to zero at 6, and showed a complex but active pattern above 8 carbons. Vegetable oils, but not beef tallow, were readily degraded. The substrate specificity of *C. deformans (S. lipolytica)* lipase studied by Muderhwa et al. (1985) was very similar to that reported by Ota et al. (1972, 1982). These findings support the contention of Muderhwa et al. (1985) that these two strains are of the same species.

3.   Mutation, Selection, and Screening Studies

This area appears to be relatively underdeveloped. One can find very little in the published literature dealing with either screening studies or strain development for increased output of yeast lipase. There undoubtedly are such studies in existence, but perhaps in closed, commercial files. This area appears ripe for exploitation using both the classical mutation and screening techniques and those of modern molecular biology. A variety of indicator assays suitable for screening lipase producers have been available since the early decades of this century. Many of the early techniques were reviewed, in brief, by Collins and Hammer (1934). In the main, the basis of these techniques is the formation of either zones of clearing in opaque media or color change zones using dye-impregnated media. Collins and Hammer (1934) described the colors produced by the interactions of Nile blue sulfate and a range of triglycerides, fatty acids, and natural and hydrogenated fats. Additionally, the consequences of the action of lipolytic bacteria were described, and their observations may serve as a useful guide to preparing screening media. Sierra (1957) described the use of water-soluble Tweens as lipase indicators. In that technique, lipolytic activity results in the formation of insoluble calcium salts of the free fatty acids, leading to halo formation. Quantitation of lipase activity by measurement of hydrolysis zone diameters has been demonstrated by Lawrence et al. (1967) and Karnetova et al. (1984). Kouker and Jaeger (1987) have similarly demonstrated quantitation of lipase activity using fluorescent haloes, detected by ultraviolet irradiation of media containing lipid and rhodamine B. Fluorescent zones developed when lipolysis occurred. Use of rhodamine B resolved problems encountered with the use of potentially bacteriostatic indicators such as Nile blue sulfate or Victoria blue and pH indicators (Kouker and Jaeger, 1987).

F. Miscellaneous Extracellular Enzymes

Increased economies in the utilization of plant biomass in microbial processes require greater efficiency in the utilization of the structural components of plant tissues. An enormous amount of effort has gone into the development of microbial systems for the degradation of cellulose, hemicellulose, pectin, and lignin (Fogarty and Kelly, 1983; Enari, 1983; Dale et al., 1985; Chynoweth and Jerger, 1985). Most of this work involves bacterial and fungal enzymes, but, increasingly, efforts to find and improve yeasts with cellulolytic and xylanolytic capabilities are being described.

1. Cellulase

Cellulase, the complex of cellulolytic proteins, is among the most extensively researched enzymatic activities known. Much of this activity stems from its vast potential in industrial applications. The complex consists of endoglucanase (1,4-β-D-glucanohydrolase, E.C. 3.2.1.4), exoglucanase (1,4-β-D-glucan cellobiohydrolase, E.C. 3.2.1.91), and β-glucosidase (β-D-glucoside glucohydrolase, E.C. 3.2.1.21). Enzymes of the cellulase complex are produced, not surprisingly, by a wide variety of microbial species. An excellent review of the sources and properties of microbial cellulases has been published by Enari (1983).

Although the ability to degrade various types of cellulose is widespread among fungal and bacterial species, it is rare among the yeasts (Enari, 1983). Dennis (1972) examined 53 strains of yeasts, in 12 species, capable of growth on cellobiose, for the ability to grow on ball-milled Whatman filter paper in minimal medium. Of the 12 species examined, only strains of *Trichosporon cutaneum* (9 of 12 examined) and *T. pullulans* (12 of 13 examined) were able to grow and produce clear zones in an opaque cellulose agar medium. No effort was made to optimize either the production of the cellulase or its activity. The other species tested were *Aureobasidium pullulans, Torulopsis fragaria, Kloeckera apiculata, Cryptococcus laurentii, C. laurentii* var. *fluorescens, C. diffluens, C. albidus, Torulopsis candida, Hansenula anomala, H. saturnus, H. silvicola, Pichia pinus, P. etchellsii, Endomyces capsularis, Debaromyces hansenii,* and *Saccharomycoides ludwigii.* Analysis of the breakdown products from a *T. cutaneum* strain, grown on either carboxymethylcellulose or ball-milled Whatman No. 1 filter paper, revealed glucose and cellobiose as the major and minor products, respectively.

Stevens and Payne (1972) conducted optimization studies for cellulase production by *T. cutaneum* and *T. pullulans* cultured on a ball-milled filter paper medium. Maximum activity, for both species, was observed at 25°C and in the pH range of 4.8–6.0. Analysis of the volumetric cellulase activity of eight strains of *T. cutaneum* and five strains of *T. pullulans* showed variation of over 15-fold in the former and almost 40-fold in the latter. Specific activity was also variable, but less dramatically so. Comparisons made under identical conditions, with the cellulase activities of *Trichoderma viride* CS12 (2,790 U/ml) and *Myrothecium verrucaria* IMI 45541 (2,560 U/ml) with the best of the *Trichosporon* strains showed that the latter had activity of the same order (*T. cutaneum* G31, 950 U/ml; *T. pullulans* C35; 2,060 U/ml). Analysis of the products of cellulolysis with two substrates, ball-milled filter paper and carboxymethylcellulose, revealed that both strains, activities produced celloboise and glucose as the major and minor products, respectively, in contrast to the results of Dennis (1972). Also observed were traces of cellotroise and cellotetraose. The individual activities of the different enzymes of the cellulase complex produced by these yeasts were not characterized.

There has recently been a single report of a truly cellulolytic strain of *Aureobasidium pullulans,* a yeastlike member of the fungi imperfecti (Gilbon et al., 1986). No characterization of the enzymatic activity was presented. This is the first positive report of cellulolytic activity for this species and contrasts with the results of Dennis (1972), who was unable to demonstrate cellulolytic activity with 10 strains of *A. pullulans.*

The cloning of a *T. viride* cellobiohydrolase in *S. cerevisiae,* yielding a hyperglycosylated protein, has been described (Shoemaker et al., 1986).

2. β-*Glucosidase*

A great deal of interest exists in obtaining yeast strains capable of efficiently converting cellobiose, a major product of enzymatic cellulolysis, to ethanol (Sims and Barnett, 1978; Blondin et al., 1983). Large numbers of yeast species have been reported to utilize cellobiose in an aerobic manner but few are able to ferment it to ethanol efficiently (Barnett, 1976; Freer and Detroy, 1983). Blondin et al. (1982,

1983) have described the purification and properties of the β-glucosidase of a yeast, *Dekkera intermedia*, that is capable of fermenting cellobiose to ethanol anaerobically. The characteristics of the *Dekkera* β-glucosidase were very similar to those of the *Saccharomyces cerevisiae* enzyme. It was intracellular and had a similar substrate spectrum, molecular weight (310,000 daltons), k_m (0.2 mM), and sensitivity to thiol-specific reagents. The *Dekkera* enzyme was synthesized under both aerobic and anaerobic conditions and was repressed by glucose. The strain fermented cellobiose to ethanol as efficiently as from glucose and at respectable levels, 75 g/liter (Blondin et al., 1983).

A screening study of 400 yeast strains capable of growth upon xylose revealed a new species, *Kluyveromyces cellobiovorans*, capable of fermenting both xylose and cellobiose to ethanol, aerobically or anaerobically (Morikawa et al., 1986). This strain produced a cellobiase with a relatively narrow pH optimum between pH 6 and 7. No activity was detected below pH 4.5. No further characterization of the cellobiase was reported.

The β-glucosidase gene from *Candida pelliculosa* var. *acetaetherius* has been identified, characterized, and cloned (Kohchi and Toh-e, 1985, 1986). The enzyme is tetrameric with a subunit size of 90,000. Transformation of *S. cerevisiae* by a plasmid carrying this gene has been shown to confer the ability to express high levels of β-glucosidase and to secrete it to the periplasmic space (Kohchi and Toh-e, 1986). Eighty percent of the activity in the transformant was located in the periplasmic space. Data on the fermentation of cellobiose to ethanol were not presented, but the transformant was able to grow on cellobiose. A successful result in cloning and expressing a heterologous β-glucosidase in *S. cerevisiae* is an excellent indication that efficient cellobiose fermentation to ethanol will be demonstrated in the near future.

3. Xylanase

Xylanase (exo-1,4-β-xylosidase, E.C. 3.2.1.37) has been found as an extracellular enzyme among a few species of yeast. Similar to the very narrow distribution of cellulolytic activity among yeast, the ability to produce xylanase is relatively rare. Biely et al. (1978) tested 95 strains of yeast in 35 genera for the ability to grow on xylan. Strains of only three genera were found to be positive for this trait. These included *Aureobasidium* (all strains tested), *Cryptococcus* (8 of 12), and *Trichosporon* (1 of 12). Analysis of the xylanase levels in liquid medium indicated that the levels of extracellular xylanase are similar for all the strains tested. In recent years, the xylanases of all three of the positive-testing genera identified by Biely et al. (1973) have been much more closely scrutinized.

The most carefully studied xylanase has been that of *C. albidus*. *C. albidus* has been shown to synthesize two distinct xylanases (Biely et al., 1980a). The secreted form is an endo-1,4-β-xylanase, which decomposes xylan to xylobiose and xylotriose. The second form is a cell-bound β-xylosidase which hydrolyzes the oligomers. The molecular weight of the extracellular enzyme was 26,000 daltons. Production of both enzymes was somewhat growth-associated, but synthesis continued well into stationary phase in liquid culture (Biely et al., 1980a). Synthesis was repressed during growth on glucose and inducible by oligoxylosides including xylobiose. Additionally, the nonmetabolizable glycoside β-methylxylopyranoside induced xylanase production (Biely et al., 1980b). The production of xylanase in response to the presence of xylobiose and other inducers was shown to be a true induction requiring their presence and not the result of a derepression mechanism. The results of Biely et al. (1980a,b), were largely confirmed by Morosoli et al. (1986), with the exception that the latter reported a molecular weight for β-xylanase of 48,000 daltons. A β-xylanase for *C. flavus* similar to that described for *C. albidus* by Biely et al. (1980a,b) was determined to have a molecular weight of 23,000–25,000 daltons (Nakanishi et al., 1984).

Yasui et al. (1984) investigated in detail the effects of a variety of inducers upon xylanase production by *C. flavus* IFO 0407. In an initial screen of various hexoses, xylo-oligosaccharides, and xylan, only the xylose-containing compounds induced xylanase formation. All of the sugars tested supported growth. A series of eight nonmetabolizable α- and β-derivatized xylosides were tested, and only the β-methyl- and β-ethylxylosides stimulated xylanase synthesis. The former compound resulted in threefold higher activity, which was directly proportional to the amount taken up by the cells. The stimulatory effect of β-methylxyloside peaked at 0.3 mg/ml. Morosoli et al. (1986) determined the sequence of the first 72 amino acids for the *C. albidus* xylanase. Active site homology with that of lysozyme was demonstrated, confirming earlier predictions of a lysozymelike mechanism by Biely et al. (1981b). Interestingly, the enzyme purified by Morosoli et al. (1986) was able to hydrolyze xylan to xylose and xylobiose, unlike the β-xylanases purified by Biely et al. (1980a) or Nakanishi et al. (1984), which had no xylobiase activity. The question of whether *Cryptococcus* β-xylanase is a glycoprotein remains unresolved. Biely et al. (1980a) alluded to unpublished observations supporting a glycoprotein nature. Nakanishi et al. (1984) were unable to detect any carbohydrate in their purified *C. flavus* enzyme.

The β-xylanase of *T. cutaneum* was purified to homogeneity (Stuttgen and Sahm, 1982). The production of characteristics were similar to those described for the β-xylanase of *C. albidus* (Biely et al., 1980a). Synthesis of the enzyme was growth-associated, glucose-repressible, and inducible by xylan and xylose. The molecular weight was estimated at about 45,000 daltons. The activity was stable at 35°C but deteriorated rapidly at higher temperatures. Maximum activity occurred at about pH 5.0. The hydrolysis product spectrum from xylan was similar to that from the *C. albidus* strain of Morosoli et al. (1986) in that xylose was a major product in addition to xylobiose and xylotriose. No covalently attached carbohydrate was observed.

Leathers (1986) reported the production of high levels of extracellular xylanase with very high specific activity from pigmentation variants of *A. pullulans*. Normally pigmented *A. pullulans* strains produced volumetric and specific xylanase activities similar to those of *C. albidus*. Volumetric activities of 2.4–7.3 U/ml and specific activities of 9.7–26.9 U/mg were reported. The culture supernatants of the color variant strains had volumetric activities of 12–373 U/ml and specific activities of 50–540 U/mg. Purification of the xylanase activity resolved a protein of approximately 20–21 kDa with a specific activity of 2,110 U/mg. Interestingly, the molecular weight of the xylanase of *C. albidus* strain chosen for comparison purposes in that study was determined to be 43 kDa. This value contrasts with the molecular weight of xylanase (23 kDa) reported for *C. albidus* by Biely et al. (1980a) and is in agreement with the value reported by Morosoli et al. (1986). The enzymatic activity was maximal at pH 4.0–4.5 and 35–50°C, values similar to those reported for the *Cryptococcus* enzyme. The *A. pullulans* color variant studied produced xylanase more slowly than the more typically pigmented strain chosen as a control. Production was otherwise similar in that it was growth-associated. The inducers of xylanase in *A. pullulans* were demonstrated to be xylose, xylobiose, and arabinose (Leathers et al., 1986). Production of the enzyme was glucose-repressible. Xylanase yields of up to 0.3 g/liter were reported for the color variant strains of *A. pullulans*, making these strains extremely potent producers of high-specific-activity extracellular xylanase (Leathers, 1986).

REFERENCES

Andrews, J., and Gilliland, R. B. (1952). Super attenuation of beer: A study of three organisms capable of causing abnormal attenuations. *J. Inst. Brew.* 58:189–196.

Augustin, J., Zenek, J., Kuckova-Kratochvilova, and Kuniak, L. (1978). Production of α-amylase by yeasts and yeast-like organisms. *Folia Microbiol.* 23:353–361.

Aunstrup, K., Andersen, O., Falch, E. A., and Nielsen, T. K. (1979). Production of microbial enzymes. In *Microbial Technology* Peppler, H. J., and Perlman, D. (Eds.). Academic Press, New York, pp. 282–309.

Azzoulay, E., Jouanneau, F., Bertrand, J. C., Janssens, J., and Lebeault, J. M. (1980). *Appl. Environ. Microbiol.* 39:41–47.

Bajon, A. M., Guiraud, J. P., and Galzy, P. (1983). Isolation of an inulinase derepressed mutant of *Pichia polymorpha* for the production of fructose. *Biotech. Bioeng.* 26 :128–133.

Bajpai, P., and Margaritis, A. (1985). Improvement of inulinase stability of calcium alginate immobilized *Kluyveromyces marxianus* cells by treatment with hardening agents. *Enz. Microb. Technol.* 7:34–36.

Bales, S., and Castillo, F. J. (1979). Production of lactase by *Candida pseudotropicalis*- grown in whey. *Appl. Environ. Microbio.* 37:1201–1205.

Ballou, C. E. (1982). Yeast cell wall and cell surface. In *The Molecular Biology of the Yeast Saccharomyces: Metabolism and Gene Expression.* Cold Spring Harbor Laboratory, Cold Spring Harbor, NY.

Bambosa, M. F. S., Silva, D. O., Pinheiro, A. J. R., Guimares, W. V., and Borges, A. C. (1985). Production of β-D-galactosidase from *Kluyveromyces fragilis* grown in cheese whey. *J. Dairy Sci.* 68:1618–1623.

Barrett, J. A. (1976). The utilization of sugar by yeasts. *Adv. Carb. Chem. Biochem.* 32 :125–234.

Belin, J. M. (1981). Identification of yeasts and yeast-like fungi, I. Taxonomy and characteristics of new species described since 1973. *Can. J. Microbiol.* 27:1235–1251.

Beluche, I., Guiraud, J. P., and Galzy, P. (1980). Inulinase activity of *Debaromyces cantarelli. Folia Microbiol.* 25:32–39.

Biely, P., Vrsanska, M., and Kratky, Z. (1980a). Xylan-degrading enzymes of the yeast *Cryptococcus albidus.* Identification and cellular location. *Eur. J. Biochem.* 108:313–321.

Biely, P., Vrsanska, M., and Kratky, Z. (1980b). Induction and inducers of endo-1,4-β-xylanase in the yeast *Cryptococcus albidus. Eur. J. Biochem.* 108:313–329.

Biely, P., Vrsanska, M., and Kratky, Z. (1981a). Substrate-binding site of endo-1,4-β-xylanase of the yeast *Cryptococcus albidus. Eur. J. Biochem.* 119:559–564.

Biely, P., Vrsanska, M., and Kratky, Z. (1981b). Mechanisms of substrate digestion by endo-1,4-β-xylanase of *Cryptococcus albidus.* Lysozyme type pattern of action. *Eur. J. Biochem.* 119:565–571.

Bierman, L., and Glantz, M. D. (1968). Isolation and characterization of β-galactosidase from *Saccharomyces lactis. Biochim. Biophys. Acta.* 167:373–377.

Blondin, B., Ratomahenina, R., Arnaud, A., and Galzy, P. (1982). A study of cellobiose fermentation by a *Dekkera* strain. *Biotech. Bioeng.* 24:2031–2037.

Blondin, B., Ratomahenina, R., Arnaud, A., and Galzy, P. (1983). Purification and properties of the β-glucosidase of a yeast capable of fermenting cellobiose to ethanol: *Dekkera intermedia* Van der Walt. *Eur. J. Appl. Microbiol. Biotechnol.* 17:1–6.

Bothast, R. J., Kurtzman, C. P., Saltarelli, M. D., and Slininger, P. J. (1986). Ethanol production by 107 strains of yeast on 5, 10 and 20% lactose. *Biotechnol. Lett.* 8:593–596.

Bourgi, J., Guiraud, J. P. and Galzy, P. (1986). Isolation of a *Kluyveromyces fragilis* derepressed mutant hyperproducer of inulinase for ethanol production from Jerusalem antichoke. *J. Ferment. Technol.* 64:239–243.

Brockerhoff, H., and Jensen, R. G. (1974). *Lipolytic Enzymes.* Academic Press, New York.

Bucke, C. (1981). Enzymes in fructose manufacture. In *Enzymes and Food Processing,* G. A. Birch, N. Blakeborough, and K. J. Parker (Eds.). Applied Science Publications, pp. 51–72.

Burgess, K., and Shaw, M. (1983). In *Industrial Enzymology,* T. Godfrey and J. Reichelt (Eds). Macmillan Publications, Surrey, U.K., pp. 269–276.

Calleja, G. B., Levy-Rick, S., Lusena, C. V., Nasim, A., Moranelli, F. (1982). Direct and quantitative conversion of starch to ethanol by the yeast *Schwanniomyes alluvius. Biotechnol. Lett.* 4:543–547.

Caputto, R. L., Leloir, L. F., and Trucco, L. E. (1948). Lactase and lactose fermentation in *Saccharomyces fragilis. Enzymologia* 12:350–355.

Carlson, M., Osmond, B. C., and Botstein, D. (1980). SUC genes of yeast: A dispersed gene family. *Cold Spring Harbor Symp. Quant. Bio.* 45:799–803.

Carlson, M., Osmond, B., and Botstein, D. (1981). Mutants of yeasts defective in sucrose utilization. *Genetics* 98:25–40.

Carlson, M., and Botstein, D. (1982). Two differentially regulated mRNA's with different 5ʻ end encode secreted and intracellular forms of yeast invertase. *Cell* 28:145–154.

Carlson, M., Osmond, B., Neigeborn, L., and Botstein, D. (1984). A suppressor of SNF-1 mutations causes constitutive high level invertase synthesis in yeast. *Genetics*107:1 9–32.

Celenza, J., and Carlson, M. (1984a). Cloning and genetic mapping of SNF-1, a gene required for expression of glucose-repressible genes in *Saccharomyces cerevisiae*. *Mol. Cell. Biol.* 4:49–53.

Celenza, J., and Carlson, M. (1984b). Structure and expression of the SNF-1 gene of *Saccharomyces cerevisiae*. *Mol Cell. Biol.* 4:54–60.

Celenza, J., and Carlson, M. (1986). A yeast gene that is essential for release from glucose repression encodes a protein kinase. *Science* 233:1175–1180.

Ceska, M. (1971a). A new approach for quantitative and semi-quantitative determinations of enzymatic activities with simple laboratory equipment. *Clin. Chim. Acta* 33:135–145.

Ceska, M. (1971b). Enzymatic catalysis in solidified media. *Eur. J. Biochem.* 22:186–192.

Chautard, P., Guiraud, J. P., and Galzy, P. (1981). Inulinase Activity of *Pichia Polymorpha*. *Acta Microbiol. Acad. Sci. Hung.* 28:245–255.

Chynoweth, D. P., and Jerger, D. E. (1985). Anaerobic digestion of woody biomass. *Dev. Indust. Microbiol.* 26:235–246.

Clementi, F., Rossi, J., Costamagna, L., and Rosi, J. (1980). Production of amylases by *Schwanniomyces castelli* and *Endomycopsis fibuligera*. *Antonie van Leeuwenhoek*46:3 99–405.

Collins, M. A., and Hammer, B. W. (1934). The action of certain bacteria on some simple tri-glycerides and natural fats, as shown by Nile Blue sulfate. *J. Bacteriol.* 27:473–485.

Dale, B. E., Henk, L. L., and Shiang, M. (1985). Fermentation of ligncellulosic materials treated by ammonia freeze-explosion. *Dev. Indust. Microbiol.* 26:223–233.

Das, S., Breunig, K. D., and Hollenberg, C. P. (1985). A positive regulatory unit is involved in the induction of the β-galactosidase gene from *Kluyveromyces lactis*. *EMBO J.*4:793–798.

Davies, R. (1964). Lactose utilization and hydrolysis in *Saccharomyces fragilis*. *J. Gen. Microbiol.* 37:81–98.

Demeulle, S., Guiraud, J. P., and Galzy, P. (1981). Study of inulase from *Debaryomyces phaffi* Capriotti. *Allg. Mikrobiol.* 21:181–189.

DeMot, R., and Verachtert, H. (1986a). Enhanced production of extracellular α-amylase and glucoamylase by amylolytic yeasts using β-cyclodextrin as carbon source. *Appl. Microbiol. Biotechnol.* 24:459–462.

DeMot, R., and Verachtert, H. (1986b). Secretion of α-amylase and multiple forms of glucoamylase by the yeast *Trichosporon pullulans*. *Can. J. Microbiol.* 32:47 51.

DeMot, R., Van Oudendijk, E., and Verachtert, H. (1985). Purification and characterization of an extracellular glucomylase from the yeast *Candida tsukabaensis* CBS 6389. *Antonie van Leeuwenhoek* 51:275–287.

DeMot, R., Andries, K., and Verachtert, H. (1984a). Comparative study of starch degradation and amylase production by ascomycetous yeast species. *Syst. Appl. Microbiol.* 5:106–118.

DeMot, R., Demeersman, M., and Verachtert, H. (1984b). Comparative study of starch degradation and amylase production by non-ascomycetous yeast species. *Syst. Appl. Microbiol.* 5:421–432.

DeMot, R., Oudendijk, E., and Verachtert, H. (1984c). Production of extracellular debranching activity by amylolytic yeasts. *Biotechnol. Lett.* 6:581–586.

DeMot, R., Van Oudendijk, E., Hougaerts, S., and Verachtert, H. (1984d). Effect of medium composition on amylase production by some starch degrading yeasts. *FEMS Microbiol. Lett.* 25:169–173.

Dennis, C. (1972). Breakdown of cellulose by yeast species. *J. Gen Microbiol.* 71:409–411.

Dhawale, M., Wilson, J., Khachatourians, G., and Ingledew, W. (1982). Improved method for the detection of starch hydrolysis. *Appl. Environ. Microbiol.* 44:747–750.

Dhawale, M., and Ingledew, W. (1983). Starch hydrolysis by derepressed mutants of *Schwanniomyces castelli. Biotech. Lett.* 5:185–190.

Dickson, R. C. (1980). Expression of a foreign eukaryotic gene in *Saccharomyces cerevisiae:* β-galactosidase from *Kluyveromyces lactis. Gene* 10:347–355.

Dickson, R. C., and Barr, K. (1983). Characterization of lactose transport in *Kluyveromyces lactis. J. Bacteriol.* 154:1245–1251.

Dickson, R. C., and Markin, J. S. (1978). Molecular cloning and expression in *E. coli* of a yeast gene coding for β-galactosidase. *Cell* 15:123–130.

Dickson, R. C., and Markin, J. S. (1980). Physiological studies of β-galactosidase induction in *Kluyveromyces lactis. J. Bacteriol.* 142:777–785.

Dickson, R. C., Dickson, L. R., and Markin, J. S. (1979). Purification and properties of an inducible β-galactosidase isolated from the yeast *Kluyveromyces fragilis. J. Bacteriol.*137:51–61.

Dickson, R. C., Sheetz, R. M., and Laey, L. R. (1981). Genetic regulation: Yeast mutants constitutive for β-galactosidase activity have an increased level of β-galactosidase messenger ribonucleic acid. *Mol. Cell. Biol.* 1:1048–1056.

Enari, T. M. (1983). Microbial Cellulose. In *Microbial Enzymes and Biotechnology,* W. M. Fogarty and C. T. Kelly (Eds.). Applied Science Publishers, New York, pp. 183–223.

Erratt, J. A., and Stewart, G. G. (1978). Genetic and biochemical studies on yeast strains able to utilize dextrins. *J. Am. Soc. Brew. Chem.* 36:151–161.

Filho, S. A., Galembeck, E. V., Faria, J. B., and Frascino, A. C. (1986). Stable yeast transformants that secrete functional α-amylase encoded by cloned mouse pancreatic DNA. *Biotechnology* 4:311–315.

Fogarty, W. M. (1983). Microbial Amylases. In *Microbial Enzymes and Biotechnology,* W. M. Fogarty (Ed.). Applied Science Publishers, London, pp. 1–92.

Fogarty, W. M., and Kelly, C. T. (1979). Starch degrading enzymes of microbial origin. *Prog. Indust. Microbiol.* 15:87–150.

Fogarty, W. M., and Kelly, C. T. (1980). Amylases, amyloglucosidases and related glucanases. In *Microbial Enzymes and Bioconversions: Economic Microbiology,* Vol. 5, A. H. Rose (Ed.). Academic Press, London, pp. 115–170.

Freer, S. N., and Detroy, R. W. (1983). Characterization of cellobiose fermentation to ethanol by yeasts. *Biotech. Bioeng.* 25:541–557.

Frommer, W., and Rauenbusch, R. (1975). Invertase from mutant strains of *Saccharomyces. carlsbergensis* Hansen. U.S. Patent 3,887,434.

Frommer, W., and Rauensbusch, E. (1973). Microbiological production of invertase. U.S. Patent 3,887,434.

Gascon, S., and Lampen, J. O. (1968). Purification of the internal invertase of yeast. *J. Biol. Chem.* 243:1567–1572.

Gascon, S., Neuman, N. P., and Lampen, J. O. (1968). Comparative study of the properties of the purified internal and external invertases from yeast. *J. Biol. Chem.* 243:1573–1577.

Gekas, V., and Lopez–Leiva, M. (1985). Hydrolysis of lactase: A literature review. *Process Biochem.* 20:2–12.

Gianetto, A., Berruti, F., Glick, B. R., and Kempton, A. G. (1986). The production of ethanol from lactose in a tubular reactor by immobilized cells of *Kluyveromyces fragilis*. *Appl. Microbiol. Biotechnol.* 24:277–281.

Gilbon, A., Huitron, C., Farias, F. G., and Ulloa, M. (1986). Scanning electronic microscopy of a true cellulolytic strain of *Aureobasidium* grown in crystalline cellulose. *Mycologia*78:804–809.

Godfrey, T. (1983). Comparison of key characteristics of industrial enzymes by type and source. In *Industrial Enzymology*, T. Godfrey and J. Reichelt (Eds.). Nature Press, Macmillan Publishers, New York, pp. 466–502.

GrootWassink, J. W. D., and Fleming, S. E. (1980). Non-specific β-fructofuranosidase (inulase) from *Kluyveromyces fragilis:* Batch and continuous fermentation, simple recovery method and some industrial properties. *Enz. Microb. Technol.* 2:45–53.

Guiraud, J. P., and Galzy, P. (1981). Enzymatic hydrolysis of plant extracts containing inulin. *Enz. Microb. Technol.* 3:305–308.

Guiraud, J. P., Viard-Gaudin, C., and Galzy, P. (1980). Etude de L'inulinase de *Candida salmenticensis* Van Uden et Buckley. *Agric. Biol. Chem.* 44:1245–1252.

Guiraud, J. P., Bermit, C., and Galzy, P. (1982). Inulinase of *Debaromyces cantarelli. Folia Microbiol.* 27:19–24.

Guy, E. J., and Bingham, E. W. (1978). Properties of β-galactosidase of *Saccharomyces lactis* in milk and milk products. *J. Dairy Sci.* 61:147–151.

Hahn-Hagerdal, B. (1985). Comparison between immobilized *Kluyveromyces fragilis* and *Saccharomyces cerevisiae* co-immobilized with β-galactosidase with respect to continuous ethanol production from whey permeate. *Biotech. Bioeng.* 27:914–916.

Halpern, M. G. (1981). *Industrial Enzymes from Microbial Sources. Recent Advances.*Noyes Data Corporation, New Jersey.

Hewitt, G. M., and GrootWassink, J. W. D. (1984). Simultaneous production of inulase and lactase in batch and continuous cultures of *Kluyveromyces fragilis. Enz. Microb. Technol.* 6:263–270.

Hoq, M. M., Yamane, T., Shimizu, S., Funada, T., and Ishida, S. (1985). Continuous hydrolysis of olive oil by lipase in microporous hydropholic membrane bioreactor using *Candida cylindracea* enzyme. *J. A. Oil Chem. Soc.* 62:1016–1021.

Hohmann, S., and Zimmerman, C. K. (1986). Cloning and expression on a multicopy vector of five invertase genes of *S. cerevisiae. Curr. Genet.* 11:217–225.

Ingle, M. B., and Boyer, E. W. (1976). Production of industrial enzymes by *Bacillus* species. In *Microbiology,* D. Schlesinger (Ed.). American Society for Microbiology, pp. 420–426.

Innis, M. A., Holland, M. J., McCabe, P. C., et al. (1985). Expression, glycosylation and secretion of an *Aspergillus* glucoamylase by *Saccharomyces cerevisiae*. *Science* 228:21–26.

Itoh, T., Suzuki, M., and Adachi, S. (1982). Production and characterization of β-galactosidase from lactose fermenting yeasts. *Agric. Biol. Chem.* 46:899–904.

Karnetova, J., Mateju, J., Rezanka, T., Prochazka, P., Nohynek, M., and Rokes, J. (1984). Estimation of lipsae activity by the diffusion plate method. *Folia Microbiol.* 29: 346–347.

Kelly, C. T., and Fogarty, W. M. (1983). Microbial β-glucosidase. *Process. Biochem.* 18: 6–12.

Kelly, C. T., Moriarty, M. E., and Fogarty, W. M. (1985). Thermostable extracellular α-amylase and β-glucosidase of *Lipomyces* starkeyi. *Appl. Microbiol. Biotechnol.* 22: 352–358.

Kilara, A. (1985). Enzyme-modified lipid food ingredients. *Proc. Biochem.* 20:35–45.

Kilara, A., and Shahani, K. (1979). *CRC Crit. Rev. Food Sci. Nutr.* 12:161.

Kim, K. C. (1975). Studies on the hydrolysis of inulin in Jerusalem antichoke by fungal inulase. *J. Korean Agric. Chem. Soc.* 18:177–182.

Kohchi, C., and Toh-e, A. (1985). Nucleotide sequence of *Candida pelliculosa* β-glucosidase gene. *Nucleic Acid Res.* 13:6273–6282.

Kohchi, C., and Toh-e, A. (1986). Cloning of *Candida pelliculosa* β-glucosidase gene and its expression in *Saccharomyces cerevisiae*. *Mol. Gen. Genet.* 208:89–94.

Kohchi, C., Hayashi, M., and Nagai, S. (1985). Purification and properties of β-galactosidase from *Candida pelliculosa* var. *acetaetherius*. *Agric. Biol. Chem.* 49:779–784.

Kouker, G., and Jaeger, K. E. (1987). Specific and sensitive plate assay for bacterial lipases. *Appl Environ. Microbiol.* 53:211–213.

Kulp, K. (1975). Carbohydrases. In *Enzymes in Food Processing,* 2d ed, G. Reed (Ed.). Academic Press, New York, pp. 55–122.

Lam, K. S., and GrootWassink, J. W. D. (1985). Efficient, non-killing extraction of β-D-fructofuranosidase (an exo-inulase) from *Kluyveromyces fragilis* at high cell density. *Enz. Microb. Technol.* 7:239–242.

Lampen, J. O. (1971). Yeast and neurospora invertases. In *The Enzymes,* 3d ed., P. Boyer (Ed.), pp. 291–305.

Lawrence, R. C., Tryer, T. F., and Reiter, B. (1967). Rapid method for the quantitative estimation of microbial lipases. *Nature* 213:1264–1265.

Leathers, T. D. (1986). Color variants of *Aureobasidium pullulans* overproduce xylanase with extremely high specific activity. *Appl. Environ. Microbiol.* 52:1026–1030.

Leathers, T. D., Kurtzman, C. P., and Detroy, R. W. (1984). Overproduction and regulation of xylanase in *Aureobasidium pullulans* and *Cryptococcus albidus. Biotech. Bioeng. Symp.* 14:225–240.

Leathers, T. D., Detroy, R. W., and Bothast, R. J. (1986). Induction and glucose repression of xylanase from a color variant strain of *Aureobasidium pullulans. Biotechnol. Lett.* 8:867–872.

Lindgren, C. C., and Lindgren, G. (1956). Eight genes controlling the presence or absence of carbohydrate fermentation in *Saccharomyces. J. Gen. Microbiol.* 15:19–28.

Linko, P., and Linko, Y. (1986). Industrial applications of immobilized cells. *CRC Crit. Rev. Biotechnol.* 1:289–338.

Luenser, S. (1982). Microbial enzymes for industrial sweetener production. *Dev. Indust. Microbiol.* 24:79–96.

Macrae, A. R. (1983). Extracellular microbial lipases. In *Microbial Enzymes and Biotechnology*, W. M. Fogarty (Ed.). Applied Science Publishers, Essex, U.K., pp. 225–250.

Mahoney, R. R., and Adamchuk, C. (1980). Effect of milk constituents in the hydrolysis of lactase from *Kluyveromyces fragilis. J. Food. Sci.* 45:962–968.

Mahoney, R. R., and Whitaker, J. R. (1978). Purification and physiochemical properties of β-galactosidase from *Kluyveromyces fragilis. J. Food Sci.* 43:584–591.

Mahoney, R. R., Nickerson, T. A., and Whitaker, J. R. (1974). Selection of strain, growth conditions and extraction procedures for optimum production of lactase from *Kluyveromyces fragilis. J. Dairy Sci.* 58:1620–1629.

Maiorella, B. L., and Castillo, F. J. (1984). Ethanol, biomass and enzyme production for whey waste abatement. *Process. Biochem.* 19:157–161.

Meaden, P., Ogden, K., Bussey, H., and Tubb, R. S. (1985). A DEX gene conferring production of extracellular amyloglucosidase on yeast. *Gene* 34:325–334.

Modena, D., Vanoni, M., England, S., and Marmur, J. (1986). Biochemical and immunological characterization of the STA-2 encoded extracellular glucoamylase from *Saccharomyces diastaticus. Arch. Biochem. Biophys.* 248:138–150.

Montenecourt, B. S., Kuo, S., and Lampen, J. O. (1973). *Saccharomyces* mutant with invertase formation resistant to repression by hexoses. *J. Bacteriol.* 114:233–238.

Moresi, M., Solinas, M. A., and Matteucci, S. (1983). Investigation on the operating variables of potato starch fermentation by *Schwanniomyces castelli. Eur. J. Appl. Microb. Biotechnol.* 18:92–99.

Morikawa, Y., Takasaw, S., Maunaga, I., and Takayama, K. (1985). Ethanol production from D-xylose and cellobiose by *Kluyveromyces cellobiovorus. Biotech. Bioeng.* 27:509–513.

Mormeneo, S., and Sentandreu, R. (1982). Regulation of invertase synthesis by glucose in *Saccharomyces cerevisiae. J. Bacteriol.* 152:14–18.

Morosoli, R., Roy, C., and Yaguchi, M. (1986). Isolation and partial primary sequence of a xylanase from the yeast *Cryptococcus albidus. Biochim. Biophys. Acta* 870:473–478.

Moulin, G., Boze, H., and Galzy, P. (1982). Amylase activity in *Pichia polymorpha. Folia Microbiol.* 27:377–381.

Muderhwa, J. M., Ratomahenina, R., Pina, M., Graille, J., and Galzy, P. (1985). Purification and properties of the lipase from the *Candida deformans* (Zach) Langevon and Guerra. *J. A. O. C. S.* 62:1031–1036.

Muderhwa, J. M., Ratomanhenina, R., Pina, M., Graille, J., and Galzy, P. (1986). Purification and properties of the lipases from *Rhodotorula pilimanae* Hedrick and Bunke. *Appl. Microbiol. Biotechnol.* 23:348–354.

Myerback, K. (1960). *The Enzymes*, 2d Ed., Vol. 4, pp. 374.

Nakamura, T., Kurokawa, T., Nakatsu, S., and Ueda, S. (1978). Crystallization and general properties of an extracellular inulase from *Aspergillus* sp. *Agric. Biol. Chem.* 52:159–166.

Nakanishi, K., Arai, H., and Yasui, T. (1984). Purification and some properties of xylanase from *Cryptococcus flavus. J. Ferment. Technol.* 62:361–369.

Negoro, H., and Kito, E. (1973). β-Fructofuranosidase from *Candida kefyr. J. Ferment. Technol.* 51:96–102.

Neigeborn, L., and Carlson, M. (1984). Genes affecting the regulation of SUC2 gene expression by glucose repression in *Saccharomyces cerevisiae. Genetics* 108:845–858.

Newberg, C., and Mandl, I. (1950). *The Enzymes* 1st Ed., Vol. 1, Part 1, p. 527.

Nijpels, H. H. (1981). Lactoses and their applications. In *Enzymes and Food Processing*, A. A. Birch, N. Blakebrough, and K. J. Parker (Eds.). Applied Science Publishers, London, pp. 89–104.

Ota, Y., and Yamada, K. (1966a). Part I: Anionic surfactants as the essential activator in the systems emulsified by polyvinyl alcohol. *Agric. Biol. Chem.* 30:351–358.

Ota, Y., and Yamada, K. (1966b). Part II. Alkaline earth metal ions as the cofactor in the shaken system containing no emulsifier. *Agric. Biol. Chem.* 30:1030–1038.

Ota, Y., Nakayima, T., and Yamada, K. (1972). On the substrate specificity of the lipase produced by *Candida paralipolytica. Agric. Biol. Chem.* 36:1895–1898.

Ota, Y., Morimoto, Y., Sugiura, T., and Minoda, Y. (1978). Soybean fraction increasing the extracellular lipase production by *Saccharomycopsis lipolytica (Candida paralipolytica) Agric. Biol. Chem.* 42:1937–1938.

Ota, Y., Gomi, K., Kato, S., Sugiura, T., and Minoda, Y. (1982). Purification and some properties of cell-bound lipase from *Saccharomyces lipolytica. Agric. Biol. Chem.* 46:2885–2893.

Ottolenghi, P. (1971). Some properties of five non-allelic β-D-fructofuranosidases (invertases) of *Saccharomyces. C. R. Trav. Lab. Carlsberg* 38:213–221.

Parviainen, P., Vaara, K., Ali-Yrrko, S., Antila, M., and Kalo, P. (1985). Changes in the triglyceride of butter fat induced by lipase and sodium methoxide catalysed inter-esterifraction reactions: Quantitative determination. *Milchwissenshaft* 41:82–85.

Pedrique, M., and Castillo, F. J. (1982). Regulation of β-D-galactosidase in *Candida pseudotropicalis. Appl. Environ. Microbiol.* 43:303–310.

Peppler, H. J. (1979). Production of yeasts and yeast products. In *Microbial Technology*, H. J. Peppler and D. Perlman (Eds.). Academic Press, New York, pp. 157–185.

Perlman, D., and Halvorson, H. (1981). Distinct repressible in RNA's for cytoplasmic and secreted yeast invertase are encoded by a single gene. *Cell* 25:525–536.

Pretorius, I. S., Chow, T., and Marmur, J. (1986a). Identification and physical characterization of yeast glucoamylase structural genes. *Mol. Gen. Genet.* 203:36–41.

Pretorius, I. S., Chow, T., Modena, D., and Marmur, J. (1986b). Molecular cloning and characterization of the STA2 glucoamylase gene of *Saccharomyces diastaticus. Mol. Gen. Genet.* 203:29–35.

Rao, M. V., and Dutton, S. M. (1978). Lactase activity of microorganisms. *Folia Microbiol.* 23:210–215.

Richardson, G. H. (1975). In *Enzymes in Food Processing*, G. Reed (Ed.). Academic Press, New York, pp. 361–395.

Russel, I., Crumplen, C. M., Jones, R. M., and Stewart, G. G. (1986). Efficiency of genetically engineered yeast in the production of ethanol from the dextrinized cassava starch. *Biotechnol. Lett.* 8:169–174.

Ruschen, S., and Winkler, U. (1972). Stimulation of the extracellular lipase activity of *Saccharomycopsis lipolytica* by hyaluronate. *FEMS Microbiol. Lett.* 14:117–121.

Sakai, T., Kou, K., Saitoh, K., and Kutsuragi, T. (1986). Use of protoplast fusion for the development of rapid starch fermenting strains of *Saccharomyces diastaticus. Agric. Biol. Chem.* 50:297–306.

Sarto, V., Marzetti, A., and Focher, B. (1986). β-D-Galactosidases immobilized on soluble matrices: Kinetics and stability. *Enz. Microb. Technol.* 7:515–520.

Sarokin, L., and Carlson, M. (1984). Upstream region required for regulated expression of the glucose repressible SUC2 gene of *Saccharomyces cerevisiae. Mol. Cell. Biol.*4: 2750–2757.

Searle, B. A., and Tubb, R. S. (1981). A rapid method for recognizing strains of yeast able to hydrolyze starch on dextrin. *FEMS Microbiol. Lett.* 111:211–212.

Shahani, K. M. (1975). Lipases and esterases. In *Enzymes in Food Processing,* G. Reed (Ed.). Academic Press, New York, pp. 181–217.

Sheetz, R. M., and Dickson, R. C. (1981). Lac4 is the structural gene for β-galactosidase in *Kluyveromyces lactis. Genetics* 98:729–745.

Shoemaker, S. P., Gelfand, D. H., Kwok, et al. (1986). Expression of *Trichoderma reesei* L27 cellulases in *Saccharomyces cerevisiae. Am. Chem. Soc., Natl. Meeting,* pp. 188 (abstract).

Sills, A. M., Zygora, P. S. J., and Stewart, G. G. (1984). Characterization of *Schwanniomyces castelli* mutants with increased productivity of amylases. *Appl. Microbiol. Biotechnol.* 20:124–128.

Sims, A. P., and Barnett, J. A. (1978). The requirement of oxygen for the utlization of maltose, cellobiose and D-galactose by certain anaerobically fermenting yeasts (Kluyver effect). *J. Gen. Microbiol.* 106:277–288.

Skogman, H. (1976). The Symba process. *Die Starke* 28:278–282.

Snyder, H. E., and Phaff, H. J. (1960). Studies on a β-fructosidase (inulinase) produced by *Saccharomyces fragilis. Antonie van Leeuwenhoek* 26:433–452.

Snyder, H. E., and Phaff, H. J. (1962). The pattern of action of inulinase from *Saccharomyces fragilis* on inulin. *J. Biol. Chem.* 237:2438–2442.

Sonawat, H. M., Agrawal, A., and Dutta, S. M. (1981). Production of β-galactosidase from *Kluyveromyces fragilis* grown on whey. *Folia Microbiol.* 26:370–376.

Spencer-Martins, I. (1982). Extracellular isoamylase produced by the yeast *Lipomyces kononenkoae. Appl. Environ. Microbiol.* 44:1253–1257.

Spencer-Martins, I., and Van Uden, N. (1977). Yields of yeast growth on starch. *Eur. J. Appl. Microbiol.* 4:29–35.

Spencer-Martins, I., and Van Uden, N. (1979). Extracellular amylolytic system of the yeast *Lipomyces kononenkoae. Eur. J. Appl. Microbiol. Biotechnol.* 6:241–250.

Sreekrishna, K., and Dickson, R. C. (1980). Construction of strains of *Saccharomyces cerevisiae* that grow on lactose. *Proc. Natl. Acad. Sci. U.S.A.* 82:7909–7913.

Steverson, E., Korus, R. A., Admassu, W., and Heimsch, R. (1984). Kinetics of the amylase system of *Saccharomyces fibuligera. Enz. Microb. Technol.* 6:549–554.

Stewart, G. G. (1984). Yeast as an industrial microorganism and as an experimental eukaryote. *Dev. Indust. Microbiol.* 25:183–193.

Stevens, B. J. H., and Payne, J. (1977). Cellulose and xylanase production by yeasts of the gene *Trichosporon. J. Gen. Microbiol.* 100:381–393.

Straathof, A. J., Kieboom, A. P., and Van Bekkum, H. (1986). Invertase-catalysed fructosyl transfer in concentrated solution of sucrose. *Carbohydr. Res.* 146:154–159.

Stuttgen, E., and Sahm, H. (1982). Purification and properties of endo-1,4-β-xylanase from *Trichosporon cutaneum. Eur. J. Appl. Microbiol. Biotechnol.* 15:93–99.

Stutzenberger, F. (1985). Regulation of cellulolytic activity. *Annu. Rep. Ferment. Proc.* 8:111–154.

Swaizgood, H. E. (1985). Immobilization of enzymes and some applications in the food industry. In *Enzymes and Immobilized Cells,* A. I. Laskin (Ed.). Benjamin Cummings, Menlo Park, CA, pp. 1–24.

Tamaki, H. (1978). Genetic studies of ability to ferment starch in *Saccharomyces:* Gene polymorphism. *Mol. Gen. Genet.* 164:205–209.

Tanaka, K., Uchiyama, T., and Ito, A. (1972). Formation of di-D-fructofuranose 1,2`:2,3-dianhydride from an extracellular inulase of *Arthrobacter ureafaciens.* II. *Biochim. Biophys. Acta* 284:248–256.

Taussig, R., and Carlson, M. (1983). Nucleotide sequence of the yeast SUC2 gene for invertase. *Nucleic Acids Res.* 11:1943–1954.

Touzi, A., Prebois, J. P., Moulin, G., Deschamps, F., and Galzy, P. (1982). Production of food yeast from starchy subtrates. *Eur. J. Appl. Microbiol. Biotechnol.* 15:232–236.

Trimble, R., and Maley, F. (1977). Subunit structure of external invertase from *Saccharomyces cerevisiae. J. Biol. Chem.* 252:4409–4412.

Tubb, R. S. (1986). Amylolytic yeasts for commercial applications. *Trends Biotechnol.* 4:98–104.

Uchiboro, T., Seino, H., and Nishitani, T. (1986). Enzymatic synthesis of carbohydrate esters of fatty acid V. Reaction products of sorbitol with fatty acids. *JAOCS* 63:463–464.

Van Uden, N., Cabeca-Silva, C., Madeira-Lopes, A., and Spencer-Martins, I. (1980). Selective isolation of derepressed mutants of an alpha-amylase yeast by the use of 2-deoxyglucose. *Biotech. Bioeng.* 22:651–654.

Velati-Bellini, A., Pedroni, P., Martegani, E., and Alberghina, L. (1986). High levels of inducible expression of cloned β-galactosidase of *Kluyveromyces lactis* in *Saccharomyces cerevisiae. Appl. Microbiol. Biotechnol.* 25:124–131.

Wendorff, W. L., Amundson, C. H., and Olson, N. F. (1970). Nutrient requirements and growth conditions for production of lactase enzyme by *Saccharomyces fragilis. J. Milk Food Technol.* 33:451–455.

Wierzbicki, L. E., and Kosikowski, F. V. (1973). Lactase potential of various microorganisms grown in whey. *J. Dairy Sci.* 56:26–32.

Williams, R. S., Trumbly, R. J., MacColl, R., Trimble, R. B., and Maley, F. (1985). Comparative properties of amplified external and internal invertase from the yeast SUC2 gene. *J. Biol. Chem.* 260:13334–13341.

Wilson, I. J., and Ingledew, W. M. (1982). Isolation and characterization of *Schwanniomyces alluvius* amylolytic enzymes. *Appl. Environ. Microbiol.* 44:301–307.

Wilson, J. J., Khachatourians, G. G., and Ingledew, W. M. (1982). *Schwanniomyces:* SCP and ethanol from starch. *Biotechnol. Lett.* 4:333–338.

Wiseman, A., and Woodward, J. (1975). Industrial yeast invertase stabilization. *Process Biochem.* 1:24–30.

Witt, I., Kronau, R., and Holzer, H. (1966). Repression von alkoholdehydrogenase, malatedehydrogenase, isocitratelyase, and malatsynthase in hefe durch glucose. *Biochim. Biophys. Acta* 118:522–537.

Workman, W. E., and Day, D. F. (1983). Purification and properties of the β-fructofuranosidase from *Kluyveromyces fragilis. FEBS Lett.* 160:16–20.

Yamashita, I., and Fukui, S. (1983). Molecular cloning of a glucoamylase producing gene in the yeast *Saccharomyces. Agric. Biol. Chem.* 47:2689–2692.

Yamashita, I., Nakamura, M., and Fukui, S. (1985a). Diversity of molecular structures in the yeast extracellular glucoamylases. *J. Gen. Appl. Microbiol.* 31:399–401.

Yamashita, I., Suzuki, K., and Fukui, S. (1985b). Nucleotide sequence of the extracellular glucoamylase gene STA1 in the yeast *Saccharomyces diastaticus. J. Bacteriol.* 161:567–573.

Yamashita, I., Maemura, T., Hatano, T., and Fukui, S. (1985c). Polymorphic extracellular glucoamylase genes and their evolutionary origin in the yeast *Saccharomyces diastaticus. J. Bacteriol.* 161:574–582.

Yamashita, I., Takano, Y., and Fukui, S. (1985d). Control of STA1 gene expression by the mating-type locus in yeasts. *J. Bacteriol.* 164:769–773.

Yamashita, I., Hatano, T., and Fukui, S. (1984). Subunit structure of glucoamylase of *Saccharomyces diastaticus. Agric. Biol. Chem.* 48:1611–1616.

Yamashita, I., Suzuki, K., and Fukui, S. (1986). Proteolytic processing of glucoamylase in the yeast *Saccharomyces diastaticus. Agric. Biol. Chem.* 50:475–482.

Yasui, T., Nhuyen, B. T., and Nakanishi, K. (1984). Inducers of xylanase production by *Cryptococcus flavis. J. Ferment. Technol.* 62:353–359.

Zemek, J., Kuniak, L., and Augustin, J. (1977). Tests of glucanase and glycosidase production by microorganisms. *Folia Microbiol.* 22:442–443.

Zimmerman, F. K., and Scheel, I. (1977). Mutants of *Saccharomyces cerevisiae* resistant to carbon catabolite repression. *Mol. Gen. Genet.* 154:75–82.

8
Yeast Strain Selection for Fuel Ethanol Production

Chandra J. Panchal
VetroGen Corporation
London, Ontario, Canada

Flavio Cesar Almeida Tavares
University of São Paulo
Piracicaba, Brazil

I. INTRODUCTION

Yeasts have been known to be involved in alcohol production for many centuries. As some of the other chapters in this book describe, the use of yeasts for production of beer, wine, and spirits has been a long-standing practice that perhaps predates recorded history. While some entrepreneurial scientist in the 1930s and 1940s began using yeasts for production of fuel-grade ethanol for use in automobiles and eventually in planes, the rapid development in oil drilling, followed by a cheap and plentiful supply of oil, retarded further growth in the fuel ethanol business. It took the oil crisis of the 1970s to rejuvenate the interest in fuel ethanol as an alternative, and a renewable source of energy. This renewed interest in ethanol still persists, in spite of the fluctuations in the world oil prices, and many countries such as Brazil have made major national efforts at converting to fuel ethanol as the major alternative to gasoline. In North America and Europe, the recent environmental crisis and the desire to phase out lead from gasolines has once again raised the profile of fuel ethanol as an octane enhancer and a lead replacer in gasolines (1,2).

While the advantages of fuel ethanol are well known, the economics of ethanol production has been a major stumbling block in the ready acceptance of ethanol as a gasoline replacer or a fuel supplement. This has prompted a great deal of research activity world wide aimed at improving the efficiency of ethanol production in general, and enhancing yeast fermentation capability in particular (3,4).

II. YEAST STRAIN SELECTION

While some research efforts have been made and are continuing to be made at using bacteria (e.g., *Zymomonas mobilis* [5,6] and *Escherichia coli* [7]) or other yeasts for ethanol production, the yeast *Saccharomyces cerevisiae* is still the primary choice for fermentation ethanol production. There have been numerous reports, however, on the improvements or attempted improvements of "naturally occurring" *S.cerevisiae* strains vis-à-vis ethanol production capability (2,8-13).

Why does this yeast need to be improved and what kind of improvements are being aimed at?

In selecting yeasts for the efficient production of ethanol for fuel (as opposed to potable ethanol), microbiologists have set out certain requirements of the yeasts (5). While these requirements gradually keep increasing as researchers learn more about the physiology and genetics of ethanol production by yeast (much remains to be learnt about these topics), by and large the following list would describe most of them.

An "ideal" yeast for fuel ethanol production should:

1. Be ethanol tolerant
2. Be osmotolerant
3. Be genetically stable
4. Be acid tolerant
5. Be thermotolerant
6. Be a rapid and efficient fermentor
7. Be easy to propagate
8. Be able to utilize a wide variety of substrates
9. Generate minimal heat during fermentation
10. Possess flocculating or nonflocculating characteristics depending upon the process requirements
11. Possess "killer" activity
12. Be derepressed for di- or polysaccharide uptake in presence of glucose
13. Be resistant to certain toxic wastes

It is safe to assume that no single yeast strain used in industry today possesses *all* the above characteristics; hence, the continued research activity in this area.

It is well known that the production of ethanol by yeasts is a polygenic phenomenon (9). The involvement of the various enzymes in the Embden-

Meyerhof-Parnas pathway has been well documented and is known to all students of biochemistry (14). From the pathway through which glucose is finally "converted" into ethanol, it has been calculated that theoretically 1 mol of glucose should yield 2 mol of ethanol, or 1 g of glucose should yield 0.51 g of ethanol and 0.49 g of carbon dioxide. In practice, however, cell growth occurs during the fermentation process and some of the carbon is diverted to biomass, thus resulting in about 0.46 g ethanol, 0.44 g CO_2, and 0.10 g biomass from 1 g of glucose (15). While this represents about 90% efficiency in carbon conversion, yields such as this are achieved in the most ideal cases. Several factors play a role in diminishing the efficiency and extent of ethanol fermentations by yeasts, the major ones being:

1. Inhibition by the end product, ethanol
2. Inhibition by certain by-products such as organic acids
3. Inhibition due to osmotic pressures resulting from high sugar concentrations
4. Inhibition by elevated temperatures
5. Inhibition of fermentation but enhancement of cell growth due to aeration/agitation
6. Inhibition of fermentation due to contamination by bacteria or other yeasts, or by high levels of certain cations (especially in industrial scale fermentations)
7. Inhibition of fermentation due to strain instability leading to formation of mutants/variants

Of all the above listed factors, the inhibitory effect of ethanol on yeast cells is considered to be by far the most significant factor in fermentation alcohol production. Not surprisingly, this has been the area of the most intense research effort. Many papers and reviews have been written on the subject, and as such the reader is referred to the list of references (8,12,13,16-19). While a great deal of information has been obtained on the various effects of ethanol on the yeast cell, there is as yet no concensus on the most critical factor(s) that plays a central role in determining the ethanol tolerance/intolerance of a yeast (or even a bacterial) cell. There are, however, certain observations that have been almost universally made by researchers working in this field.

These can be itemized as:

1. The role of lipid in the cell membrane is very important in ethanol tolerance—unsaturated lipids enhance ethanol tolerance (and membrane fluidity), whereas saturated lipids diminish ethanol tolerance (and make membranes more rigid).
2. Initial aeration of fermentation broth is necessary for efficient ethanol production, but detrimental during the fermentation, as it allows for better cell growth as opposed to ethanol production.
3. A factor (or factors) in the mitochondrial cell membrane is important for determining ethanol tolerance in yeast cells.

4. Rapid fermentations, due to high inoculum rates or high temperatures, generally result in higher cell death rates at the end of the fermentations.
5. Cells producing ethanol at faster rates and to high levels are not necessarily more tolerant to ethanol which has been added to the fermentation media as a "narcotic."
6. Certain nutrients and lipids enhance ethanol tolerance and ethanol productivity by the yeasts when added to the broth.
7. Small-size yeast cells are generally more ethanol tolerant (and faster producers of ethanol) than large-size yeast cells.
8. While the evidence is not totally convincing, it is widely believed that yeast cells are genetically predisposed to be either ethanol tolerant or intolerant.
9. Ethanol tolerance, osmotolerance, and thermotolerance are closely linked and related to the structure of the cell membrane.

A subject of contention, however, has been the diffusion of ethanol across the cell membrane as being a rate-limiting step. It has been shown in several instances (20-22) that under certain fermentation conditions, the rate of production of ethanol can exceed the rate of diffusion of the ethanol across the cell membrane. This can thus lead to an accumulation of ethanol inside the cell, which has the tendency to cause severe damage to the various cellular components, including enzymes, lipids, and membrane proteins (18,23-25). Thus, the ability of the yeast cell to excrete the ethanol rapidly would play a major role in determining its ethanol tolerance, and thus its survivability.

Smaller yeast cells with a larger surface to volume ratio would thus be able to excrete ethanol faster than larger yeast cells which have a smaller surface area to volume ratio. Cells with a highly unsaturated lipid component in the cell membrane would thus be able to excrete the ethanol rapidly (as well as facilitate the uptake of sugars into the cell [25]). It should, however, be stressed that by and large the transport of ethanol across the cell membrane is very rapid and only in certain circumstances does intracellular ethanol accumulation occur (e.g., rapid fermentations, with high osmotic pressures and with high temperatures).

III.   INHIBITORY EFFECTS OF ORGANIC ACIDS

Recently, Viegas et al.(26) reported on the inhibitory effects of the organic acids octanoic acid and decanoic acid on the growth of *Saccharomyces cerevisiae*. When these acids were added to growth media with a sublethal concentration of ethanol that was sufficient to solubilize the acids when in concentrations of up to 16 mg/liter for octanoic acid and 8 mg/liter for decanoic acid (the range present in wines), the specific growth rate of *S.cerevisiae* decreased exponentially. Thus, when 6% (vol/vol) ethanol was added to the growth medium, the growth rate was reduced to 50% of the maximum specific growth rate at 30°C; simultaneous addition of 8 mg/liter of octanoic acid led to an additional growth inhibition of 26.2% at pH 3.8 and 33.1% at pH 3.0 (26). Similarly, addition of 4 mg/liter of decanoic acid to medium already containing 6% ethanol led to a further decline in the growth rate of 24.6% at pH 3.8 and 45.3% at pH 3.0. Since growth inhibition was more drastic for lower pHs, this suggested that the undissociated form of the acid was the toxic form. Undissolved acids (e.g., when ethanol concentration was below 0.6%) did not have any inhibitory effect.

Besides the inhibitory effect of the acids on the growth rate, Viegas et al. (26) also detected a substantial decrease in the overall biomass yield of the yeast strain in the presence of the acids. They summarized that on a molar basis decanoic acid was more toxic than octanoic acid, and that these two acids were much more toxic than ethanol, relating directly to their solubilities in lipids. Octanoic and decanoic acids have been proposed to enter the yeast cell across the plasma membrane by passive diffusion of the undissociated molecules, which are readily soluble in membrane phospholipids (26). Their entry into the membrane would then decrease the hydrophobic lipid-lipid and lipid-protein interactions, leading to disturbance of the spatial organization of the membrane, resulting in an increase in permeability of the membrane and an adverse effect on the membrane-bound enzymes.

The involvement of membrane-bound adenosine triphosphatase (ATPase) for maintaining the intracellular levels of ATP as well as pH during alcoholic fermentations has been purported to be very important for cell viability; ethanol has been shown to affect the activity of this enzyme, and it is likely that the organic acids also affect it negatively. However, the inhibitory effect of the organic acids at such low concentrations suggests that more attention be paid to monitor their levels during fermentations. Viegas et al. (26), however, indicate that the levels of the acids found in the broth are dependent on the formulation of the growth media as well as the strain. An opportunity thus exists for a rational approach to yeast strain selection for ethanol production using the inherent ability of the strain to produce organic acids as a criterion!

IV. EFFECT OF OSMOTIC PRESSURE

It is a well known fact that when yeasts are subjected to fermentations in media containing high concentrations of sugars (>15% w/v), the efficiency of ethanol production drops (20,22). Thus, fermentations by *Saccharomyces cerevisiae* in high osmolarity media are very sluggish. Osmotolerant yeasts such as *Saccharomyces rouxii* can grow well in media containing high concentrations of sugars, but they produce very low amounts of ethanol.

While high osmolarity (and subsequently low water activities) has been used to "preserve" foods and confectioneries for hundreds of years, the effects of osmotic pressure on bacterial yeast cells have been investigated only in recent years. The effects on cell membranes and membrane-associated proteins have been elucidated (18,25). The effect, however, of osmotic pressure on the efficiency of ethanol production by yeasts is not quite well understood. In an attempt to understand the phenomenon, Panchal and Stewart (20) and D'Amore et al. (22), from the same laboratory, studied the effect of osmotic pressure on fermentation by a brewing yeast in media containing varying concentrations of sugar. Table 1 summarizes some of the results obtained by the group (22) using an ale brewing strain. The results clearly indicate that there was a decrease in both the growth rate as well as the fermentation rate with an increase in glucose concentration. It was also found that a decrease in the efficiency of ethanol production (measured as % theoretical ethanol yield) occurred with an increase in glucose concentration. It was previously also shown that an increase in osmotic pressure led to an increase in the concentration of intracellular ethanol (20), which would then have detrimental effects on intracellular enzymes involved in ethanol production. Such inhibitory effects of ethanol on enzymes of the Embden-Meyerhof pathway have been described before (27,28). Similar results were also obtained when fermentations were carried out with a lager brewing strain, *Saccharomyces uvarum* in a sucrose-yeast nitrogen base medium containing a fixed amount of sucrose and variable concentrations of sorbitol (a carbohydrate that is taken up by the yeast but not metabolized), to impart osmotic pressure (20). As the osmotic pressure increased, the yeast viability and fermentative ability decreased due to accumulation of high levels of intracellular ethanol early in the fermentation. Other reports confirm these results and indicate that an intracellular ethanol level of 0.20 - 0.25 $\times 10^{-6}$ mg per viable cell is the upper limit before yeast cell viability is negatively affected (8,29).

It has, however, been shown that exogenously added unsaturated lipids (e.g., linoleic acid) and nutrients (peptone-yeast extract) to fermentation broth alleviate to a great degree the inhibitory effects of osmotic pressure on the fermentation capacity of yeasts (5,24). This would imply that osmotic pressure on yeast cells leads to a depletion of essential nutrients for cell growth and maintenance, and

Table 1. Effect of Glucose Concentration on *Saccharomyces cerevisiae* Growth and Fermentation

Glucose (g/liter)	Growth rate (mg dry wt/ml, hr)	Fermentation (mol ethanol/ml, hr)	% Theoretical ethanol yield
100	0.33	54.0	93.5
200	0.24	52.7	66.4
300	0.11	42.5	59.0
400	0.03	14.2	23.6

Source: Adapted from Ref. 22.

perhaps this is due to the cell membrane becoming more "rigid," thus affecting the entry of these nutrients inside the cell. Addition of unsaturated lipid such as linoleic acid increases the fluidity of the cell membrane, allowing it to become more permeable to nutrients as well as permitting the intracellular ethanol to diffuse out of the cell much faster.

It has also been shown that elevated temperatures (above 40°C) also cause similar effects on the yeast cell membrane (30). Casey and Ingledew (31), studying the performance of brewer's yeast in high gravity (27% carbohydrate) fermentations, concluded that the limiting factors in such fermentations were the lipid content of the medium (wort) and the total amount of assimilable nitrogen in the medium. Supplementation of wort with 1% yeast extract, 40 ppm ergosterol, and 0.4% Tween 80 (a source of oleic acid) led to dramatic reductions in fermentation times and increased ethanol concentrations in the medium with high viability of the cells maintained. They indicate that under such medium conditions there was a significant increase in cell biomass, which they believe led to the improved performance of the yeast in such high osmolarity medium.

Improvements in ethanol fermentations under high osmotic pressures can also be made by sequential addition of the substrate to the fermentation broth. By doing this the yeast cells are not subjected to an osmotic "shock" with high substrate concentrations (22,32). It has been shown that yeast cell viability remains high when such fermentations are carried out using the "fed-batch" approach.

V. EFFECTS OF ELEVATED TEMPERATURE

It has been known for a long time that temperatures in excess of 40°C adversely affect yeast cell growth (30). Yeasts that grow at 40°C and above are considered to be thermotolerant. There are a few reports in the literature on yeasts that have the ability to propagate as well as ferment at these temperatures. Table 2 lists some of the yeasts reported to be thermotolerant. It should, however, be recognized that the growth of the yeasts at such temperatures is not only a function of the genetic make-up of the yeast, but also the medium composition as well as the fermentation conditions. It is, nonetheless, safe to state that there are very few yeasts that can grow and ferment sugars to produce ethanol at high temperatures (in excess of 40°C). Why is this the case?

Several reviews have been written describing the effects of high temperature on yeast cell physiology and the reader is referred to them (30,33,34). In general, at high temperatures, a decline in the growth rate leads to an overall decrease in biomass, resulting in a decrease in cellular protein, ribonucleic acid (RNA), deoxyribonucleic acid (DNA), and free amino acids. In addition, high temperatures induce rigidity of the cell membrane (similar to the effect of osmotic pressure), resulting in reduced permeability of solutes and essential nutrients into the cell. Also, at high temperatures there is a marked decrease in the respiratory activity of

Table 2. Yeast Strains Capable of Fermentation at >40°C

Yeast	Maximum ethanol produced (%w/v)
Kluyveromyces fragilis YKL1	4.0
Kyuyveromyces marxianus NCYC 587	5.03
Candida lusitaniae Y-5394	4.57
Candida psuedotropicalis YCa9	6.87
Candida tropicalis NCYC 405	3.43
Saccharomyces cerevisiae ATCC 4132	4.39
Saccharomyces cerevisiae YSa86	6.38
Saccharomyces cerevisiae NP3	7.58
*Saccharomyces cerevisiae*Y24	10.27
Saccharomyces sp. 1400	7.00
Saccharomyces uvarum inulyticus	7.82
Schizosaccharomyces pombe YSC 3	2.15
Hansenula polymorha ATCC 4516	1.80

Source: Adapted from Ref. 30.

yeasts (33). It has been shown that high temperatures induce respiratory deficiency in yeasts, resulting in formation of petites, a phenomenon also known to occur with ethanol. These results imply the role played by mitochondria in thermotolerance (as well as ethanol tolerance).

A possible factor that may play a crucial role in determining the thermotolerance of yeast is the presence (or absence) of heat shock proteins (HSPs). It has been found that a brief exposure to high temperatures induces a set of specific HSPs which are presumed to confer resistance against thermal damage in the yeast cells (34). Several classes of heat shock proteins have been described, and many are also associated with other stress-causing factors such as ethanol and osmotic pressure. Watson and Cavicchioli (35) reported that heat shock in *S.cerevisiae* not only induced HSPs and thermotolerance, but also imparted a certain level of ethanol tolerance in the yeasts. However, the yeast was not able to ferment very well at high temperatures.

The association of HSPs with the acquisition of thermotolerance led many researchers to clone the HSP genes and reintroduce them into yeasts to acquire the desired characteristics. So far this approach has not met with much success. Finkelstein and Strausberg (36) cloned a gene for an abundantly synthesized HSP in yeast. However introduction of this gene into a mesophilic yeast did not impart any degree of thermotolerance. Lindquist (37) reported similar results with a different HSP clone. This suggests that perhaps a complex regulation of these genes exists in yeast (and possibly other living forms where HSPs have been identified [38]). It is generally believed that HSPs are not new proteins, but constitutively expressed at low levels in the yeast. Under stressful conditions, which include heat, osmotic shock, ethanol, and nutrient depletion as well as depletion of specific metal ions (38), the HSPs are expressed at much higher levels. Since they are, however, genetically controlled, they would be susceptible to some degree of genetic manipulation, and present opportunities for strain improvement programs.

VI. STRAIN IMPROVEMENT PROGRAMS

An effective yeast strain improvement program must be well targeted as well as be flexible as to the types of techniques that can be applied to achieve the aims. These include the traditional genetic techniques as well as the more recent recombinant DNA methods. Some of the methods that have been used for obtaining improved ethanol producing yeasts are:

1. Natural crosses
   Polycross breeding
   Recurrent selections
   Selection after self-crossings
2. Hybridizations
3. Rare matings
4. Mutations
5. Spheroplast fusions
6. Recombinant DNA technology

It is safe to say that in industrial fermentations for the production of ethanol, little attention has been paid to the yeast strain, since the availability of dried baker's yeast is widespread, facilitating the use of a large inocula required for rapid fermentations. It is, however, becoming increasingly apparent that for economic as well as operational reasons, the use of a yeast possessing most, if not all, of the characteristics outlined earlier above would definitely be advantageous. The overall aim, of course, would be to increase the "productivity" of the yeast, such that the economics become attractive. Since it is known that productivity (both general and specific) depends on cell mass concentration, velocity of product formation, and on cellular efficiency at maximum product formation, and considering the complexity of the characteristics involved in productivity and yield, the selection criteria and appropriate evaluation of the traits under investigation have made breeding of new yeast strains a complicated matter (2).

The adoption of plant breeding techniques to yeast breeding has been reported before by Tavares et al. (2). Yeasts are amenable to such breeding methods for several reasons: (a) they are amenable to selection at any ploidy level, (b) they carry out discrete matings between single cells, and these can be analyzed, (c) they have the advantage of tetrad analysis, (d) they have a high level of recombination frequency, and (e) large numbers of individual yeast cells as well as yeast populations can be handled with ease.

Tavares et al. (2) developed a technique to handle large numbers of strains and cells similar to the polycross test and other methods used in plant breeding. The first step in such a breeding program was to perform a mass mating among strains followed by sporulation, and subsequently heat treatment to kill nonsporing cells and stop further mass matings. The cells were then plated on nutrient plates, and large colonies were selected for further evaluation by growth under high temperature and on selective sugar sources. From about 50 colonies, 20% were generally selected and a second qualitative evaluation made to select those possessing better fermentation characteristics than the original parental strain. Upon final selection, they were further sporulated and mass mated again to start another selection round. An alternative approach, avoiding plating, was also used where the cells after mass mating were subjected to specific fermentation conditions, sporulated, heat treated, and again mass mated followed by further fermentations. Table 3 shows some of the results obtained after three rounds of selection with four different strains. As can be seen, after every round of selection, there was a significant improvement in ethanol production by the tester strain when compared to the original parent. In these experiments, the best strain obtained was 39% better than the standard Fleischmann yeast used (2).

Another method that has been successfully adapted to yeast breeding program was selection after self-crossing to choose superior recombinant haploid strains and hybrids. In this scheme, a superior hybrid for fermentation was selected, sporulated, and the spores heat treated. The cells were then plated, and the large isolates (about 10%) were selected for evaluation in fermentations and the best 20% were subjected to further rounds of sporulation and mass matings. While gains in ethanol productivity were obtained, they were slightly lower than those obtained with the polycross method (2).

Table 3. Evaluation of Selected Strains After Three Rounds of Polycross Selection

Tester/Parental (T/P)	Initial %(w/v) ethanol	Final %(w/v) ethanol
M 300 A (P)	8.08	9.42
Fleischmann (P)	8.72	9.76
FI-A (T)	7.80	9.16
IZ-671 (T)	7.46	10.51
IZ-672 (T)	6.92	10.28

Source: Adapted from Ref. 2.

Inbreeding has been used to obtain strains of decreased genetic variability. Another method is to perform back-crossing or recurrent selection. These latter techniques are useful in obtaining isogenic strains of different mating types, appropriate strains for DNA transformations, or strains for conducting quantitative genetic studies. Typically in these programs, two haploid strains are mated, followed by sporulation and isolation of the single spores by micromanipulation. The spores are then scored for the appropriate mating types. In such experiments involving two haploid parental strains, Tavares et al. (2) randomly mated an α-mating–type segregant with one of the recurrent parents (M 304-2C). The new hybrid was then sporulated and a second back-cross round started. After four back crosses the hybrids were tested for ethanol production. As can be seen in Table 4, the hybrids produced more ethanol than one of the parents, but slightly less than the other parent. However, Tavares et al. (39) carried out further experiments in which the diploid formed from the two parents was sporulated and the asci dissected to obtain spores of appropriate mating types; these spores were then used for both intramating and intermating with other spores. The hybrids were then statistically analyzed for both carry over of polygenic characters as well as ethanol production.

Tavares et al. (39) state that such a method gives access to statistically controlled fermentation trials where parental, hybrid, and segregant generations of haploid and diploid nature are present and simultaneously evaluated for ethanol production. Such a biometrical approach helped the partitioning of genetic variance for haploid lines derived from monosporic colonies of individual asci. Sister spore matings in every ascus provided the diploid intrameiotic lines which can then be analyzed separately. They found that for the haploid segregant lines, additive effects were responsible for 30.5% of the total genetic variance, whereas 69.5% was epistatic in nature. With such a genetic variance, they estimated the heritability

Table 4. Ethanol Production in Back-Cross Breeding

Strains	Ethanol (%w/v)	Ethanol productivity (g/l/hr)	% Variation in successive fermentations
P1 M 304-2C	10.79	5.96	0.80
P2 × 2180-1B	10.34	5.46	4.69
M × I	10.48	5.50	3.25
M × II	10.31	5.15	3.24
M × IV	10.55	5.72	2.12
M × V	10.60	5.55	1.85

Mean ethanol production value given for parental, hybrid, and back crosses.
Source: From Ref.2.

value (h, representing the total amount of genetic traits inherited by the progeny) to be 80.3%. In similar studies, the h value for variances among asci was 79%, whereas h for variance within asci was 75.8%. These results thus suggest differences in efficiencies for selection procedures which should be kept in mind when conducting a genetic breeding program. Using these approaches, Tavares et al. (39) have selected various hybrids which have shown enhanced ethanol-producing capabilities, and some are currently being used in industrial-scale fermentations in Brazil.

Using somewhat similar approaches, Del Castillo selected hybrids between homothallic and heterothallic yeast strains to study heredity of ethanol tolerance and production (40). He found that the most tolerant spores derived did not generally produce ethanol well, implying that the tolerance to ethanol and the capacity to produce high levels of ethanol were two genetically independent phenomena. Furthermore, no ethanol-sensitive strains were able to produce high levels of ethanol. From his results, Del Castillo (40) suggested that at least four distinct genes were involved in ethanol tolerance.

In an attempt to see the effect of yeast ploidy on fermentation capacity, Dilorio et al. (41) constructed a series of isogenic diploids, triploids, and tetraploids from two haploid yeasts which differed only in the mating type. When subjected to various anaylses, including ethanol productivity, Dilorio et al. (41) found an almost direct relationship between an increase in ethanol productivity and increased ploidy (Table 5). They also found that (a) cell mass and protein content increased in direct proportion to cell ploidy; (b) cell growth rates of the four strains were identical and independent of ploidy; (c) the specific activities of two representative native enzymes, alcohol dehydrogenase and tryptophan synthetase, remained constant, independent of ploidy, but per cell enzyme activities increased in direct proportion to ploidy; and (d) the efficiency of ethanol production per unit cell mass was greater in cells of higher ploidy. These results were different from those

Table 5. Specific Ethanol Production Rates for Constant Cell Mass Batch Fermentations in Exponential Phase

Ploidy	Specific ethanol production rates (g/g/hr)		
	2% glucose	5% glucose	10% glucose
N	0.99	1.13	0.78
2N	1.22	1.30	1.18
3N	1.35	1.75	1.15
4N	1.85	2.60	1.34

Source: Adapted from Ref. 41.

reported earlier by Takagi et al. (42), who found no significant ploidy-dependent differences in fermentation rates among diploid, triploid, and tetraploid strains of an inbred, homothallic ploidy series. These differences could be due to strain variations and/or differences in measurement methods and resolutions.

Natural selection of ethanol-producing/tolerant strains by continuous recycling in the presence of ethanol has been used over many years by people using yeasts industrially. In earlier studies, Ismail and Ali (43) found no significant increase in ethanol tolerance of yeasts subjected to several rounds of growth in the presence of increasing concentrations of ethanol. Brown and Oliver (44) and Oliver (45), however, devised a different technique in which they monitored the rate of fermentation accurately using an infrared detector to analyze for carbon dioxide evolution. The signal for CO_2 evolution was fed to a potentiometric controller which regulated a pump that added ethanol to the fermenter at the time the culture produced high levels of CO_2. After about 1 month, the amount of ethanol added to the culture had more than doubled from 2 to 4.7% (w/v). However, cells exposed to this level of ethanol were found to ferment sugars at twice the rate of the original parent in the presence of 10% (w/v) ethanol. Although as yet no industrially useful strains have been obtained from such a feedback system, the potential of obtaining one is high, and the system could be applicable for isolating yeast variants with other desired characteristics, such as thermotolerance and osmotolerance.

VII. MUTATION AND SELECTION

Many of the studies employing mutation selection have been carried out in the context of improving yeasts involved in brewing or baking. Some of these studies have been referred to in other chapters in this book. Nonetheless, it is worth mentioning that yeasts that have been genetically modified using mutagenesis can and do have usefulness in the fuel ethanol fermentations. Mutations have been induced in industrial yeasts using N-methyl-N-nitrosoguanidine (NTG) and ultraviolet light (46) as well as isolating mutants by natural selection. Mutants of use in fermentations have been those that are derepressed for glucose; i.e., those that can utilize other carbon sources besides glucose in the presence of high concentrations of glucose. Such mutants can substantially reduce fermentation times when the substrate contains disaccharides (e.g., maltose, sucrose), as well as glucose, since with normal yeasts, the disaccharides would be utilized only when the glucose has all been consumed (due to the repressing effect of glucose on disaccharide utilization). Such mutants have been isolated in brewing yeasts and the fermentations improved as a result (47).

Mutant yeasts have also been used for isolating hybrids with desired characteristics. Thus, hybrid brewing yeasts have been obtained by "rare-mating" (48), involving crosses between auxotrophic haploid and diploid laboratory yeast strains and respiratory-deficient (49) or antibiotic-resistant mitochondrial mutants

(50) or spore clones (51) of brewing yeasts. It was mentioned above that one of the desirable characteristics in an industrial ethanol-producing yeast is the possession of the "killer," or zymocidal, factor which would effectively kill all wild yeasts during the fermentation (if these wild yeasts were not killer yeasts themselves). The technique of cytoduction has been used to transfer the linear double-stranded RNA of *S.cerevisiae*, which codes for the killer toxin, into a brewing yeast by a specialized form of rare-mating involving the *kar1* mutation in the donor strain. This mutation prevents karyogamy (or nuclear DNA transfer), but allows cytoplasmic DNA and RNA transfer into the hybrid yeast. Effective killer brewing yeasts have been obtained in this manner (52), and such procedures are being attempted to obtain killer fuel ethanol-producing yeasts. Some reports in the early 1970s had indicated that respiratory-deficient (RD) mutants were more efficient in ethanol production than respiratory-sufficient yeasts, since very little biomass was produced by these mutants (53). While theoretically appealing, these results have not been reproducible by many of the groups attempting to isolate such mutants. Invariably the RD mutants were found to be slow fermentors in addition to being less efficient than the parental strains, suggesting that it is difficult to totally uncouple respiration from fermentation.

VIII. SPHEROPLAST (OR PROTOPLAST) FUSION

The technique of spheroplast fusion has received a great deal of attention for the genetic improvement of industrial yeast and bacterial strains where the classic breeding techniques have not been possible to use. Spheroplast fusion disregards the mating type characteristic of the yeast. Briefly, it involves the use of two parental strains that are genetically marked with complementary auxotrophies. The cell walls of the yeasts are removed enzymatically, the resulting spheroplasts, which are osmotically fragile, are washed and fused in equal numbers in the presence of a fusogen (e.g., polyethylene glycol). The mixture is then plated on regeneration agar plates, and fusion products isolated on selective media plates.

Using the spheroplast fusion technique, Stewart et al. (12) obtained stable fusion products of the polyploid brewing yeast *Saccharomyces uvarum*, strain 21 and a genetically constructed diploid *Saccharomyces diastaticus*, strain 1384. One of the fusion products, strain 1400, was found to have poor brewing performance but enhanced ethanol-producing capability. The yeast 1400 was found to have an accelerated fermentation capability in defined media (with glucose substrate) as well as in whole corn mash media (54). In addition, it was found to be more osmotolerant and more thermotolerant than either of the parental yeast strains (54,55). Since one of the parents (*S.diastaticus*) also had glucoamylase-producing capability, strain 1400 was found to secrete glucoamylase into the media, enabling it to utilize dextrins in the corn mash substrate. Large-scale fermentations have been carried out with this yeast, and it is expected that it will have significant potential for industrial ethanol production.

Spheroplast fusions have also been used to isolate brewing (and other industrial) yeasts with increased osmotolerance (56). Fusion products have been obtained that ferment high concentrations of sugars better than the parental strains. Several groups have reported isolation of fusion yeast products with enhanced ethanol-producing capabilities, and the importance as well as usefulness of this simple, yet effective, technique has been established.

IX. RECOMBINANT DNA TECHNOLOGY

As mentioned before (see Chaps. 4,6), the use of recombinant DNA technology in yeasts effectively started in 1978, when the first transformation experiments in yeast were described by Hinnen et al. (57). Since that time the gene-cloning systems in yeast have been well developed to the point where recombinant yeasts have been used industrially to produce pharmaceutically important products (58,59). The yeast host-vector system is now considered to be a favored system in many research institutes and biotechnology companies for the production of heterologous proteins. Advances have also been made with recombinant yeasts in brewing and baking research, although progress has been slow in using recombinant yeasts for the traditional brewing and baking processes, primarily due to "percieved" ethical reasons (see Chap. 4). However, the use of recombinant DNA technology for the isolation of improved ethanol-fermenting yeasts has almost been nonexistent.

The only area where there has been major activity has been in the attempts to clone cellulase genes into fermenting yeasts. Cellulose and hemicellulose, which comprise more than 70% of plant biomass and are the two most abundant organic compounds in the biosphere, occur in the wastes from agriculture and forestry as well as from vegetable and fruit processing and municipal waste treatment (60). Recycling of these wastes to produce useful products such as ethanol and methane makes not only economic sense, but also contributes to a cleaner environment. Upon enzymatic or acid hydrolysis, plant biomass yields a mixture of sugars consisting of mostly cellobiose, xylose, glucose, together with mannose, galactose, and arabinose. Except for one report (60), microorganisms that are able to ferment all the above sugars to produce high concentrations of ethanol have not been found.

Since *S.cerevisiae* has been the organism of choice for most ethanol fermentations, research activity has mostly centered around imparting this yeast with characteristics allowing it to ferment cellulose directly or indirectly (after acid or enzymatic hydrolysis). Thus, the cellobiohydrolase genes from the fungus *Trichoderma reesei* and the bacterium *Cellulomonas fimi* have been cloned and expressed in yeast (61,62), and attempts have been made to impart xylose-fermenting capability to *S.cerevisiae*. Ethanol-fermenting yeasts have also been transformed with glucoamylase and α-amylase genes, thus enabling them to utilize starch directly (63). While these developments have given the ethanol-fermenting yeasts added enzymatic and substrate-utilizing capabilities, they have done little to make them better fermentors. Nonetheless, the capability to utilize cellulosic materials directly, and to ferment them to produce ethanol efficiently, would give yeasts a tremendous capability and make the process of producing fuel ethanol by fermentation a very attractive and economically viable proposition, since a major cost factor in ethanol production is attributed to the substrate.

It is, thus, quite apparent that in the forseeable future, yeasts will be the choice organisms for fuel ethanol fermentations, and hence can be expected to be the focus of attention for genetic improvement programs in this field. The world fermentation ethanol industry is a huge one by any standards, and with legitimate attention being given to the protection of the environment, cleaner burning fuels such as ethanol are becoming more and more attractive. A small improvement, therefore, in the efficiency of ethanol production by a genetically modified yeast could be economically significant and perhaps crucial!

REFERENCES

1. Murtagh, J.E. Fuel ethanol production—The U.S. experience. Proc. Biochem. 21:61 (1986).
2. Tavares, F.C.A., and Echeverrigaray, S. Yeast breeding for fuel production. In: *Biological Research on Industrial Yeasts*. Vol. I. Stewart, G.G., Russell, I., Klein, R.D., and Hiebsch, R.R. (Eds.). CRC Press, Boca Raton, Florida, 1987, p. 59.
3. Lyons, T.P. Ethanol production in developed countries. Proc. Biochem. 18:18 (1983).
4. Keim, C.R. Technology and economics of fermentation alcohol—an update. Enzyme Microb. Technol. 5:103 (1985).
5. Stewart, G.G., Panchal, C.J., Russell, I., and Sills, A.M. Biology of ethanol-producing microorganisms. Crit. Rev. Biotechnol. 1:161 (1984).
6. Rogers, P.L., Lee, K.J., Skotniki, M.L., and Tribe, D.E. Ethanol production by *Zymomonas mobilis*. In: *Advances in Biochemical Engineering*. Vol. 23. Feichter, A. (Ed.). Springer-Verlag, New York, 1980, p. 37.
7. Ingram, L.O., Conway, T., Clark, D.P., Sewell, G.W., and Preston, J.F. Genetic engineering of ethanol production in *Escherichia coli*. Appl. Environ. Microbiol. 53:2420 (1987).
8. D'Amore, T., and Stewart, G.G. Ethanol tolerance of yeast. Enzyme Microbial. Technol. 9:322 (1987).
9. Ho, N.W.Y. Yeast alcohol tolerance and recombinant DNA for improved alcohol production process. In: *Annual Reports on Fermentation Processes*. Vol. 4. Tsao, G. (Ed.) Academic Press, New York, 1980, p. 235.
10. Haraldson, A., and Bjoirling, T. Yeast strains for concentrated substrates. Eur. J. Appl. Microbiol. Biotechnol. 13:34 (1981).
11. Hacking, A.J., Taylor, T.W.F., and Hanas, C.M. Selection of yeast able to produce ethanol from glucose at 40°C. Appl. Microbiol. Biotechnol. 19:361 (1984).

12. Stewart, G.G., Russell, I., and Panchal, C.J. The genetics of alcohol metabolism in yeasts. Brew. Dist. Int. 12:23 (1982).

13. Panchal, C.J., Russell, I., Sills, A.M., and Stewart, G.G. Genetic manipulation of brewing and related yeast strains. Food Technol. 38:99 (1984).

14. Lehninger, A. *Biochemistry*. Worth, New York, 1975.

15. Lyons, T.P. Industrial uses of yeast in the production of fuel ethanol. In: *Developments in Industrial Microbiology*. Vol. 25. 1984, p. 231.

16. Navarro, J.M. Fermentation alcoolique. Influence des conditions de culture sur l'inhibition par l'ethanol. Cell. Molec. Biol. 26:241 (1980).

17. Casey, G.P., and Ingledew, W.M. Reevaluation of alcohol synthesis and tolerance in brewers yeast. Am. Soc. Brew. Chem. J. 43:25 (1985).

18. Van Uden, N. Ethanol toxicity and ethanol tolerance in yeasts. In: *Annual Report on Fermentation Processing*. Vol. 8. Tsao, G. (Ed.). Academic Press, New York, 1985, p. 11.

19. Casey, G.P., and Ingledew, W.M. Ethanol tolerance in yeasts. Crit. Rev. Microbiol. 13:219 (1986).

20. Panchal, C.J., and Stewart, G.G. The effect of osmotic pressure on the production and excretion of ethanol and glycerol by a brewing yeast strain. J. Inst. Brew. 86:207 (1980).

21. Nagodawithana, T.W., Castellano, C., and Steinkraus, K.H. Influence of the rate of ethanol production and accumulation on the viability of *Saccharomyces cerevisiae* in rapid fermentation. Appl. Environ. Microbiol. 31:158 (1976).

22. D'Amore, T., Panchal, C.J., Russell, I., and Stewart, G.G. Osmotic pressure effects and intracellular accumulation of ethanol in yeast during fermentation. J. Indust. Microbiol. 2:365 (1988).

23. Leao, C., and van Uden, N. Effects of ethanol and other alkanols on the glucose transport system of *Saccharomyces cerevisiae*. Biotechnol. Bioeng. 24:2601 (1982).

24. Leao, C., and van Uden, N. Effects of ethanol and other alkanols on passive proton influx in the yeast *Saccharomyces cerevisiae*. Biochem. Biophys. Acta 774:43 (1984).

25. Beavan, M.J., Charpentier, C., and Rose, A.H. Production and tolerance of ethanol in relation to phospholipid fatty acyl composition in *Saccharomyces cerevisiae* NCYC 431. J. Gen. Microbiol. 128:1447 (1982).

26. Viegas, C.A., Rosa, M.F., Sa-correia, I., and Novaio, J.M. Inhibition of yeast growth by octanoic and decanoic acids produced during ethanolic fermentation. Appl. Environ. Microbiol. 55:21 (1989).

27. Nagodawithana, T.K., Whitt, J.T., and Cutaia, A.J. Study of the feedback effect of ethanol on selected enzymes of the glycolytic pathway. J. Am. Soc. Brew. Chem. 35:179 (1977).

28. Millar, D.G., Griffiths-Smith, K., Algar, E., and Scopes, R.K. Activity and stability of glycolytic enzymes in the presence of ethanol. Biotechnol. Lett. 4:601 (1982).

29. Dasari, G., Roddick, F., Connor, M.A., and Pamment, N.B. Factors affecting the estimates of intracellular ethanol concentrations. Biotechnol. Lett. 5:715 (1983).

30. Slapack, G.E., Russell, I., and Stewart G.G. Thermophilic bacteria and thermotolerant yeasts for ethanol production. CRC Press, Boca Raton, Florida. (1988).

31. Casey, G.P., and Ingledew, W.M. High gravity brewing: influence of pitching rate and wort gravity on early yeast viability. J. Am. Soc. Brew. Chem. 41:148 (1983).

32. Panchal, C.J., and Stewart, G.G. Ethanol production by a highly flocculent brewing

yeast strain. Dev. Indust. Microbiol. 22:711 (1981).

33. Van Uden, N. Effects of ethanol on the temperature relations of viability and growth in yeast. CRC Crit. Rev. Biotechnol. 1:263 (1984).

34. Van Uden, N. Temperature profiles of yeast. In: *Advances in Microbial Physiology*. Vol. 25. Rose, A.H., and Tempest, D.W. (Eds.). Academic Press, London, 1984, p. 195.

35. Watson, K., and Cavicchioli, R. Acquisition of ethanol tolerance in yeast cells by heat shock. Biotechnol. Lett. 5:683 (1983).

36. Finkelstein, D.B., and Strausberg, S. Identification and expression of a cloned yeast heat shock gene. J. Biol. Chem. 258:1908 (1983).

37. Lindquist, S. Heatshock proteins in yeast. Int. Yeast Genet. Meet. Proceed. Banff, Alberta, 1986.

38. Tanguay, R.M. Genetic regulation during heat shock and function of heat shock proteins: A review. Can. J. Biochem. Cell. Biol. 61:387 (1983).

39. Tavares, F.C.A., Kido, E.A., and Vencovsky, R. Biometrical analysis of ethanol production in *Saccharomyces cerevisiae*. Can. J. Microbiol. (submitted).

40. Del Castillo, A.L. Ethanol tolerance in yeast. Curr. Microbiol. 12:41 (1985).

41. Dilorio, A.A., Weathers, P.J., and Campbell, D.A. Comparative enzyme and ethanol production in an isogenic yeast ploidy series. Curr. Genet. 12:9 (1987).

42. Takagi, A., Harashima, S., and Oshima, Y. Effect of ploidy on fermentation by yeasts. Appl. Environm. Microbiol. 45:1034 (1983).

43. Ismail. A.A., and Ali, A.M.M. Selection of high ethanol yielding *Saccharomyces*. I Ethanol tolerance and the effect of training *Saccharomyces cerevisiae*. Hansen. Fol. Microbiol. 16:346 (1981).

44. Brown, S.W., and Oliver, S.G. Isolation of ethanol tolerant mutants of yeast by continuous selection. Eur. J. Appl. Microbiol. Biotechnol. 16:119 (1982).

45. Oliver, S.G. Biological limits to ethanol production. Chem. Ind. 12:425 (1984).

46. Molzahn, S.W. A new approach to the application of genetics to brewing yeasts. J. Am. Soc. Brew. Chem. 35:54 (1977).

47. Stewart, G.G., Jones, R., and Russell, I. The use of derepressed yeast mutants in the fermentation of brewery wort. Proc. Eur. Brew. Conv. 20th Congress, Helsinki. 1985, p. 243.

48. Gunge, N., and Nakatani, Y. Genetic mechanisms of rare matings of the yeast *Saccharomyces cerevisiae* heterozygous for mating type. Genetics 70:41 (1972).

49. Tubb, R.S., Searle, B.A., Goodey, A.R., and Brown, A.J.P. Rare mating and transformation for construction of novel brewing yeasts. Proc. Eur. Brew. Conv. 18th Congress, Copenhagen. 1981, p. 487.

50. Spencer, J.F.T., and Spencer, D.M. The use of mitochondrial mutants in the isolation of hybrids involving industrial yeast strains. Mol. Gen. Genet. 177:355 (1980).

51. Kielland-Brandt, M.C., Gjermansen, C., Nilsson-Tilgren, T., Peterson, J.G.L., Holberg, S., and Sigsgaard, P. Genetics of a lager production yeast. MBAA Tech. Quat. 18:185 (1981).

52. Young, T.W. Brewing yeasts with anticontaminant properties. Proc. Eur. Brew. Conv. 19th Congress, London. 1983, p. 129.

53. Bacila, M., and Horii, J. Improving the efficency of alcohol production with respiration-deficient yeast mutants. Trends Bio. Sci., March 1979, p. 59.

54. Panchal, C.J., Harbison, A., Russell, I., and Stewart, G.G. Ethanol production by

genetically modified strains of *Saccharomyces*. Biotechnol. Lett. 4:33 (1982).

55. Panchal, C.J., Peacock, L., and Stewart, G.G. Increased osmotolerance of genetically modified ethanol producing strains of Saccharomyces sp. Biotechnol. Lett. 4:639 (1982).

56. Legmann, R., and Margalith, P. Ethanol formation by hybrid yeasts. Appl. Environm. Microbiol. 23:198 (1986).

57. Hinnen, A., Hicks, J.B., and Fink, G.R. Transformation of yeast. Proc. Natl. Acad. Sci. U.S.A. 75:1929 (1978).

58. Ratner, M. Protein expression in yeast. BioTechnology. 7:112 (1989).

59. King, D.J., Walton, E.F., and Yarranton, G.T. The production of proteins and peptides from *Saccharomyces cerevisiae*. In: *Molecular and Cellular Biology of Yeasts*. Walton, E.F., and Yarranton, G.T. (Eds.). Van Nostrand, New York, 1989, p. 107.

60. Wu, J.F., Lastick, S.M., and Updegraff, D.M. Ethanol production from sugars derived from plant biomass by a novel fungus. Nature 321:887 (1986).

61. VanArsdell, J.N., Kwok, S., Schweickart, V.L., Ladner, M.M., Gelfand, D.H., and Innes, M.A. Cloning, characterization and expression in *Saccharomyces cerevisiae* of endoglucanase I from *Trichoderma reesei*. BioTechnology 5:60 (1987).

62. Wang, W.K.R., Currey, C., Parekh, R.S., Parekh, S.R., Wayman, M., Davies, R.W., Kilburn, D.G., and Skipper, N. Wood hydrolysis by *Cellulomonas fimi* endoglucanase and exoglucanase co-expressed as secreted enzymes in *Saccharomyces cerevisiae*. BioTechnology 6:713 (1988).

63. Cole, G.E., McCabe, P.C., Inlow, D., Gelfand, D.H., Ben-Bassat, A., and Innis, M.A. Stable expression of *Aspergillus awamori* glucoamylase in distiller's yeast. Bio-Technology 6:417 (1988).

9

Transformation and Cloning Systems in Non-*Saccharomyces* Yeasts

Ronald D. Klein and Phillip G. Zaworski
The Upjohn Company
Kalamazoo, Michigan

I. INTRODUCTION

A major emphasis of molecular biology research in eukaryotes has been the detailed analysis and characterization of the yeast *Saccharomyces cerevisiae*. This past decade has witnessed an explosion of data regarding this organism and its application to our understanding of the complex physiological and biochemical nature of the eukaryotic world. The advances that have been made have resulted primarily from three factors: a detailed understanding of the genetics of the organism, its extensive biochemical characterization, and recent developments in recombinant DNA technology. Preliminary work has also been conducted in more complex fungi, primarily species of *Aspergillus* (1–3). The motivation for detailed study of these organisms has been both academic and applied, with an increasing emphasis being placed on the latter (4,5). Emphasis on the applied aspects of recombinant DNA technology has prompted the expansion of basic research into a wide variety of yeasts and fungi previously little known outside of specialized industries (6).

The term "yeast" has become almost synonymous with *S. cerevisiae*, which is but one of hundreds of yeast that offer the investigator an infinite variety of biochemical phenomena for serious study. In addition to their serving as models in our analysis of eukaryotic biology, the biochemical diversity of the yeasts has made them important as biocatalysts for industrial purposes (7,8). Industrial application of yeast include biomass conversions to potable and fuel-grade alcohols (9) and the production of single cell protein (SCP) as well as being sources of a number of purified enzymes. Yeasts have also been used as biocatalysts for the production of fine chemicals including amino acids, vitamins, citric acid, and a variety of complex oils and fats (10,11). In this review, we will discuss the recent expansion of recombinant DNA methodologies into alternative yeasts (here defined as members of non-*Saccharomyces* genera), the industrial and basic research importance of these developments, and the future directions of current research. The investigation of alternative yeasts represent the beginning of a revolution in biotechnology that will play a major role in shaping our exploration and exploitation of the fungal world.

The advances in recombinant DNA technology have had a major effect on the biotechnology industry. This technology offers promise for the direct genetic manipulation of organisms that have been the mainstay of the fermentation and food industry. The application of recombinant DNA techniques to the alternative yeasts has lagged behind the rapid advances made in the common laboratory strains of *S. cerevisiae*. This has been due to a number of factors, including the lack of a sexual cycle in many species, aneuploidy of these strains, and the limited number of investigators working on any one species. The lack of detailed biochemical and genetic data can be attributed to these and other factors. Without a detailed biochemical and genetic profile the application of recombinant DNA technology to any given species has been, until recently, severely restricted. Although it is clear that differences exist between the alternative yeasts and *S. cerevisiae* that necessitate modifications of existing recombinant DNA protocols, the wealth of knowledge gained in the development of *S. cerevisiae* provides an insight into what we can expect to encounter in the development of transformation and cloning systems in the non-*Saccharomyces* yeasts.

A. Protoplast Fusion and Transformation

The use of polyethylene glycol (PEG) as a fusant for plant cell and bacterial protoplasts (12) created a tremendous interest in its application to yeasts that were impaired in mating or spore formation or that lacked a sexual cycle. Although these included many of the industrial strains of *Saccharomyces*, this new technology provided at last a means for the exploitation and development of numerous industrial yeasts through intergenic as well as intragenic fusions (13). The role of protoplast fusion in commercially important microorganisms has been reviewed elsewhere (12,14,15).

The technique of intergenic protoplast fusion greatly expanded the range of possibilities for strain improvement (16). Even today, as techniques for the transformation of yeast with purified DNA become more widely used, there are some arguments favoring the use of intergenic and intragenic fusion in modifying alternate yeasts over transformation. Recombinant DNA techniques require vectors, transformation systems, and well-defined genetic or selectable markers. Many of these requirements are not yet available in the alternative yeast. In addition, in many cases, it may be necessary to transfer several (potentially unlinked) genes in a given pathway from one organism to another (17) — genes and pathways that may not even be well characterized. Despite the positive aspects of protoplast fusion, there are several drawbacks including instability of hybrids, the cotransfer of undesired characteristics, and the bias that exists in the transfer of chromosomes from one of the donor cells (17).

Interest in developing a transformation system in *S. cerevisiae* began soon after it became evident that foreign DNA could be introduced into bacteria. Attempts in the early 1960s to modify *S. cerevisiae* by transformations centered on obtaining strains of *Saccharomyces* that were able to utilize different sugars (18). In 1978, Hinnen et al. (19) reported the successful and reproducible transformation of protoplasts of a *S. cerevisiae leu*2 strain with a plasmid consisting of the bacterial vector Col E1 and the *S. cerevisiae* LEU2 gene. At the same time J. D. Beggs (20) reported the transformation of a *leu*2 strain of *S. cerevisiae* using an endogenous 2-μm plasmid carrying the LEU2 gene. Both reports employed a protoplasting procedure utilizing the fusant PEG.

These initial yeast transformation procedures were based on developments in protoplast fusion techniques and involved the coprecipitation of yeast protoplasts that had been treated with calcium ions and transforming DNA by PEG. Methods for the generation of protoplasts in a number of yeast genera entailed enzymatic treatment with snail enzyme or Zymolyase 500 in the presence of dithiothreitol or 2-mercaptoethanol and an isoosmotic stabilizer such as $MgSO_4$, mannitol, KCl, or sorbitol (21). The adaptation of these procedures for the reproducible transformation of yeasts required the optimization of factors such as time of cell harvest, reaction times, and choice and concentration of an isoosmotic stabilizer.

Coprecipitation of protoplasts with purified DNA is only one of several methods used to introduce DNA into yeast. Foreign DNA has been transformed into yeast by mixing bacterial and yeast protoplasts in the presence of PEG and selecting for a specific yeast marker (22). Attempts at transferring plasmids into yeast protoplasts by encapsulating them in liposomes has been reported, and, though effective, it did not markedly increase the transformation efficiency (23). The key feature of transforming yeast is the necessity to reproducibly permeabilize the cell membrane.

As early as 1975 Kimura and Morita (24) observed that treatment of yeast with Triton X-100 permitted the incorporation of various extracellular mononucleotides into intact yeast cells. This observation was quickly followed by attempts to introduce DNA into intact yeast cells using transformation protocols similar to those used for *E. coli*. These involved treating cells with various detergents and/or $CaCl_2$ followed by the addition of PEG. Although such procedures saved considerable time, since they did not require the generation and regeneration of protoplasts, the transformation efficiency in general was quite low (25). More recent studies have focused on optimizing the conditions essential for increasing the efficiency of yeast transformation without removal of the cell wall. These procedures have been widely accepted since they are far simpler than the existing protoplasting methods and give nearly comparable efficiencies (26–28).

Table 1 summarizes the essential features of the most commonly employed transformation protocols for both *Saccharomyces* and non-*Saccharomyces* yeasts.

B. Cloning Vectors

The advances that have been made in the development of vectors in the yeast *S. cerevisiae* have been due to the exploitation of the endogenous 2-μm plasmid (29) and the isolation of DNA sequences capable of permitting the autonomous replication of plasmids that contain them (30). The 2-μm plasmid is 6300 base pairs (bp) in length and contains one origin of replication in addition to four replication functions (REP1, REP2, REP3, and FLP), which are responsible for the plasmid's stability and copy number of 50–100 copies per cell (31,32). The structural features and organization of the 2-μm plasmid are summarized in Figure 1 and the accompanying legend.

In yeast there are three classes of cloning vectors that are defined by function and the source of replicating sequences. The most stable autonomously replicating vectors contain defined replication origins from the centromeric regions of *S. cerevisiae* chromosomes or CEN sequences (33). In addition to centromeric origins of replication, these sequences apparently include spindle attachment sites permitting the plasmids to segregate as "minichromosomes" (34). These plasmids are maintained at one copy per cell and are very stable even in the absence of selective pressure.

The second class of vectors carry sequences that confer the property of autonomous replication, which consist of ARSs (autonomously replicating sequences) from a variety of sources or the origin of replication from the 2-μm plasmid. The former are referred to as YRp vectors; the latter, YEp vectors. In *S. cerevisiae* an ARS consensus sequence has been defined as, 5′ TTTTATGTTTA 3′ (35). It is of interest that sequences that confer ARS activity in *S. cerevisiae* do not necessarily show the same functions in other yeasts and vice versa, even though these sequences are nearly homologous with the *S. cerevisiae* ARS consensus sequence.

The last class of vectors are those that carry sequences that permit integration into the host cell's chromosome. These integrating or YIp vectors are detected by scoring for a selectable marker *or* the loss of a given phenotype resulting from the disruption of the target gene by integration event. Confirmation of integration is

Table 1. Generalized Yeast Transformation Protocols

Spheroplasting	Alkali cation treatment
1. Grow 200 ml culture in YEPD to mid-log phase (2×10^7 cells/ml; $A_{550}=$ 0.8).	1. Grow 100 ml culture in YEPD to late log phase (A_{550} ~6).
2. Spin 5,000 rpm, 5 min. Wash once with sterile H_2O, and respin. Resuspend in 20 ml SED. Incubate 10 min at 30°C.	2. Spin 5,000 rpm, 5 min. Wash with sterile TE. Respin. Resuspend in TE at density of 2×10^8 cells/ml.
3. Spin 5,000 rpm, 5 min. Wash with 20 ml sorb. Respin. Resuspend in 20 ml SCE.	3. Remove 0.5-ml aliquot of resuspended cells. Add equal volume of 0.2 M lithium acetate. Incubate 60 min at 30°C with gentle agitation.
4. Add 0.2 ml glusulase or Zymolyase 500. Incubate at 30°C with gentle mixing. Assay spheroplasting by diluting treated cells into a drop of 5% SDS on a microscope slide. Observe ghosts at 400× phase contrast.	4. Remove 0.1 ml of lithium-treated cells. Add plasmid DNA (up to 10 µg in 15 µl of TE buffer). Incubate 30 min at 30°C.
5. Spin 2,000 rpm, 3 min. Resuspend in 20 ml sorb and respin. Resuspend in 20 ml STC and respin. Resuspend in 0.1 ml STC.	5. Add equal volume of 70% PEG 4000. Vortex to mix. Incubate 60 min at 30°C.
6. Divide into 50- to 100-µl aliquots in 1.5-ml Eppendorf tubes. Add DNA (0.1 to 10 µg) in 1 to 10 µl of TE buffer. Incubate at room temperature for 15 min.	6. Heat shock in 42°C bath for 5 min.
7. Add 1.0 ml 20% PEG to each tube. Leave at room temp for 15 min. Spin 2,000 rpm, 3 min. Resuspend in 100 µl STC.	7. Cool to room temperature. Wash cells twice by pelleting at 5,000 rpm for 5 min and resuspension in sterile water.
8. Add cells to melted, isoosmotic top agar at 45°C and spread on selective media.	8. Resuspend in water and plate directly on selective media.

Summary and comparison of the steps involved in yeast transformation by spheroplasting and by treatment of intact cells with alkali cations. Abbreviations: YEPD: complete media consisting of 1% yeast extract, 2% peptone and 2% glucose; SORB: 1 M sorbitol; SED: 1 M sorbitol; 25 mM EDTA, 50 mM dithiothretol; SCE: 1 M sorbitol, 100 mM sodium citrate pH 5.8, 10 mM EDTA; PEG 20–40%; PEG 4,000, 10 mM TRIS HCl pH 7.5, 10 mM $CaCl_2$; STC: 1 M sorbitol, 10 mM TRIS pH 7.5, 10 mM $CaCl_2$ (19,20,26,27); TE buffer: 10 mm TRIS pH 7.5, 0.1 mM EDTA.

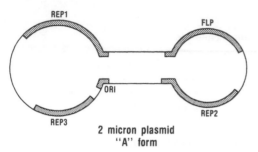

2 micron plasmid
"A" form

Figure 1. The 2-μm circle is a 6,318-bp double-stranded DNA plasmid found in the nucleoplasm of most strains of *S. cerevisiae*. Salient structural features are shown. Parallel lines represent two inverted repeats of 599 bp. High-frequency recombination occurs at these sites resulting in the 2-μm circle being found in two isomeric forms which differ in the orientation of a specific region resulting from this recombination event. ORI represents the region of DNA that has been demonstrated to function as an origin of DNA replication both in vitro and in vivo. Regions labeled REP1 and REP2 both encode *trans*-acting proteins required for high copy number maintenance and proper segregation of the plasmid. REP3 is also required for proper copy number and segregation, but functions only in *cis* and is the site through which REP1 and REP2 act. FLP encodes a site-specific recombinase required for high-frequency recombination within the inverted repeat regions.

usually by Southern hybridization analysis of isolated genomic DNA. These vectors are stable once integrated since the foreign DNA is inherited as an integral part of the host cell's chromosome.

The efficiency of transformation depends on the particular vector: CEN and YIp vectors transform at frequencies of 1–10 transformants per microgram of purified plasmid DNA, whereas the YRp and YEp plasmids give efficiencies of 10^2–10^4 transformants per microgram of purified plasmid DNA. The essential features of these plasmids are summarized in Figure 2.

C. Selectable Markers

Perhaps the most difficult component to acquire in the development of a transformation and cloning system in the non-*Saccharomyces* yeast is a stable selectable marker. The absence of extensive genetic characterization and the problems associated with hidden mutations affecting production strains after undergoing standard mutagenesis have hampered development of genetically modified industrial yeasts.

Several new approaches have been employed to provide useful markers or defined mutants to serve as the basis for a cloning system. The use of positive

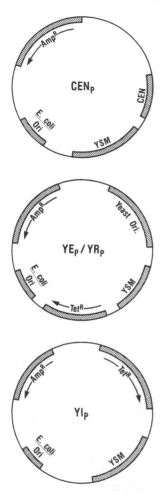

Figure 2. Diagrammatic representation of the components required in three types of yeast plasmid vectors. In all cases, the fragment labeled *E. coli* Ori, represents an origin of replication which allows the plasmid to be propagated in *E. coli*. Amp^R and Tet^R code for resistance to ampicillin and tetracycline, respectively, and are selectable markers in *E. coli*. The fragment labeled YSM represents a yeast-selectable marker, usually a gene complementing an auxotrophic mutant such as URA3, LEU2, TRP1, etc. In the CENp, the fragment labeled CENp represents a yeast centromere which includes a yeast origin of replication in addition to sequences allowing proper segregation. In YEp/YRp, the fragment labeled Yeast Ori represents an ARS sequence which allows plasmid replication in yeast. A yeast origin of replication is lacking in the YIP plasmid vector.

selection for defined mutants has been successful, with attention focusing on mutants for which complementing genes are readily available. Success has been achieved using α-aminoadipate for *lys*2 (36) and 5-fluoro-orotic-acid (FOA) for *ura*3 and *ura*5 (37–39).

One of the most fruitful methodologies has been the isolation of specific genes from other organisms by complementation in *S. cerevisiae.* This entails the construction of cDNA or genomic DNA libraries in a suitable *S. cerevisiae* expression vector, followed by the transformation and selection for the desired gene in an appropriate *S. cerevisiae* mutant. The work of McKnight and McConaughy (40) illustrates the power of this technique in their construction of cDNA pools in which specific yeast mRNAs were represented in medium and low abundance. The investigators were able to clone ADC1, URA3, HIS3, and ASP5 by complementation of appropriate *S. cerevisiae* strains. Their technique has been successful in isolating genes from other yeast and fungi including the ADH gene from *Aspergillus nidulans* (41).

Once a specific gene has been cloned and characterized it becomes possible to delete or disrupt the gene in vitro and to reintroduce the altered gene into the host by homologous recombination, thus generating a specific mutation at the corresponding chromosomal locus (42). In the case of polyploid strains, further manipulations may be required such as mating and standard tetrad analysis; where no sexual cycle is evident, rare mating and cell fusion techniques may have to be employed to generate the appropriate host strain. With a selectable marker and defined mutant, it is possible to isolate ARSs and begin construction of a defined vector. One advantage of gene disruption is the maintenance of the isogenic nature of a particular strain; this may be important if one is dealing with a production strain whose genetic background must be maintained. The process of gene disruption is illustrated in Figure 3.

Where it may be difficult to acquire defined mutants of a specific gene by complementation, there are a number of genetic markers that have been used to select for transformants in a variety of genera. These markers include genes that code for resistance to the aminoglycosides G418 (43) and hygromycin (44), and genes conferring resistance to the toxic effect of heavy metals such as the CUP1 gene, which confers resistance to copper toxicity (45). However, little work has been reported utilizing either hygromycin or G418 selection in alternative yeasts. CUP1 selection has been used successfully in a number of *Saccharomyces* species that are of commercial interest, but little work using this selection system has been reported in other genera.

The most widely publicized use of cloning systems is the expression of heterologous genes, the proteins of which are of commercial value. Numerous genes that encode proteins of commercial interest have been cloned and expressed in *S. cerevisiae.* These include human epidermal growth factor (46), hepatitis B surface antigen (47–50), calf chymosin (51), calf prochymosin (52), human

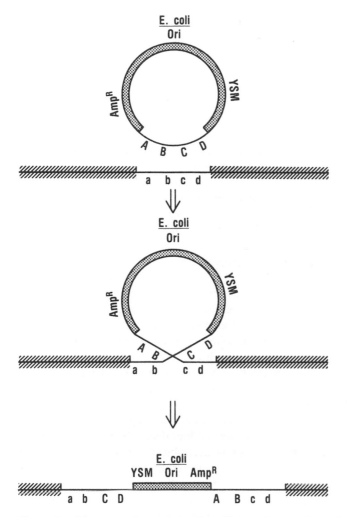

Figure 3. Diagrammatic representation of integrative transformation in yeast. The host is transformed with the donor plasmid. This plasmid carries a bacterial origin of replication (*E. coli* ori) and ampicillin resistance marker (Amp^R) which allow its propagation and selection in *E. coli*. In addition, it carries a marker selectable in yeast (YSM) and sequences homologous to a region of the host genome ("abcd") here designated on the plasmid as ABCD. A single recombination event between the homologous region ABCD carried on the plasmid, and abcd in the host genome results in the integration of the plasmid DNA flanked by recombined ABCD sequences. Recombinants may be detected by selection for the integrated selectable marker or by loss of the phenotype encoded by the target gene.

α-interferon (53–55,61) tissue plasminogen activator (56,57), β-endorphin (58), calcitonin (58), human insulin (59), human arterial natriuretic peptide (60), and α-1=antitrypsin (62).

The expression of such genes requires the presence of promoters, termination sequences, and regulatory elements. Several of the alternate yeast systems have progressed to this level of sophistication. The use of these sequences for the analysis of biological problems and the general nature of such sequences have been reviewed elsewhere (63–65). Although much publicity has centered on the expression of foreign genes of commercial interest, there are numerous examples of the use of cloning and transformation systems to modify industrial microorganisms for a variety of purposes.

The modification of the bacterium *Methylophilus methylotrophus* by scientists at the Imperial Chemical Industries (ICI) to make the organisms more energy efficient is a case in point (66,67). The gene coding for methanol dehydrogenase (Mdh) was replaced by the gene coding for alcohol dehydrogenase (Adh) from *Bacillus stearothermophilus*. The *M. methylotrophus* enzyme uses cytochrome C as the electron acceptor in the oxidation of methanol, whereas the *Bacillus* enzyme uses NAD(P). In the latter case the NAD(P)H produced may be used directly for biosynthesis or in other physiological processes, yielding three times as much ATP as that produced using the endogenous enzyme (Mdh). The energy conversion efficiency is significantly higher with carbon conversion improved from 20% to 30%. Similar applications could be employed in yeast to engineer specific enzymes, remove catabolite repression, or modify an organism's ability to utilize alternative and less expensive carbon sources. An excellent example of the application of recombinant DNA technology for yeast strain improvement is reported by Hanley and Yocum (68), who cloned the regulatory and structural genes for lactose utilization from *K. lactis* and introduced them into a strain of *S. cerevisiae*. The resulting recombinant grew anaerobically on lactose as well as the parental *Kluyveromyces* strain but had both superior osmotolerance and higher ethanol tolerance. Another example is the construction of a *Saccharomyces* strain capable of starch fermentation by the introduction of a mouse pancreatic α-amylase gene by DNA transformation and subsequent crossing with strains expressing endogenous maltase and glucoamylase activities. The recombinant yeast could grow on starch as its sole carbon source with a conversion efficiency of starch to ethanol of 95% (69).

The application of these techniques to the alternative yeasts is very promising since it permits the isolation, manipulation, and construction of strains for which there is virtually no known genetic information. The resulting strains can be modified at specific loci and otherwise remain isogenic. When these techniques are combined with recent advances in protein purification, protein sequencing, oligonucleotide synthesis, cloning by hybridization procedures, polymerase-chain reactional strategics (PCR), and protocols that enrich for low abundance messenger

RNAs by substractive hybridization, the permutations and applications not only are broad in scope, but have begun to redefine our approach to industrial yeasts.

One problem in the assembly of a diverse literature concerned with a large number of diverse genera and species is taxonomy. The problem has been elegantly stated elsewhere (71). We have attempted to follow Barnett et al. (72), except in cases that would be confusing owing to historical concerns. Questions regarding taxonomy are discussed by Dr. Kurtzman elsewhere in this volume. In addition, regarding the use of abbreviations, we have employed those cited by the authors to avoid confusion upon consulting specific references. Thus the reader will find, for example, three abbreviations for *Kanamycin* resistance: Kn, Kan, and Km. These examples are rare, and fortunately provide little variety (or challenge) for the reader.

In the following sections, we will discuss specific applications of recombinant DNA technology to a variety of alternative yeasts.

II. *Candida*

Members of the genus *Candida* represent a diverse group of economically important yeasts. Members of this genus are used in the production of industrially important enzymes, fine chemicals, and single cell protein (SCP); in addition, several species are human pathogens. Many of the *Candida* species are capable of degrading starch to its constituent components, primarily through the action of α-amylase. Members of the genus possessing α-amylase activity are *C. utilis* (72), *C. homilentomas*, and *C. silvanorum* (7). *C. tsukukaensis* has been shown to possess α-amylase in addition to a glucoamylase-like activity (7). These amylolytic activities make these species attractive candidates for biomass conversion. In addition to the amylolytic enzymes, several of the *Candida* species can degrade cellobiose by means of β-glucosidase; *C. wickerhamii* and *C. pelliculosa* are the best-known members of this group (72,73).

The metabolic versatility of the *Candida* is also illustrated by the use of various species to convert inexpensive substrates into SCP or alcohols by single or associative fermentations. For example, *C. utilis* in the Symba process (75,76) is grown in mixed culture with *Saccharomycopsis fibuligera* on waste streams from potato processing. This process is effective in lowering the biological oxygen demand (BOD) of the waste stream in addition to producing food yeast. SCP is also produced by growth of *C. utilis* on ethanol and waste liquors from paper processing (76) whereas *C. lipolytica* has been grown for SCP on n-alkanes (77). Attempts to improve the ethanol yields from xylose using *C. shehatae* have been reported (78,79). Inulin is becoming an increasingly important substrate for SCP production, and a number of direct fermentations have been conducted using *C. kefyr, C. pseudotropicalis,* and *C. macedoniensis* (80). In addition to inulin lactose in whey has been converted to SCP and ethanol by *C. pseudotropicalis* (81).

Several members of the *Candida* genus have been used for direct bioconversions of a variety of substrates in the production of fine chemicals (8). In general, *C. albicans* has been used for the oxidation of hydroxyls (7) and the production of 6-aminopenicillanic acid from penicillin by the action of penicillin acylase (82). One attraction of the *Candida* species as "biocatalysts" in the fine-chemical industry is the extracellular accumulation of a variety of metabolic products. Examples of efficient bioconversions with the extracellular accumulation of product include the production of ATP by *C. boidinii* (83), 1,2-hexadecanediol by *C. lipolytica* (8), 1-decene and 1-decanol by *C. rugosa*, tryptophan and long-chain fatty acids by *C. tropicalis*, citrate by *C. lipolytica (84)*, isocitrate by *C. zeylanvides* (8), riboflavin by *C. scatti* (8), vitamin B_6 by *C. albicans,* and fumaric acid by *C. hydrocarbofumarica* (8). These compounds are produced from a variety of hydrocarbon substrates.

Several of the species of the genus *Candida* are endogenous opportunistic pathogens. These include *C. albicans,* the most common cause of invasive candidiasis, and less common species such as *C. tropicalis, C. parapsilosis, C. krusii, C. globrata,* and *C. pseudotropicalis* (85,86).

A. C. albicans

C. albicans is the most common human fungal pathogen, and infections are extremely difficult to treat. The organism is implicated in the morbidity and mortality of immunocompromised individuals. The frequency of superficial and invasive candidiasis infections has increased markedly in the past few years, probably owing to increasing use of immunosuppressive therapy, broad-spectrum antibacterial antibiotics, and parenteral feeding (87–90). Thus at risk are those individuals undergoing chemotherapy. Disseminated fungal infections are the major cause of morbidity and mortality in patients with leukemia and a variety of immunodeficiency diseases. Concern with fungal pathogens will increase as the average population grows older, immunosuppressive therapies become more widespread, and the number of *Candida* species resistant to standard antimycotic agents increases.

The development of useful antimycotic agents and effective therapies requires an understanding of the biochemistry and physiology of pathogen/host-pathogen interactions as well as the genetics of the organism. The organism is dimorphic, displaying both budding and hyphal forms, and natural isolates have been shown to be diploid (91,92). To date no sexual cycle has been described, and the isolation of recessive mutations has thus been extremely difficult; however, adenine-requiring mutants have been extensively characterized. Spheroplast fusion has been employed to construct recombinant strains (93,94) and analyze complementation between independent mutants (90). These "parasexual" procedures have resulted in the assignment of at least 14 mutations to five linkage groups (90). The nuclear events following spheroplast fusion have been shown to be variable and the strain characteristics unstable. With these difficulties a number of laboratories have taken an alternative route to classical analysis by employing recombinant DNA technology.

The isolation of *C. albicans* genes by complementation in *S. cerevisiae* has been successful. Rosenbluth et al. (95) have cloned size-fractionated *Sau*3A restriction fragments of genomic DNA into a standard *S. cerevisiae* vector, YEp13, and isolated sequences that complement *trp*1 and *his*3 mutations. The source of the cloned DNA was confirmed to be *C. albicans* by Southern hybridization analysis. The complementing genetic elements have been shown to be plasmid mediated by cotransforming the selectable marker LEU2 with either TRP or HIS and demonstrating segregational loss of the marker and complemented phenotypes in the absence of selective pressure. Useful host cell mutations corresponding to these cloned genes can be generated by standard mutagenesis protocols and identified by complementation of the specific mutations generated by gene disruption techniques. The availability of a mutant host cell and the complementing genes provides the means for isolating ARSs and thus the tools to begin developing a gene cloning system.

Rather than focus on developing autonomously replicating vectors, Kurtz et al. (96) have successfully introduced foreign DNA into the chromosome of *C. albicans* utilizing a number of integrating vectors. A library was constructed from a *Sau*3A partial digest of genomic DNA and cloned into YEp13. This library was then used to isolate a DNA sequence that complemented an *ade*2 mutant of *S. cerevisiae*. A plasmid, pMK3, was constructed using a DNA fragment that carried the adenine-complementing activity and was used to transform an *ade*2 mutant of *C. albicans*. The strain hOG300 (*ade*2, *pro, met*) was transformed by a modification of the procedure used for *Neurospora crassa* which included a dimethyl sulfoxide and a heat shock step, and by standard yeast spheroplasting procedures (as outlined in the Introduction) used for *S. cerevisiae;* the two gave similar results. Resulting adenine prototrophs were found to be 100% stable when grown in the presence of adenine. Since the plasmid pMK3 was believed not to contain an ARS, the investigators suspected that the plasmid had integrated at the ADE2 locus. Southern hybridization analysis confirmed this hypothesis.

These plasmids provide a means to introduce foreign DNA stably into *Candida albicans* at a specific locus. It was also possible to recover the integrated plasmids by isolating genomic DNA, cleaving with *Bam*HI, ligating the products, and transforming *E. coli* and selecting for amplicillin resistance. Plasmids recovered from the ampicillin-resistant *E. coli* were indistinguishable from the parental plasmid. Figure 4 shows the scheme for integrative transformation and a map of the resulting modifications at the ADE2 locus based on detailed Southern hybridization analysis. The *Bam*HI sites that were essential for releasing the intact

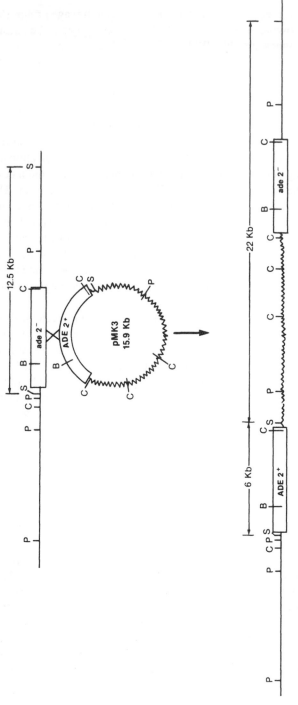

Figure 4. Model for integrative transformation in *C. albicans*. The predicted structure of the ADE2 region after a single homologous recombination event between pMK3 and one chromosomal homologue is shown. Fragments expected for a single recombination event for *Sph*1-digested DNA were 6.0 and 22 kb. Similar results were obtained with *Pvu*II. Host DNA digested with *Sph*1 yielded a 12.5-kb fragment. Key: Wavy lines, YEp13 vector sequences; boxed, ADE2 sequences; horizontal line, flanking host chromosome sequences. Abbreviations: P, *Pvu*II; C, *Cla*I; S, *Sph*I; B, *Bam*HI. (From 96, courtesy of Dr. D. R. Kirsch.)

plasmid were shown to be present in both the wild-type and mutant genes. The integration event occurred at only one chromosomal homologue; tandem copies were not found. It should be noted that a detailed characterization of the *C. albicans*ADE2 gene was not necessary (i.e., nucleotide sequencing and detailed restriction enzyme mapping *or* deletion analysis). The investigators were able to go directly from an isolated DNA fragment that conferred adenine prototrophy in *S. cerevisiae* to the transformation of their *ade2* strain of *C. albicans*.

Specific *ura3* strains of *C. albicans* were generated by directed mutation using the ADE2 gene. The URA 3 gene from *C. albicans* was cloned by complementation of a *ura3* mutation in *S. cerevisiae* and a *pyrf* mutation in *E. coli* (196). A mutated URA3 was created by inserting the ADE2 gene into the coding region of the wild-type URA3 gene (197). The mutated URA3 gene was then introduced into *ade2* strains of *C. albicans* and *ade+* transformants selected. Southern hybridization analysis showed the *ade+* transformants to be disrupted or deleted at one of the URA3 homologues; none of the *ade+* transformants were found to be homozygous for *ura3*. Several *ura3* strains were generated by treating the disrupted transformants with UV light (to increase mitotic recombination) followed by replica plating. Of the resulting homozygous *ura3* strains, all were found to be nonreverting owing to the partial deletion of the URA3 loci. This was the first report of the isolation of a specific mutant of *C. albicans* by directed mutagenesis.

In addition to the in vitro mutagenesis of URA3, Kirsch and co-workers isolated an ARS from a RsaI *C. albicans* genomic library cloned into a pBR322 vector containing the ADE2 gene (198). A 350-bp RsaI fragment was found to be responsible for ARS activity, permitting transformation efficiencies of from 100 to 1,000 *ade+* transformants per microgram of plasmid DNA. Of interest was the fact that the introduced DNA was extrachromosomal and of high molecular mass, consisting of oligomers of the original plasmid.

The isolation of the ADE2 gene, an *ade2* auxotrophic host, and the development of a successful procedure for introducing foreign DNA provides the means for isolating ARSs and developing a plasmid-based vector system useful for a detailed analysis of the organism. This is the first demonstration of a transformation system for an asexual, diploid microorganism and the first introduction of foreign DNA into a yeast lacking a sexual cycle. These results provide some of the basic materials and protocols essential for the development of a recombinant DNA system essential to the study of the molecular biology of the pathogen *C. albicans*.

The role of molecular biology in addressing the problem of treating candidiasis and alleviating an increasing degree of human suffering will be manifold (85,88,97,98). The best-tolerated drug by the patient in the treatment of candidiasis is 5-flurocytosine, which generates resistant strains of *Candida* at a high frequency. Other antimycotic agents that are effective are poorly tolerated and often exhibit serious side effects. What role could molecular biology play? One possible application is in the development of new antimycotics. The currently available drugs act against the plasma membrane, and since *Candida* is a eukaryote, its plasma membrane is similar in molecular content and structure to that of the human host. Shepard et al. (9) have suggested that cell wall biosynthesis be targeted, since the presence of chitin and β-glucans clearly indicates a macromolecular biosynthetic pathway different from that of the human host. The elucidation of these pathways, via mutant generation and the construction of tester strains could be accomplished by applying recombinant DNA techniques. The goals of such an approach would be to elucidate metabolic pathways that would be the targets of new drugs effective at specific stages in cell wall biosynthesis, and the development of strains of *Candida* suitable for screening for antimycotic agents. Additional areas for applying recombinant DNA technology would include the characterization of cell wall modifications responsible for adhesion, human-fungal interactions, and the cloning of "pathogenic factors" by new techniques such as subtractive cDNA cloning.

B.   *C. maltosa*

C. maltosa is able to utilize n-alkanes as its sole carbon source and, as is the case with many of the *Candida*, is not easily amenable to conventional genetic techniques. Studies of gene regulation and the analysis of metabolic pathways necessitate the development of recombinant DNA and cell fusion systems in order to acquire critical genetic data and generate useful strains.

Kunze et al. (99,100) have demonstrated that foreign DNA can be introduced into *C. maltosa*. A plasmid, pYe(ARG4)411, composed of pBR322 and the ARG4 gene from *S. cerevisiae*, was introduced into an arg4 mutant of *C. maltosa* (G344) by a modification of the transformation method of Iimura et al. (25) (as summarized in the Introduction). Cells were grown on enriched medium, washed, and incubated in 0.2 M $CaCl_2$, followed by the addition of TRIS HCl and 10 mM $MgCl_2$ (pH 7.4) that contained from 0.5 to 10 μg of plasmid DNA. The cells were incubated at 0°C for 20 min, heat-shocked at 37°C for 5 min, and plated directly on minimal medium lacking arginine. The transformation efficiency using this protocol was equivalent to that of the *S. cerevisiae* control at 10^3 arg+ transformants per microgram of plasmid DNA. The introduction of foreign DNA was demonstrated by hybridization of pBR322 to DNA fragments generated by the digestion of genomic DNA with BglII. The hybridized DNA fragments were identical to those expected from a BglII digest of pYe(ARG4)411. Additional evidence for the successful transformation of this yeast was obtained by the presence of argininosuccinate lyase and β-lactamase activity in all of the arg+ transformants.

When cells containing the plasmid were grown under nonselective conditions, loss of the ARG4 phenotype could be detected. Since the investigators were able to recover plasmids from lysates of the ARG4 transformants, they conclude that the plasmid pYe(ARG4)411 has the ability to replicate autonomously in the yeast *C. maltosa*. The levels of the specific activity of argininosuccinate lyase when compared to an *S. cerevisiae* control suggested that the plasmid was present in from one to three copies per cell, assuming the gene is transcribed and its product functions identically as in the *S. cerevisiae* control. The overall conclusion from this work is that the plasmid pYe(ARG4)411 contains an ARS that can be introduced and maintained in *Candida maltosa*.

A gene library from *C. maltosa* has also been constructed by Kawamura et al. (101) in the *E. coli* vector pBR322 and used in cloning a gene corresponding to the *S. cerevisiae* LEU2 by complementation of an *E. coli leu*B mutation. A plasmid designated pCMK3 from this library containing an 11-kb insert was found to complement the *leu*B mutation. Southern hybridization analysis confirmed that the DNA insert was from the yeast *C. maltosa*. The fragment was subcloned into pRC3, and the resulting plasmid, pCMK31, was used to complement a *leu*2 mutant of *S. cerevisiae* (Fig. 5). This vector contains the TRP1 gene as a marker and was used to transform a *leu*2, *trp*1 mutant of *S. cerevisiae*. All *trp*⁺ transformants were found to be *leu*⁺. A second vector was constructed, pCMK32, which had the fragment containing the complementing activity cloned in the opposite orientation. Of interest was the observation that the LEU2 gene from *C. maltosa* was expressed in either orientation in *S. cerevisiae,* but in only one orientation in *E. coli,* which suggested that a sequence of *C. maltosa* DNA was functioning as a promoter in *S. cerevisiae,* but not in *E. coli.*

The investigators further analyzed the DNA insert to determine if it contained an ARS function in *S. cerevisiae.* Of several subcloned fragments one 800-bp segment was found to function as an ARS in *S. cerevisiae.* Plasmids containing this fragment in addition to those containing the 2-μm plasmid's origin of replication were not active in *C. maltosa.* In order to isolate an ARS functional in *C. maltosa,* the investigators transformed a number of *leu* mutants of the organism with a genomic DNA library constructed using the *S. cerevisiae* vector YEp13 (carrying the *S. cerevisiae* LEU2 gene) by both the spheroplasting and lithium acetate methods. Plasmid DNA was recovered from the *leu*⁺ transformants, and a 6.3-kb segment was demonstrated to contain an ARS. Subsequent analysis localized the ARS activity to a 3.8-kb *Bam*HI fragment. This fragment of DNA was used in the construction of two novel vectors that will be useful for cloning in *C. maltosa*; Figure 6 shows the structure of these vectors: pTRA1 and pTRA11. Both vectors transform *C. maltosa* at an efficiency of 10^3 transformants per microgram of purified plasmid DNA using the spheroplast method and 4×10^2 transformants per microgram using the lithium acetate method.

Southern hybridization analysis showed that both pTRA1 and pTRA11 are capable of autonomous replication in both *C. maltosa* and *S. cerevisiae* (102). These vectors in combination with mutants that are deficient in the metabolism

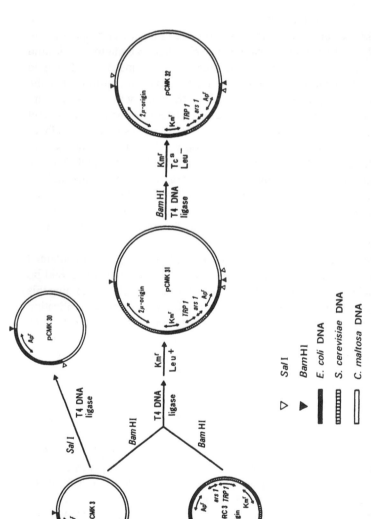

Figure 5. Construction of plasmids used in the cloning of a LEU gene and ARS from *C. maltosa*. pCMK3 was constructed from the *E. coli* vector pBR322 and *C. maltosa* DNA; both were digested with *Bam*HI. *E. coli* C600 (pCMK3) colonies were selected for *Leu*⁺ phenotype. A small region was removed from pCMK3 to construct pCMK30. pCMK31 was constructed from pCMK3 and a subcloning vector, pRC3, and selected for *Leu*⁺ phenotype after transforming *E. coli* C600 or *S. cerevisiae* SHY3. pCMK32 was constructed by self-ligation of a pCMK31 *Bam*HI digest and selection of *E. coli* C600 transformants with the *Leu*⁺ phenotype a few days after streaking on leucine-deficient minimal medium. As indicated, pCMK32 is in the opposite orientation of pCMK31. (From 101, courtesy of Dr. M. Takagi.)

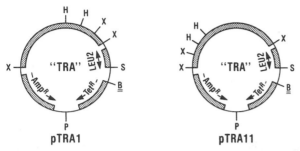

Figure 6. The structure of pTRAI and pTRAII. The plasmids are 11 kb in size with a unique *Bam*HI cloning site. The plasmids differ in the orientation of the TRA (transforming) region. The heavy lines indicate the 3.8-kb TRA fragment from *C. maltosa,* the thin lines the *S. cerevisiae* cloning vector YEp13. Both plasmids transform with equal efficiencies. Abbreviations: B, *Bam*HI; S, *Sal*I; X, *Xho*I; H, *Hind*III; P, *Pst*I. (From 102, courtesy of Dr. M. Takagi.)

of n-alkanes will provide a means for analyzing the regulatory system and genes involved in n-alkane uptake and catabolism, in addition to providing a new approach to improving this industrially important yeast.

C. *C. pelliculosa*

C. pelliculosa is noted for its efficient growth on cellobiose, making it a candidate for large-scale SCP production from plant biomass. The enzyme responsible for cellobiose degradation, β-glucosidase, has been cloned from this organism and expressed in *S. cerevisiae* (103), and its nucleotide sequence has been determined (74). This enzyme, aside from its obvious use in the conversion of cellobiose to glucose, may be of value as a model protien in the characterization and isolation of regulatory DNA sequences (e.g., activation sequences, enhancers, and promoters) in this yeast. The enzyme is easily detected on indicator plates using the chromogenic agent X-glu (5-bromo-4-chloro-3-indolyl-β-glucopyranoside) (103) and can be conveniently assayed and quantitated using the chromogen paranitrophenyl (PNPG) (104). β-Glucosidase can be used in the same manner that the *E. coli* β-galactosidase gene has been used in the analysis of a wide variety of biological phenomena in both *E. coli* and *S. cerevisiae* by gene fusions (65). The major advantage of the *C. pelliculosa* enzyme is that it is derived from a *Candida* yeast and thus problems with codon bias, mRNA structure, and posttranslational processing may be more reasonably approached and accounted for than using the *E. coli* enzyme.

D. C. tropicalis

This is one of the more unusual yeasts in the genus *Candida* because of its assimilation of fatty acids and alkanes (105) and its use in the production of tryptophan from paraffins. The metabolic diversity of this yeast makes it an attractive candidate for the production of SCP from industrial wastes and for the biotransformation of materials in the production of fine chemicals. *C. tropicalis* can be induced to produce peroxisomes when grown on n-alkanes (106). Peroxisomes are subcellular compartments considered attractive models for understanding organelle biogenesis, intracellular transport, and protein sorting and targeting, and are currently under investigation by a number of laboratories (107,108). Research in this area has been hampered by difficulty in obtaining mutations in the peroxisomal proteins and their regulatory elements. A number of n-alkane inducible proteins have been cloned by Kamiryo and Okazaki (109). DNA libraries composed of genomic DNA were probed using cDNA prepared from mRNA isolated from cells grown on oleic acid or glucose. By using these two pools of probes the investigators were able to identify 102 clones that were from mRNAs induced by growth on oleic acid.

The DNA from these clones was isolated and further characterized. Seven coding regions were found distributed in clusters of three regions of genomic DNA; of these seven, five were assigned to inducible peroxisomal proteins by characterization of peptides produced by in vitro translation experiments. Proteolytic peptide mapping of the in vitro translation products of the cloned genes permitted identifying them as PXP-4 and PXP-5. The remaining three were tentatively identified as PXP-12, PXP-11, and PXP-5.

A similar approach was taken by Lazarow et al. (110) in the cloning of cDNA coding for catalase and other peroxisomal proteins. A cDNA library was constructed from polyA$^+$-mRNAs isolated from n-alkane-grown cells. The library was screened by DNA hybridization using cDNA from n-alkane- and glucose-grown cells as probes. Clones that were shown to be positive using the n-alkane-induced probes were analyzed by hybridization to mRNAs whose cell free translation products comigrated with proteins purified from *C. tropicalis* peroxisomes. Two enzymes, catalase and acyl-CoA oxidase, were identified in addition to a number of other peroxisomal proteins. Possession of these genes provides the opportunity of using gene disruption techniques to generate specific mutations and to use their sequences as probes in Northern hybridization analysis to study gene induction. Future goals would be the use of this information to specifically modify the organism to increase its energy efficiency or to develop a host vector gene expression system for the production of commercially valuable proteins.

E. C. utilis

The species *C. utilis* is one of the most extensively used industrial yeasts. This organism has been employed for the production of a variety of amino acids and other fine chemicals (8) in addition to SCP from a wide assortment of carbon sources. The organism's ability to grow efficiently on xylose makes it attractive for large-scale biomass conversion. These qualities have made *C. utilis* a prime candidate for the development of a cloning and transformation system with the goal of improving its industrial utility (111,112). Despite the widespread use of *C. utilis* there are a few reports of the isolation of auxotrophic mutants, and thus the general state of the organism's genetic characterization is limited. There have been reports of the isolation of DNA sequences from *C. utilis* that confer ARS activity in the yeast *S. cerevisiae*.

Figure 7 shows the scheme employed by Hsu et al. (113) for the isolation of a *Sau*3A fragment from a genomic library from *C. utilis* that confers ARS activity in the yeast *S. cerevisiae*. *C. utilis* genomic DNA was cloned into the integration vector, YIp5, and used to transform a strain of *S. cerevisiae* from *ura*⁻ to *ura*⁺. A plasmid designated pHMR22 was successfully recovered from the transformants, and a DNA fragment was found to confer the ARS activity in *S. cerevisiae* characterized by subcloning. A *Hind*III-*Bam*HI subfragment was cloned into YRp5 (producing the plasmid pHMR23) and shown to possess ARS activity, indicating that the single *Bam*HI site in the *C. utilis* insert could be used as a cloning site. The cloned ARS showed limited homology to *S. cerevisiae* genomic DNA. Although further work on the ARS sequences in *C. utilis* is not reported, the investigators point out the advantages of using pHMR22 as a cloning vector in *S. cerevisiae* being its small size (6.6 kb) and the number of unique restriction sites (*Bam*HI, *Sal*I, *Hind*III, *Eco*RI, and *Pvu*II).

One of the problems of cloning into *C. utilis* is the lack of a suitable selectable marker; two laboratories have reported on the use of aminoglycoside G-418 in attempts to transform *C. utilis* (111,113). In a preliminary report Tsao and co-workers (111) have reported on the construction of a number of vectors containing kanamycin resistance (*kan*ʳ) which confers resistance to G418. The resulting plasmid pLC11 contains *kan*ʳ, *amp*ʳ, the *E. coli* origin of replication, and an ARS from *C. utilis* genomic DNA functional in *S. cerevisiae*. This plasmid was used to transform *C. utilis* and to select for G-418 resistant clones. In this preliminary report the authors cite the lack of stability of the resistant phenotype in the absence of selective pressure is an indication that G-418 resistance is plasmid-mediated. The investigators discuss the results of hybridization experiments in identifying specific pLC11 sequences in the G-418 clones, and a more rigorous detailed report is expected.

Hsu and Reddy (114) have also constructed a vector that is capable of transforming a number of genera including *Candida*. The vector, pHR40 (Fig. 8)

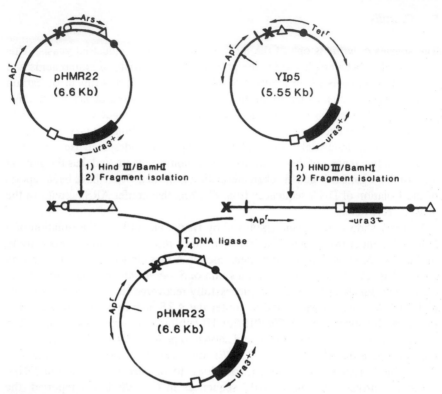

Figure 7. Subcloning analysis of the *ars* fragment from *C. utilis* of pHMR22. Plasmid pHMR22 is YIp5 plus a 1.05-kb *Mbo*I restriction fragment of *C. utilis* which includes an ARS. A new plasmid, pHMR23, was constructed by using a 1.35-kb *Hind*III–*Bam*HI fragment of pHMR22 and a 5.25-kb *Hind*III–*Bam*HI fragment of YIp5. YIp5 is used for the stable integration of DNA into the *S. cerevisiae* genome (see Introduction). Restriction enzyme sites are as follows: *Eco*RI (+), *Hind*III (×), *Bam*HI (Δ), *Pst*1 (↑), *Sal*I (•), *Pvu*II (□), and *Sau*3A (○). Not all *Sau*3A sites are shown. (From 113, courtesy of Dr. C. A. Reddy.)

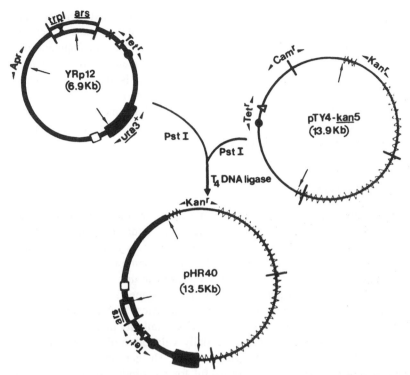

Figure 8. Construction of a chimeric plasmid used in the transformation of *Candida utilis*. YRp12 is a *S. cerevisiae* vector carrying the URA3, TRP1, and ARS1 from *S. cerevisiae*. The plasmid pTY4-Kan5 carries the Kan^r from the transposon Tn601. YRp12 DNA is shown by the dark lines; the *S. cerevisiae ars*1 is indicated, and DNA from pTY4 carry Kan^r is shown as wavy lines. The various restruction enzyme cleavage sites are as designated in the legend to Figure 7. (From 114, courtesy of Dr. C. A. Reddy.)

carries the URA3 gene and an ARS from *S. cerevisiae* and *kan^r* from *E. coli*. In a preliminary report the investigators discuss the successful transformation of a number of genera of yeasts using a modification of the lithium chloride procedure (26). One of the advantages of the G-418-based systems is the elimination of the need for auxotrophic markers and subsequent stable auxotrophs. A more detailed report of this work is awaited.

A further extension of the work in *C. utilis* has involved the isolation of the LEU2 gene by complementation in *S. cerevisiae* and *E. coli* (112). In this case *C. utilis* genomic DNA was digested partially with MboI and cloned into the yeast vector YRp12. The coding sequence for the *C. utilis* gene was shown to reside within a 2.43-kb fragment confirmed to be from *C. utilis* by Southern hybridization analysis. The *C. utilis* LEU2 gene was shown to be expressed at high efficiency in both *E. coli* and *S. cerevisiae*. With such a marker gene it is now critical that a proper mutant be generated in *C. utilis* to permit the development of a well-defined vector system. It may be possible to generate such a mutant by disrupting the chromosomal loci (assuming the organism is diploid) by means of integrative transformation of a modified LEU2 gene. The investigators would be forced to score for a rare event by replica plating, but, once characterized, a well-defined mutant with the complementing gene cloned will greatly facilitate the development of a *C. utilis* cloning system.

III. KLUYVEROMYCES

Members of this genus have been used for decades in the food industry most notably for supplying invertase used in the manufacture of jams and lactase (β-galactosidase) used for the production of sweet syrups from lactose found in whey and milk. *K. fragilis* and *K. lactis* are the major commercial sources for these enzymes. In addition to these two enzymes, members of this genus are used in the production of SCP from whey and inulin. The enzyme β-galactosidase is the product of the LAC4 gene and the key enzyme in lactose metabolism. There is a great deal of interest in this enzyme because of its similarity to the *E. coli* enzyme (115,116). These factors make it a potentially valuable tool for studying gene expression, gene regulation, and protein secretion. The gene could also be part of the development of a cloning system useful for the production of commercial proteins. Members of this genus show an interesting versatility, ranging from being sources of SCP for human consumption to successful attempts in the direct fermentation of inulin to fuel grade alcohol (80).

Progress in the development of a cloning technology for the exploitation of this genus has been advanced by the discovery of several different types of endogenous plasmids.

A. K. drosophilarium

Chen and co-workers have reported the discovery of a circular DNA plasmid in *K. drosophilarium* that is similar in structure to the 2-μm plasmid isolated from *S. cerevisiae* (117). This plasmid, pKD1, is a 1.6-μm circle of 4,757 base pairs that comprises roughly 2% of the total cellular DNA (118). Like 2-μm plasmids, this plasmid has two isomeric forms generated by internal recombination at the inverted repeats, and two identical terminal repeats. The plasmid has been sequenced and contains three open reading frames (ORFs). These ORFs have been designated A, B, and C and may be equivalent to the 2-μm plasmid's genes FLP, REP1, and REP2. The amino acid sequence of gene A, as deduced from the nucleotide sequence of pKD1, shows significant homology to the 2-μm plasmid's FLP gene product. The sequence homology at the nucleotide level is limited, supporting earlier reports of the inability to hybridize pKD1 DNA to 2-μm plasmid DNA even under low stringency conditions (118). The 5' ends of genes A, B, and C were confirmed by S1 mapping following Northern hybridization analysis of total *K. drosophilarium* DNA using cloned pKD1 sequences as probes. The latter experiment revealed three distinct species of mRNA, which correspond to sequences containing the ORFs.

The origin of replication for this plasmid was located by subcloning a number of fragments in a vector containing the URA3 gene from *S. cerevisiae* and necessary pBR322 sequences for propagation and selection in *E. coli*. The URA3 gene of *S. cerevisiae* complements the *ura*A mutation in *K. lactis*, so both *S. cerevisiae* and *K. lactis* were used as recipient hosts in this study. The origin of replication for pKD1 was located in a region near one of the inverted repeats, as is the case with the 2-μm plasmid. These plasmids, however, were unstable. Initial attempts to clone foreign DNA into pKD1 were not successful owing to the choice of cloning sites interfering with plasmid stability. A systematic study was undertaken to locate regions that could serve as cloning sites in pKD1 that would not affect plasmid stability. Again, the URA3 gene from *S. cerevisiae* was used and was cloned into a variety of restriction sites in pKD1. The resulting plasmids were tested for stability in the *K. lactis ura* A mutant. An *Eco*RI cloning site was found between genes A and B that did not affect plasmid stability; interestingly, cloning URA3 into gene A did not have a marked effect on plasmid stability either.

Like the 2-μm plasmid from *S. cerevisiae*, the plasmid pKD1 requires that its replication function for maximum stability. To be maintained stably it is necessary to introduce the plasmid into a *K. drosophilarium* strain that contains an endogenous pKD1 plasmid or to introduce the entire plasmid into the host cell. The latter has been accomplished using *K. lactis* as the recipient host, showing that pKD1 can be maintained without selective pressure. The transformation efficiency for this plasmid into *K. lactis* is comparable to that of the 2-μm-based plasmids in *S. cerevisiae*. Transformation procedures for *K. lactis* will be discussed below.

The plasmid pKD1, like the plasmids pSR1 and pSB3 from species of *Zygosaccharomyces* (discussed below), share remarkable similarity with the 2-μm plasmid from *S. cerevisiae*. Though different in size, they have a similar number of large ORFs (pKD1, pSR1, and pSB3 have three whereas the 2-μm plasmid has four). All have a pair of inverted repeats, as illustrated by the cruxiform structure in Figure 1, and all four are found in isomeric forms owing to intramolecular recombination at their inverted repeats. Like the 2-μm plasmid from *S. cerevisiae*, these plasmids will be of immense value in the development of a cloning system in these industrially important yeasts.

B. K. lactis

K. lactis was the first member of this genus to be transformed. This was accomplished by Das and Hollenberg (119) using a 2-μm-based vector and a modification of the protoplasting protocol reported by Beggs (20). The vector PTY75-LAC4 containing the 2-μm origin of replication from S. cerevisiae, kanamycin resistance (Kn^r) and LAC4 (encoding β-galactosidase) from K. lactis, was used to transform K. lactis. Cells were plated onto medium containing 200 μg/ml G-418. Of the G-418-resistant transformants less than 5% proved to be lac$^+$. The transformation efficiency was extremely low (4 transformants per μg of DNA). This low transformation efficiency was found to be due to inefficient protoplast regeneration. Several isosmotic stabilizers were screened, and the replacement of sorbitol with 0.6 M KCl increased the regeneration efficiency by a factor of 3. The optimized conditions permitted the direct selection of transformants on lactose using LAC4 as a selectable marker. Using the modified protocol, all of the lac$^+$ clones were shown to be transformants by the presence of the Kn^r marker when replica plated onto medium containing G-418. The low transformation frequency was not due to chromosomal integration, since 50% of the transformants lost both markers within 10 generations when grown under nonselective conditions. Southern hybridization analyses of undigested minilysates of the yeast lac$^+$ transformants probed with the bacterial portion of the vector showed the presence of the recombinant plasmid. The Southern hybridization result and the ability to recover the intact plasmid in E. coli confirmed that PTY75-LAC4 was replicating autonomously.

The investigators also demonstrated that K. lactis could be transformed with LAC4 cloned into YRp7 (which contains the S. cerevisiae TRP1 and ARS1). This plasmid was found to integrate into the host chromosome. This series of experiments established the conditions for transforming K. lactis and demonstrated the utility of a number of selection schemes.

To isolate a useful ARS from K. lactis, a suitable trp1 mutant was obtained by treating wild-type K. lactis with ethyl methane sulfonate (EMS) and scoring for trp auxotrophy. The specific identity of the mutants was confirmed by complementing with a plasmid, pL4, consisting of the S. cerevisiae TRP1 (containing ARS1) and the K. lactis LAC4. K. lactis ARSs (KARS) were isolated by cloning genomic XhoI fragments into pL4 and scoring for increased transformation efficiencies. Several KARSs were found, and their plasmids were shown to be autonomously replicating by Southern blot analysis. Copy numbers were estimated to be from one to five per cell. All of the plasmids were unstable under nonselective conditions, with both markers lost in a few generations.

Another source of material in the development of a transformation system for the Kluyveromyces has been the killer plasmids, the structure and biology of which have been reviewed elsewhere (120–124). The two killer plasmids isolated from

K. lactis are k1 (8.8 kb) and k2 (13.4 kb). These plasmids confer on the host strain an ability to kill sensitive cells by the production of a secreted protein toxin. They may serve as a model for protein secretion as well as a source of material for the construction of a vector system.

De Louvencourt et al. (125) constructed a series of plasmids (Fig. 9) using the URA3 gene from *S. cerevisiae* and segments of the plasmid k1, including a k1 plasmid containing an internal deletion of 2.9 kb. Since k1 requires the presence of k2 to replicate, the investigators were required to clone into a *K. lactis* strain containing k2. The cells were transformed by the protoplast method of Hinnen et al. (19), and several different constructions had varying degrees of stability. All of the resulting plasmids were shown to be autonomously replicating by Southern analysis and the recovery of intact plasmids in *E. coli*. The copy number of the plasmids was lower than that for the endogenous killer plasmids, and none of the plasmids had integrated into the host chromosome. Two of the plasmids, pL3 (7.3

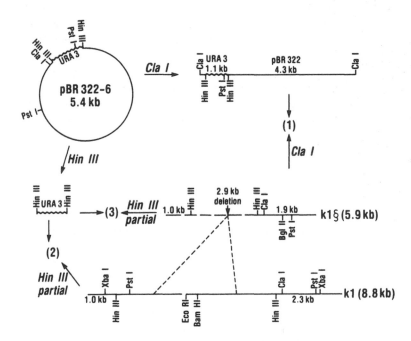

Figure 9. Construction of k1-*URA*3 hybrid plasmids. The source of the *URA*3 gene is the plasmid pBR322-6, whose structure is shown at the top. The structures of the linear plasmids k1 and k1§ are shown at the bottom. Ligation mixtures (1), (2), and (3), were prepared according to the indicated schemes. A 2.9-kb deletion of wild-type k1 is shown. The resulting plasmids pL₃, pL₄, and pL₈ were from ligation reactions 1, 2, and 3, respectively, as described in the text (125).

kb, from reaction 1; Fig. 9) and pL4 (4.4 kb from reaction 2; Fig. 9), were maintained at a frequency of 35% ura$^+$ colonies in selective-minimal medium for 100 generations. Another plasmid, pL8 (from reaction 3; Fig. 9), was maintained at 50%. The percentage of ura$^+$ colonies dropped to 2.6, 1.0, and 5.5 for pL3, pL4, and pL8, respectively, when grown in minimal media plus uracil.

Thompson and Oliver (126) have characterized two ARSs from the k1 plasmid that have activity in both *K. lactis* and *S. cerevisiae*. Both ARSs were subcloned from a *Sau*3A digest of k1. A 700-bp fragment (Kla1) was shown to be unstable in both hosts when the cells were grown under nonselective conditions. The second ARS (Kla2) was created in vitro by the ligation of two *Pst*1-*Sau*3A fragments from opposite termini of k1. The resulting fragment can be divided into two subfragments by cleavage at the internal Pst1 site. The 410-bp fragment Kla2A has ARS activity in *K. lactis* and only a 10-bp exact homology with Kla1. The investigators suggest that this sequence, 5' TCATAATATA '3, may be the consensus sequence necessary for ARS activity in *K. lactis*. The second subfragment, Kla2B, is 659 bp in length and has ARS activity only in *S. cerevisiae*. This fragment has a perfect base homology with the *S. cerevisiae* ARS consensus sequence (5' TTTTATGTTTA '3). In addition, Kla1 and Kl2B both contain 10 out of 11 bp matches to the *S. cerevisiae* consensus sequence. These results provide a refinement of our understanding of the *K. lactis* ARS and point out the clear differences in the sequence requirements for ARS activity between *K. lactis* and *S. cerevisiae*. Although these plasmids are unstable and their copy numbers are low, this work demonstrates that the killer plasmids can be used as sources for ARSs.

C. *K. fragilis*

K. fragilis is of interest since it is capable of fermenting lactose and other carbon sources over a wider temperature range than *K. lactis* (127) and can very efficiently secrete large proteins (128). The feasibility of strain modification by the introduction of foreign genes into *K. fragilis* was demonstrated by Farahnak et al. (129) in a series of experiments fusing selected strains of *K. fragilis* and *S. cerevisiae* with the goal of producing a stable hybrid capable of fermenting lactose and with an increased tolerance to higher concentrations of ethanol. Standard fusion protocols were used, and several fusants were found to ferment lactose and produce ethanol at rates and final yields that were significantly higher than those of the parental strains. Unfortunately, these fusants proved to be unstable. This work suggested that a direct approach involving cloning or specific genetic modifications in the host organism could result in a useful and more stable production strain.

Early attempts to develop a transformation system for *K. fragilis* were conducted by Hollenberg and co-workers (130). Since specific genetic markers were not available, G-418 selection was employed using Kn^r and ARS sequences from *K. lactis* (KARS2). KARS2 and the functional TRP1 gene of *S. cerevisiae* were cloned into the *S. cerevisiae* vector YRp7, yielding the plasmid pER27 (Fig. 10). The plasmid used to transform *K. fragilis* was constructed from the vector pK21 (the source of amp^r and Kn^r) and pEK27 (source of TRP1 and KARS2). The resulting plasmid pLG2 was recovered from *E. coli* that had been transformed with the ligation mixture. The plasmid pLG2 was then used to transform *S. cerevisiae*, *K. lactis*, and *K. fragilis* using the alkali metal ion method of Ito et al. (27). The frequency of G-418-resistant transformants was low, approximately 25 per microgram of DNA for *K. fragilis* and *K. lactis*.

Southern hybridization was used to demonstrate that the transformants carried autonomously replicating plasmids, and this was verified further by the recovery of intact plasmids in *E. coli*. This work demonstrated that the plasmid pGL2 can replicate freely in both *K. lactis* and *K. fragilis*. The low frequency of transformation was suggested to be due to the marker under selection. For example, transformation of *K. lactis* yielded over 1,500 trp$^+$ yeasts per 10 µg of pGL2 DNA, sixfold greater than recovered using G-418. These results indicate that a suitable plasmid and transformation procedure is now available for these two species of *Kluyveromyces*.

There has been one reported attempt at utilizing killer plasmids from *K. lactis* in *K. fragilis* (131). This study involved the direct transfer of the killer plasmids from *K. lactis* to *K. fragilis* by cell fusion. The *K. lactis* strain was auxotrophic for methionine and harbored both killer plasmids; the *K. fragilis* strain was a nonkiller lacking both plasmids. The resulting fusants, after selection, had the same phenotype as the *K. fragilis* parent but with an acquired killer activity. This demonstrated that the killer plasmids can be transferred to and function in *K. fragilis*, providing yet another potential source of materials that could be used to create a viable recombinant DNA technology in this yeast.

IV. METHOTROPHIC YEAST

In the past decade there has been an increasing interest in the development of new technologies for the exploitation of methanol as a feedstock in the production of fine chemicals (132,133) and as an alternative fuel (134). The renewed emphasis on methanol is due to its low price, worldwide surplus, and new techniques in large-scale organic synthesis. For many countries this is critical since methanol can be derived from a variety of sources (e.g., coal), thus diminishing dependency on crude oil as a fuel and chemical feedstock.

Of considerable interest are the methotrophic microorganisms that possess the capacity to utilize methanol as a sole carbon and energy source. These organisms, including members of several yeast genera, provide a means of converting

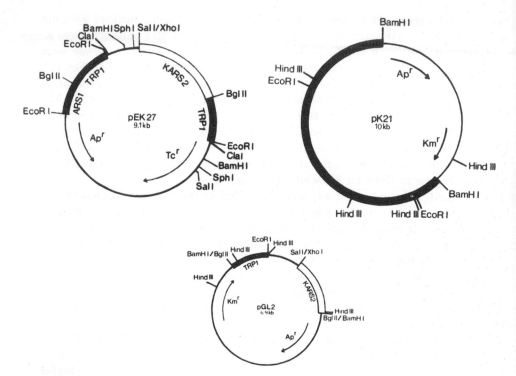

Figure 10. Construction of a *K. lactis* and *K. fragilis* vector, pGL2. A 3.3-kb *Bgl*II fragment containing the *TRP*1 gene of *S. cerevisiae* and the functional *KARS*2 sequence was subcloned from plasmid pKARS2 into the *Bgl*II site of the *S. cerevisiae* vector YRp7 as described in the text. The resulting plasmid pEK27 offers the possibility of using the restriction sites *Sph*I, *Bam*HI, *Cla*I, *Eco*RI, or *Bgl*II for inserting the *KARS*2 *TRP*1-containing fragment into another plasmid. To construct plasmid pGL2, pK21 DNA digested with *Bam*HI and pEK27 DNA digested with *Bgl*II were mixed and ligated. *E. coli* JA300 (*thr leu*B6 *thi thy*A *trp*C1117 *hsr*$_k$ *hsm*$_k$ Str^r) was transformed to Km^r. The Km^r transformants were replica-plated on minimal M9 medium containing ampicillin (100 μm/ml) and supplemented with threonine, leucine, thiamine, and thymine. DNA samples from the Ap^r Trp$^+$ colonies were analyzed. pGL2 was isolated from one such Ap^r Km^r Trp$^+$ colony. (From 130, courtesy of Dr. C. P. Hollenberg.)

methanol into SCP and potentially other useful chemicals. As discussed by Dijkhuizen et al. (134), methanol has several advantages as a feedstock for biotechnological processes: methanol is completely miscible with water, and easy to store and transport; in aerobic processes higher rates and yields are seen for biomass conversion than with other inexpensive substrates. The metabolism of the methylotrophic yeasts has been intensively studied, and a great deal is known regarding the biochemistry and physiology of methanol metabolism. In order to obtain a more detailed picture of regulation and induction of pathways involved in methanol utilization for the purposes of modifying critical stages in these pathways, development of cloning and transformation systems for these yeasts is required. In addition, the large percent of the total protein produced by methotrophic yeasts residing in key enzymes in the methanol catabolic pathway makes the strains attractive candidates for the development of expression systems for heterologous genes of commercial interest. In *H. polymorpha*, methanol oxidase (MOX) and dihydroxyacetone synthetase (DHS) can account for more than 33% of the total cellular protein when the cells are grown in the presence of methanol (135).

The principal steps in methanol assimilation and the compartmentalization of methanol metabolism are presented in Figure 11. The enzymes involved in the initial steps of methanol metabolism are found in intracellular organelles (peroxisomes), which are synthesized in response to growth in the presence of methanol (136,137). The ability of yeast to use methanol as a sole carbon source has been demonstrated for members of the genera *Candida, Torulopsis, Pichia,* and *Hansenula* (138). Of these four genera, significant progress has been made in species belonging to the latter two—namely *Pichia pastoris* and *Hansenula polymorpha*. The basic biochemistry, physiology, and molecular biology of these yeasts has been reviewed recently (199).

A. *H. polymorpha*

The demonstration that the two major proteins in methanol metabolism, methanol oxidase (MOX) and dihydroxyacetone synthetase, are regulated at the level of transcription was presented by Roggenkamp et al. (139) by the in vitro translation of *Hansenula polymorpha* mRNA. Analysis of the products from the in vitro translation of polyA$^+$-mRNA in rabbit reticulocyte lysates from repressed (2% glucose), derepressed (0.5% glucose) and induced cells (1% methanol) revealed the following: under derepressed conditions, MOX was shown to be the predominant protein, under induced conditions both MOX and DHAS were prominent, and under repressed conditions no polypeptides corresponding to MOX and DHAS could be detected. These data demonstrated that regulation of these genes was at the level of mRNA transcription, but that the regulation of methanol

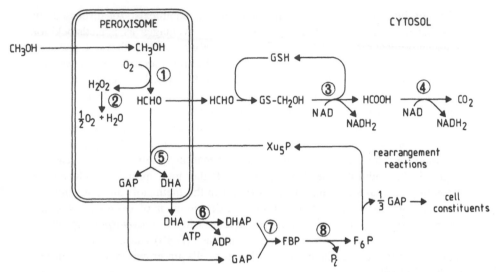

Figure 11. Compartmentalization of methanol metabolism in the methotrophic yeast, *H. polymorpha:* 1) alcohol oxidase; 2) catalase; 3) formaldehyde dehydrogenase; 4) formate dehydrogenase; 5) dihydroxyacetone synthase; 6) dihydroxyacetone kinase; 7) fructose 1,6-biphosphate aldolase; 8) fructose 1,6-biphosphate phosphatase. Abbreviations: GAP, glyceraldehyde 3-phosphate; DHA, dihydroxyacetone; DHAP, dihydroxyacetone phosphate; FBP, fructose 1,6-biphosphate; F6P, fructose 6-phosphate; Xu5P, xylulose 5-phosphate; GSH, reduced gluthathione. (From Douma et al., 1985, *Archives of Microbiology,* 143:242, courtesy of Dr. W. Harder.)

assimilation (DHAS) differed from its dissimilation (MOX). In addition, electrophoretic studies of in vivo and in vitro translation products (139,140) show that there is no detectable signal sequence for MOX's translocation into peroxisomes. These results were confirmed by protein sequence data derived from an analysis of the nucleotide sequence of the cloned gene, revealing that MOX is indeed not synthesized as a precursor (141).

The genes for both MOX and DHAS have been cloned and their nucleotide sequences determined (141,142). Since both of these proteins are found in abundance and are coordinately induced by growth in methanol, it would be expected that a comparison of regions flanking each structural gene would reveal common features. Indeed, this was found to be the case regarding the placement of the TATA box as well as several regions of dyad symmetry. The role of these features in the regulation of these genes will require a systematic analysis at the molecular level and in vitro modifications followed by reintroduction into the host organism by means of a transformation system. Two laboratories have reported the successful transformation of the yeast *Hansenula polymorpha*.

Three classes of plasmids have been constructed by Tikhomirova et al. (143) based on the source of replicator sequences, which were an ARS fragment from the *S. cerevisiae* 2-μm plasmid and mitrochondrial DNA (mtDNA) from *C. utilis*, chromosomal DNA from *H. polymorpha*, and ARSs from *H. polymorpha* mitochondria. The selectable marker in all cases was the LEU2 gene subcloned from the *S. cerevisiae* vector, YEp13. Foreign DNA was introduced into the host cell by standard protoplasting followed by treatment with a 20% PEG solution.

The plasmid pL2, containing an ARS sequence from *C. utilis* mtDNA, transformed *H. polymorpha* strain *leu356* (*leu2*) from *leu2* to *leu2*+ at 10³transformants per microgram of DNA. Autonomous replication of the plasmid was confirmed by Southern hybridization analysis and recovery into *E. coli*. Of interest was the observation that after replating several times there was a significant increase in the molecular weight of the plasmids suggesting polymerization, a phenomenon also reported for plasmids introduced by transformation into *S. pombe* (144) (see below). Plasmids containing *H. polymorpha* mitochondrial and chromosomal DNA were able to transform the host cell at a frequency of 5×10^2 transformants per microgram of DNA, a lower frequency than pL2; however, these plasmids were found to be more stable under nonselective growth conditions than pL2. The result of this work shows that plasmids that contain no *H. polymorpha* DNA in addition to those that do, not only can be used to introduce foreign DNA into the cell but can replicate antonomously. Of interest is the fact that owing to the composition of the plasmids that do not contain *H. polymorpha* DNA, chromosomal integration by homologous recombination is unlikely.

Hollenberg and co-workers (145) have developed a high-frequency transformation protocol based on complementation of a defined *H. polymorpha* (ura3) strain using the URA3 gene of *S. cerevisiae*. Two basic types of vectors have been developed: integration vectors based on the *S. cerevisiae* plasmid YIp5, and autonomously replicating plasmids containing the ARS1 from the *S. cerevisiae* plasmid YRp17 or similar sequences from *H. polymorpha*.

Appropriate mutants of *H. polymorpha* were selected for growth on 5-fluoroorotic acid following EMS treatment and nystatin enrichment. Mutants in orotidine-5′-phosphate decarboxylase, corresponding to the *ura3* mutation in *S. cerevisiae*, were characterized by enzyme analysis and designated *odc1*. Two ARSs (HARS1 and HARS2) were isolated following transformation of *odc1* mutants with YIp5 containing *H. polymorpha* chromosomal DNA. Plasmids containing HARS1, HARS2, and ARS1 from *S. cerevisiae* (YRp17) were shown to replicate autonomously in *H. polymorpha* following transformation using the LiSO₄ method of Ito et al. (27). The frequency of transformation using HARS1 was approximately 10-fold greater than ARS1. Although ARS1 from *S. cerevisiae* functioned in *H. polymorpha*, neither HARS1 nor HARS2 permitted autonomous replication in *S. cerevisiae*. However, some cloned sequences of *H. polymorpha* DNA were shown to function as ARSs in *S. cerevisiae* but did not have ARS properties in *H. polymorpha*.

Because of the lack of detectable homology between the *S. cerevisiae* URA3 and the corresponding *H. polymorpha* gene, chromosomal integration occurred at a low frequency. Selection for integration of the URA3 gene into the chromosome was performed by enriching for stable ura^+ transformants. This was accomplished by growth under nonselective conditions for 100 generations and isolating the resulting ura^+ yeast. Characterization of the ura^+ clones by Southern hybridization and assaying levels of plasmid encoding β-lactamase showed a range in the copy number of the integrated plasmid from a single copy of YIp5 to the integration of 75 copies of an HARS2-based plasmid. This is in contrast to the copy numbers of the plasmids containing ARSs; ARS1 (YRp17), HARS1, and HARS2 were shown to be 5, 40, and 25 copies per cell, respectively. These results are quite interesting since they suggest that gene integration can be employed not only for stable expression, but also to achieve significant levels of amplification of homologous and heterologous genes.

B. *P. pastoris*

Work in the methotrophic yeast has progressed rapidly in the cloning of methanol-regulated genes, development of a transformation system, the construction of *lacZ* fusions useful in dissecting the molecular events associated with methanol metabolism, and the development of *P. pastoris* as a system for the large-scale production of heterologous proteins on an industrial scale.

Alcohol oxidase (AOX) and two regulatable genes, p40 and p76, were cloned from cDNA libraries prepared from mRNA isolated from cells grown on methanol (146). Although the function of p40 is unknown, p76 has been identified as dihydroxyacetone synthetase (DHAS). The libraries were probed with polyA$^+$-mRNA isolateld from *P. pastoris* grown on methanol or ethanol. The clones were identified by analysis of protein products from the in vitro translation of mRNA that hybridized to the cDNA clones. Alcohol oxidase was identified by the direct comparison of the amino acid sequences derived from the nucleotide sequences of the gene and the first 18 amino acids of the amino terminus of the purified enzyme. Of interest is the observation that a comparison of the protein sequence derived from the cDNA clone and that of peroxisomal packaged AOX indicates the absence of a signal peptide, as is the case with MOX from *H. polymorpha*.

Studies of gene regulation indicate that p40 mRNA is made constitutively, but amplified in the presence of methanol. The mRNA for p40 is also present in cells grown on ethanol. AOX and DHAS are strongly regulated by methanol, and their mRNA is not detected in cells grown on ethanol. To further pursue an analysis of the regulatory region of AO and DHAS, gene fusions with the β-galactosidase gene of *E. coli* were made and expression studies conducted. The goal of these studies included the identification of regulatory regions, an analysis of the mechanism of regulation, and a comparison of the relative strength of the AO and DHAS promoters. The resulting data provide an indication of the utility of this system for the expression of foreign genes and an insight into understanding the mechanism of methanol regulation (147).

To ensure stable expression of DHAS-*lac*Z and AOX-*lac*Z fusions, plasmids containing these inserts were integrated into the *P. pastoris* chromosome at the HIS4 locus. The shifting of cells containing the AOX-*lac*Z fusion from growth on glucose (repressed) to carbon starvation (depressed) showed a significant increase in the levels of AO and *lac*Z. β-Galactosidase reached from 3–4% of the maximal level seen in methanol. As has been reported for *S. cerevisiae* (148), the levels of the heterologous protein (*lac*Z) were ~5 times lower than the native protein, AOX, in both carbon-starved and methanol-grown cells. However, *lac*Z expression is clearly under control of the AOX regulatory region. The DHAS-*lac*Z fusion is repressed in glucose-grown cells but expressed at high levels in methanol-grown cells. The DHAS-*lac*Z fusion is not activated in response to carbon starvation. The high levels of AOX- and DHAS-driven expression of *lac*Z occur only in response to methanol, suggesting that an induction component such as an activator protein is involved.

These data exhibit initial steps in the development of a commercially viable gene expression system in a non-*Saccharomyces* yeast. Recently, β-galactosidase expression was found to exceed 10% of the soluble protein of *P. pastoris* using modifications of the AOX and DHAS *lac*Z fusion constructions. This is comparable to several *S. cerevisiae* expression systems with the added advantage of "tight" regulation. The scale-up process would entail growth of *P. pastoris* on an inexpensive substrate, cells would be allowed to deplete their carbon source, and methanol would then be added to induce the appropriate promoter.

A transformation system for *P. pastoris* has been developed by Cregg and co-workers (149) and is based on the use of the HIS4 gene from *P. pastoris* and a modification of the protoplast transformation procedure. The HIS4 gene was isolated by complementation in *S. cerevisiae* on a 6.0-kb fragment and was used in turn to complement a well-defined *his*4 mutant of *P. pastoris*. Recombinant plasmids composed of the *P. pastoris* HIS4 gene in the *S. cerevisiae* plasmid *yep*13 transformed the strain *P. pastoris* GS115 (*his*4) at a high frequency. Plasmids could be recovered in *E. coli* following transformation with total yeast DNA from cells grown for approximately 10 generations. This result and detailed Southern hybridization analysis indicated the plasmids were able to replicate autonomously. However, in all cases after 25 generations, plasmids had integrated into the host chromosome. More detailed analysis localized the *his*4 gene to a 2.7-kb Sau3A fragment. This smaller fragment was used in the construction of other vectors.

Two efficient ARSs were isolated by cloning *P. pastoris* DNA into the HIS4-based plasmids and isolating transformants. Plasmids from these were isolated in *E. coli* and reintroduced into *P. pastoris*. Those that exhibited higher transformation efficiencies were chosen for further analysis. In order to demonstrate that the *Pichia* ARSs (PARS) were confined to subfragments and not dependent on sequences in the original plasmids, two unique vectors were constructed. These vectors, pYM4 and pYM3 (Fig. 12) contained the HIS4 genes from either *Pichia pastoris* or *S. cerevisiae*. *Taq*I fragments containing either PARS1 (Fig. 13A) or PARS2 (Fig. 13B) were inserted into the *Cla*I cloning site of both pYM4 and pYM3. It was demonstrated that either PARS fragment confered ARS activity when the plasmids were transformed into the *P. pastoris his*4 strain, GS115. It is interesting to note that these two PARSs were sequenced (Fig. 13A,B) and shown to have significant homology to the *S. cerevisiae* ARS consensus sequence, but did not function in *S. cerevisiae* despite this homology (Fig. 13C). In addition the 9.4-kb fragment containing the *S. cerevisiae* HIS4 gene was found to contain an ARS functional in *P. pastoris*, an activity that does not occur in *S. cerevisiae*. Thus it was observed that DNA fragments that did have ARS activity in *S. cerevisiae* did not have ARS activity in *P. pastoris*, and vice versa. Figure 13 shows the structural features of PARS1 and PARS2; the homologous inverted repeat sequences shown in Figure 13C, present in both, may represent a *Pichia pastoris* ARS consensus sequence.

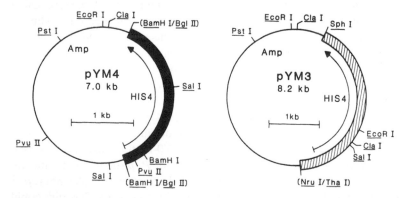

Figure 12. Restriction enzyme maps of pYM4 and pYM3. The plasmids are composed of pBR322 sequences (thin lines) and either a 2.7-kb fragment which contains the *P. pastoria* HIS4 gene (thick line) or a 3.8-kb fragment which contains the *S. cerevisiae* HIS4 gene (hatched line). Amp, ampicillin. (From 149, courtesy of Dr. J. M. Cregg.)

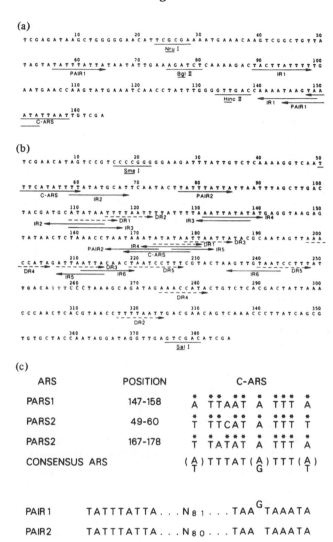

Figure 13. (a) DNA sequence of the 164-bp *Taq*I fragment which contains PARS1. (b) DNA sequence of the 385-bp *Taq*I fragment which contains PARS2. The position and direction of sequences with homology to the *S. cerevisiae* consensus ARS (C-ARS), of the PARS inverted repeat (PAIR) sequences, and of other direct repeat (DR) and inverted repeat (IR) sequences are denoted by the arrows. (c) DNA sequences of interest from PARS1 and PARS2. *P. pastoris* nucleotide positions from the PARSs which match the *S. cerevisiae* consensus ARS are denoted with asterisks. PAIR1 and PAIR2 are the PARS inverted repeat sequences from PARS1 and PARS2, respectively. (From 149, courtesy of Dr. J. M. Cregg.)

One interesting observation in this study was the high frequency of chromosomal integration of the *P. pastoris*–based vectors. The plasmids containing PARS2 transformed the host strain of *P. pastoris* at a high frequency, replicated autonomously for 10 to 25 generations before integrating. Plasmids having *S. cerevisiae* ARSs and the *P. pastoris* HIS4 transformed at a low frequency and also had a high level of chromosomal integration. It was demonstrated that *P. pastoris* plasmids that contain sequences homologous to the *P. pastoris* genome integrate at a frequency much higher than is seen in *S. cerevisiae*. In addition, the *P. pastoris* strains containing ARS-based plasmids grow at a higher rate than comparable *S. cerevisiae* strains with a rate of loss less than that seen for *S. cerevisiae* strains when cells are grown under nonselective conditions. The investigators suggest that these phenomena can be explained by the higher homologous recombination frequency in *P. pastoris*. Plasmids with homology to the *P. pastoris* genome will integrate frequently.

Cregg et al. suggested the construction of a vector that does not contain *P. pastoris* DNA. Such a vector could consist of the 9.4-kb *Pst*1 fragment containing the HIS4 gene from *S. cerevisiae*; not only does this gene function in *P. pastoris,* it also contains a PARS. The efficiency of HIS4 expression is low, so the cell would be forced to maintain multiple copies of the gene (or a plasmid containing the gene) under selective conditions. This could provide a means for gene amplification perhaps increasing the levels of heterologous gene expression from that seen for the integration of single copies of the plasmid and heterologous gene.

Integration vectors have been used successfully in producing high levels of heterologous proteins in *P. pastoris* using the methanol-inducible expression system (200). A vector, pGS102 containing the SUC2 gene from *S. cerevisiae* under control of the AOX1 promoter of *P. pastoris*, was integrated into the HIS4 and AOX1 locus of an aox1 *his*4 mutant. Under continuous fermentation conditions in the presence of methanol, approximately 80–90% of the total secreted protein was found to be the SUC2 gene product, invertase. After 100 hr of growth, following induction, the enzyme was found at 2.5 g/liter of culture medium. The final cell density after 256 hr was 40 g dry weight per liter with the level of the secreted invertase still approximately 2.5 g/liter. Invertase is normally secreted into the periplasmic space of *S. cerevisiae* and is hyperglycosylated, with a molecular mass ranging from 100 to greater than 140 kDa. The deglycosylated form of the enzyme from *S. cerevisiae* is 56 kDa. The enzyme secreted into the medium by *P. pastoris* was found to be more homgeneous, with a molecular mass ranging from 85 to 90 kDa. This material was reduced to 58 kDa by treatment with Endo-H. Finding a yeast system that secretes high-molecular-mass proteins into the culture medium at significant levels is important since it will facilitate recovery of commercially valuable heterologous gene products. It is especially critical that the material be active and not contain large amounts of unwanted carbohydrate, which has been found to be highly antigenic.

Invertase has also been used as a dominant selectable marker in the transformation of *P. pastoris*. *P. pastoris* cannot use sucrose as a sole carbon source; thus a vector consisting of the SUC2 gene from *S. cerevisiae* and PARS1 was introduced into a *suc⁻* strain of *P. pastoris*. Transformants were allowed to regenerate on dextrose before being screened for a *suc⁺* phenotype on sucrose-containing medium. Even under selective conditions the introduced plasmids were highly unstable owing to cross-feeding since invertase is secreted by the transformants. Integrated transformants, however, were found to be completely stable for the *suc⁺* phenotype after 20 generations on selective and nonselective media (200).

The *P. pastoris* transformation and cloning system is the most advanced of the non-*Saccharomyces* gene expression systems having coupled significant developments in high cell density fermentation and the biochemistry and molecular biology of methyltrophic yeasts. Recently, the production of commercial products was reported, using this system: hepatitis B suface antigen, epidermal growth factor, streptokinase and tissue plaminogen activator (201, 209).

V. *RHODOSPORIDIUM TORULOIDES*

The major interest in the basidiomycete, *R. toruloides,* is its ability to utilize phenylalanine as a source of carbon, nitrogen, and energy. The enzyme, phenylalanine ammonium lyase (PAL), catalyzes the initial step in phenylalanine catabolism. PAL is a valuable industrial enzyme used in the synthesis of 2-phenylalanine from *trans*-cinnamic acid (150) and medically has a potential role in the diagnosis and treatment of phenylketonuria (151).

Analysis of the proteins produced by in vitro translation of mRNA from *R. toruloides* grown under a variety of physiological conditions shows that glucose, ammonia, and phenylalanine regulate PAL synthesis by adjusting the level of functional mRNA. This regulation occurs at the level of transcription (152). The gene encoding PAL was cloned from a gene bank of *R. toruloides* DNA screened using cDNA synthesized from partially purified PAL mRNA (153). In addition to using the gene to characterize the biochemical properties of the enzyme, it was also used as a selectable marker for the development of a cloning system in the organism (154).

Two sets of vectors were constructed with the goal of developing a transformation system in *R. toruloides* that contained the PAL gene by itself or the PAL gene and the LEU2 gene from *S. cerevisiae*. *R. toruloides* was transformed using a modified protoplasting procedure using crude cell wall-degrading extracts from *P. lilacinus*. A *pal⁻* strain of *R. toruloides* was used as the recipient in initial experiments with a plasmid (pHG2) containing PAL. Selection was for growth on phenylalanine and the resulting transformants characterized for stability and the presence of foreign DNA by Southern hybridization. Of the first 20 *pal⁺*transformants, 16 contained the PAL gene on an autonomously replicating plasmid, whereas the remaining four had plasmid DNA integrated into their chromosomes. These observations were confirmed by the successful isolation of plasmids from the "unstable" yeast strains in *E. coli* confirming the presence of autonomously replicating plasmids. Plasmids were recovered from the stable transformants only after digestion with restriction enzymes followed by ligation; extracts of total DNA that was not digested with restriction enzymes did not result in successful transformation of *E. coli*. These studies suggested that an ARS was present in the PAL insert, which was subsequently demonstrated to be the case. The ARS was shown to be located downstream from the PAL gene.

In addition a vector was constructed by cloning the *S. cerevisiae* LEU2 into pHG2 to generate a new plasmid pHG8. The plasmid pHG8 was used to transform a *leu2, pal⁻* of *R. toruloides*. All of the resulting plasmids with either selection were found to be both *leu⁺* and *pal⁺*. Further experiments revealed that 2-μm-based plasmids containing the LEU2 gene were not able to transform *R. toruloides*.

VI. SCHIZOSACCHAROMYCES POMBE

The fission yeast *Schizosaccharomyces pombe* has been used by industry primarily for the production of single-cell protein from biomass. Recently *S. pombe* has received much attention for the expression of heterologous gene products and as a model system for the study of eukaryotic cell biology. This organism has been well characterized both genetically and physiologically and is thus well suited for molecular biology research. Although this yeast and *S. cerevisiae* are both members of the Saccharomycetaceae of the ascomycetous yeast, they differ significantly in aspects of cell division, morphology, and chromosome number (155). *S. pombe* reproduces through binary fission and contains three chromosomes, in contrast to *S. cerevisiae*'s budding form of reproduction and higher chromosome number (17). In addition, Sipiczki et al. (156) have demonstrated that their isoaccepting tRNA species have an average difference of 25%.

Since the first reported transformation of *S. pombe* in 1981 by Beach and Nurse (157), work on the biochemistry and molecular biology of *S. pombe* has increased dramatically, and with this increase have come additional data confirming differences between *Saccharomyces* and *Schizosaccharomyces* yeasts. *S. pombe* has been shown to be capable of excising intervening sequences (introns) from transcripts of higher eukaryotic genes (158). Studies of heterologous gene expression in *S. cerevisiae* have shown that this organism is incapable of performing this function (159,160). Cell cycle differences between the two organisms reveal that *S. pombe* very closely resembles higher eukaryotic cells in that there occur a distinct G2 phase and visible condensation of the three chromosomes during mitosis, whereas *S. cerevisiae* does not (161). RAS genes have been isolated from *S. pombe* and *S. cerevisiae*, and although both contain sequences homologous to mammalian RAS genes, they have different activities. The RAS genes from *S. cerevisiae* function in the activation of adenylate cyclase (162) whereas the RAS gene from *S. pombe* does not (163). Although there are many genetic and physiological differences between these two yeasts, the application of technology developed in *S. cerevisiae* has been successfully applied in the study of *S. pombe*.

Beach and Nurse (157) first reported the transformation of *S. pombe* in 1981. Their protocol has remained largely unchanged since that time. The *S. pombe* strain used in this initial study was *leu*1, which was complemented by a plasmid carrying the LEU2 gene (B-isopropylmalate dehydrogenase) of *S. cerevisiae*. The vector pJDB248, carrying the *S. cerevisiae* LEU2 gene, the 2-μm origin of replication, and antibiotic resistance genes from the bacterial plasmid pMB9, was used to transform the *S. pombe* mutant from *leu⁻* to *leu⁺* at a frequency of 10⁴ transformants per microgram of DNA. The plasmid pJDB248 is inherently unstable in *S. cerevisiae,* and the same was found to be true in *S. pombe*. The investigators transformed their *leu*1 host strain with a derivative of this plasmid where the pMB9 DNA had been replaced by sequences from pBR322 and the 2-μm origin of replication removed. This new plasmid, pDB248 (Fig. 14), was shown to transform *S. pombe* at the same frequency as the parental plasmid, and had the same level of instability. Despite the instability problem the plasmid pDB248 is a useful cloning vector since it contains several unique cloning sites including *Pst*I, *Hind*III, *Bam*HI, and *Hpa*1. In addition, the transformation efficiency suggested that a functional ARS was present in the remaining 2-μm plasmid DNA sequences.

To address the instability problem, several *S. cerevisiae* ARSs were tested in LEU2 containing plasmids which transformed *S. pombe* at a low frequency,

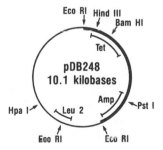

Figure 14. Plasmid map of pDB248. pDB248 contains pBR322 and the 4.4-kb *Eco*RI fragment of pJDB248. The *Eco*RI sites and unique *Pst*I, *Hind*III, *Bam*HI, and *Hpa*I sites are shown. Amp, region coding for ampicillin resistance; Tet, region coding for tetracycline resistance. (Reprinted by permission from *Nature,* 290 (5802), p. 141, copyright © 1981, Macmillan Journals Limited [157].)

indicating that they were not recognized as functional ARSs. A genomic library was prepared from *S. pombe* consisting of partial *Pst*I digested DNA cloned into the plasmid, pBR325, containing the *S. cerevisiae* LEU2 gene. Several plasmids were isolated from this library and shown to have the ability to replicate in *S. pombe*. When cells containing these plasmids were grown in the absence of selective pressure the portion of *leu*+ cells in the cultures ranged from 35% to 95%. Despite this range of stabilities, nine of the plasmids were able to transform *S. pombe* at a 10-fold greater efficiency that the original 2-μm-based plasmids.

Yamamoto et al. (164) have cloned the gene coding for aspartate transcarbamylase from an *S. pombe* genomic library (generated by a partial *Hind*III digest) in pBR322, by complementation of an *E. coli* strain carrying a pyrB (*ura*-) mutation. Two plasmids, pFYM2 and pFYM3, were isolated from *ura*+ transformants. Restriction mapping and subsequent retransformation of subcloned fragments revealed that two contiguous *Hind*III fragments of 2.7 and 1.5 kb were responsible for the complementing activity. The fragments were shown to be from *S. pombe* by Southern hybridization analysis and contained sequences that functioned as promoters in *E. coli*. The *Hind*III fragments containing the pyrB complementating activity were cloned into pDB248 and used to transform *leu*1-, *ura*1, *leu*1- double, and *ura*5 mutants of *S. pombe*. No *ura*5 complementation was observed, whereas the *leu*1 *ura*1 strain was transformed at a high frequency, thus identifying the specific enzyme activity encoded by the *S. pombe* URA1 gene. Sakaguchi and Yamamoto (165) also transformed the *leu*1 *ura*1 strain with a number of derivatives of the original pFYM plasmids which did not contain sequences having known ARS activity in *S. pombe*. All of the plasmids that contained only the URA1 gene transformed *S. pombe* with a high efficiency, suggesting that an ARS was located within or near the URA1 gene. Autonomous replication was demonstrated by Southern hybridization analysis of the *ura*+ transformants, and interestingly, several of the plasmids that were shown to be episomal had lost all of their pBR322 sequences.

Of significance was the finding that these plasmids were present in polymeric form primarily as decamers; in fact the investigators could not find any monomer or dimers. The polymers were believed to have been formed by some mechanism that promoted reciprocal recombination. This observation was complicated by the comigration of the polymeric forms with genomic DNA on agarose gels. Thus Southern hybridization analysis of intact genomic DNA initially suggested that the plasmids had integrated into the genome in multiple copies. However, detailed restriction enzyme digests and Southern hybridization analysis demonstrated that these DNAs were indeed episomal in nature.

The polymerization of DNA transformed into the cell has been used to develop a high-frequency cotransformation system for introducing nonreplicating circular DNA into *S. pombe* (144). This system requires that the *ura*1 complementing plasmids (in this case pFYM2 and pFYM225) provide not only a marker gene but also an origin of replication. The nonreplicating DNA must be on a vector that has sequences in common with the complementing plasmid. Since virtually all of the commonly used cloning vectors have the same *E. coli* sequences, practically any plasmid containing the desired DNA can be used. Sakai and co-workers (144) show that cotransformation takes place very efficiently and that plasmids efficiently recombine at sites of common DNA sequence forming heteropolymers. This method obviates the necessity of subcloning DNA fragments and permits introducing potentially large DNA molecules into *S. pombe.* One can speculate as to the reason for such a high frequency of recombination. Sakai et al. (144) argue that the reason for polymer formation is directly related to efficient plasmid replication. For example, a plasmid with a sequence that has a probability of being propagated 20% of the time will be lost from a population of cells in a few generations if it is in monomeric form. However, if that plasmid is converted to a tandem 20mer, the number of functional replication origins will increase proportionally and the intact molecule will complete a round of replication with a nearly fivefold higher probability at each cell cycle. Thus an intact plasmid will be available to be transferred to each daughter cell.

In addition to circular plasmids, *S. pombe* has been successfully transformed with linear plasmids (166). These molecules consist of the 2-μm origin of replication, the *S. cerevisiae* LEU2 gene, and teleomeres from *Tetrahymena pyriformi.* Plasmids are maintained under selective pressure as autonomously replicating elements as demonstrated by Southern hybridization analysis at an average of 85 copies per cell. This number reflects the necessity of *S. pombe* to have multiple copies of the inefficiently expressed LEU2 gene. Although evidence of recombination was found in some of the transformants, the extensive polymerization reported by Sakaguchi and Yamamoto (165) was not observed.

New vectors have been reported both for integrating DNA directly into the *S. pombe* chromosome and for autonomous replication (167). These include a series of shuttle vectors that are functional in *E. coli, S. cerevisiae,* and *S. pombe.* The original vector used to transform *S. pombe* was pDB248, which is not well suited for many recombinant DNA manipulations such as the construction of gene libraries. Wright and co-workers (167) have designed a new series of vectors more suitable for the isolation of genes and the introduction of foreign DNA into *S. pombe.* These vectors are shown in Figure 15. The vectors designed for autonomous replication consist of modified versions of the parental plasmid pDB262 (Fig. 15A). This plasmid was constructed from the *E. coli* vector TR262 by inserting a 5.2-kb *Hind*III-*Xho*I fragment from pDB248 (Fig. 14) containing the 2-μm ARS (functional in *S. pombe*) and the LEU2 gene from *S. cerevisiae* into a unique *Pvu*II site. Direct selection of DNA inserts is possible by cloning into restriction enzyme sites that inactivate the λ CI repressor gene product and allow expression of tetracycline resistance. The plasmid pDB262 transforms *S. pombe* with a frequency up to 3×10^4 transformants per microgram of purified plasmid

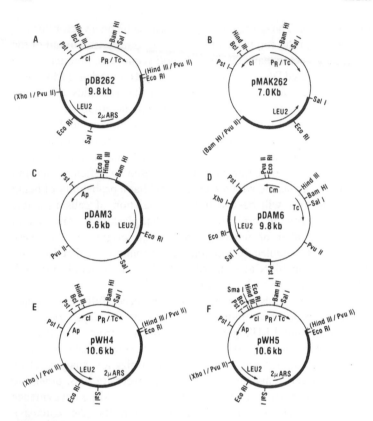

Figure 15. Restrictioin maps of *S. pombe* plasmid vectors. The thick lines show the fragments that were inserted into preexisting vectors (thin lines): key sites are shown. The gene designations are as follows: cI is bacteriophage λ cI repressor gene; P_R/Tc is tetracycline resistance gene under control of the λ P_R promoter, Tc is tetracycline resistance gene; Ap is ampicillin resistance gene; Cm is chloramphenicol resistance gene; LEU is LEU2 gene from *S. cerevisiae*; 2μ ARS is *S. cerevisiae* 2μ plasmid origin of replication. The unique cloning sites in pDB262 are *Hind*III and *Bcl*1 (which can accept *Bam*HI, *Bgl*II, *Sau*3A, *Mbo*I, and *Xho*II cohesive ends) (167).

DNA, similar to pDB248. This plasmid was modified by restoring the β-lactamase gene and several unique cloning sites to produce pWH4 (Fig. 15E) and pWH5 (Fig. 15F). Removal of the 5-kb SalI fragment containing the 2-μm ARS results in a vector that can integrate into the host chromosome. Additional integrating vectors pDAM3 and pDAM6 are also shown (Fig. 15C,D). The site of integration dependents on the particular sequences cloned into the vectors. The plasmid pMAK262 (Fig. 15B) has been used for the isolation of ARSs, and the integrating vectors have been used to map the ADE1 locus. The availability of vectors with useful cloning sites will be of great value; characterization of these vectors is ongoing.

Wright et al. (168) have isolated ARSs from S. pombe using the plasmid pDAM3 which consists of the S. cerevisiae LEU2 gene in the plasmid pBR322. Transforming S. pombe with this construct, the investigators found that the plasmid integrated into the host cell's chromosome, and yet they were able to recover intact plasmids. It was postulated that since the plasmid did not initially contain any S. pombe sequences, subsequently through some nonhomologous integration event followed by aberrant excision the plasmid had removed flanking chromosomal sequences which fortuitously contained an ARS function, thus permitting extrachromosomal replication. The presence of S. pombe sequences was shown by Southern hybridization analysis. These DNA sequences increased transformation efficiencies from less than 10 transformants per microgram of DNA to more than 10,000. The isolation of such sequences and their obvious utility in transformation systems will greatly enhance research efforts in S. pombe.

Though still in the developmental stages, there are several examples of the application of recombinant DNA technology to the improvement of S. pombe and the understanding of its molecular biology. Toda et al. (169) have been able to map ribosomal RNA (rRNA) genes to chromosome III of S. pombe by constructing vectors containing RNA genes and a selectable marker. DNA fragments containing sequences coding the rRNA genes were subcloned into YIp33, which has the S. cerevisiae LEU2 gene but does not have an origin of replication for S. pombe. Figure 16 shows the structure of the YIp plasmids containing the single-unit repeat of the S. pombe rDNA in addition to three subfragments. These plasmids were used to transform a leu1 strain of S. pombe by the method of Beach and Nurse (157), and leu+ auxotrophs were selected. Two of the plasmids, YIp 10.4 and YIp 7.3, were shown to contain an ARS functional in S. pombe. Both of these plasmids were unstable when grown in nonselective medium. Only stable transformants grown under nonselective conditions were assayed for integration of the plasmid into the host chromosome. Since the LEU2 gene from S. cerevisiae is not homologous with the same gene in S. pombe, it was expected and demonstrated that integration had occurred at the rRNA loci. The integration site was localized to chromosome III by the linkage of LEU2 and the chromosomal marker ADE5.

Ueng et al. (170) recently isolated the xylose isomerase gene from E. coli and

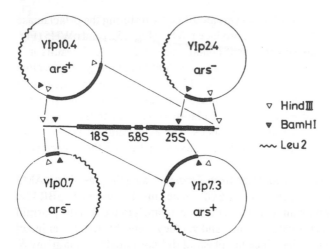

Figure 16. Structures of YIp plasmids containing *S. pombe* rDNA sequences. The 10.4-kb repeat unit of rDNA is represented by the middle line. The positions of 18S, 5.8S, and 25S rRNA genes are indicated. Restriction sites of *Hind*III, *Bam*HI, and *Hind*III–*Bam*HI in the YIp33 vector were employed for the cloning of 10.4-, 7.3-, and 2.4- or 0.7-kb rDNAs, respectively. Inserted rDNA sequences are shown by the thick lines. (From 169, courtesy of Dr. T. Toda.)

expressed the gene in *S. pombe*. Xylose is the major degradation product of both the enzymatic and acid hydrolysis of cellulose. *S. pombe* has a high ethanol tolerance and a rapid fermentation rate with a high end-product yield. The goal of introducing xylose isomerase into *S. pombe* was to obtain an organism that could easily ferment xylose to ethanol. Plasmids containing the gene were successfully introduced into *S. pombe* and expressed. Several transformants showed a fivefold greater yield of xylose converted to ethanol than the untransformed parental organism. The above summarizes some of the initial efforts at applying recombinant DNA technology to the analysis of the biochemistry and physiology of *S. pombe* for both basic research and industrial applications.

VII. YARROWIA (FORMERLY *SACCHAROMYCOPSIS*) *LIPOLYTICA*

The yeast *Y. lipolytica* has been used in the industrial production of 1,2-hexadecanedial from hexadecane (8), citric acid from decane (8), SCP, and biotin (171). Industrial interest has stimulated considerable work in the genetics of this yeast in attempts to elucidate a variety of biosynthetic pathways. Recently Bassel and Mortimer have reported the isolation of mutants that demonstrate that n-alkane uptake is inducible (172). *Y. lipolytica* is also capable of the efficient secretion of a number of enzymes that are of commercial interest including acid proteases and an exocellular alkaline protease.

One approach to increase the metabolic range of this yeast has been through the generation of intergenic hybrids by protoplast fusion of *Y. lipolytica* with *K. lactis.* In an extensive study, Groves and Oliver (17) were forced to use physiological and molecular biological analyses in the characterization of hybrids because of the nature of the resulting intergenic fusions. The investigators found that several of the fusants had carbon assimilation patterns intermediate between the two parents with one fusant, for example, able to use galactose, glucose, and n-decane. All of the fusants had growth rates considerably lower than the parental strains, and although uninucleate, the *Y. lipolytica* parent made the largest contribution to the fusant's genome as determined by DNA buoyant densities. In this rigorous study the investigators report that they were successful in achieving complementation between the two parental strains and demonstrated that the preferential recruitment of genes from one strain is more selective in these types of experiments than previously thought. It is clear that undesirable traits are incorporated along with desirable ones, and it may be possible that such traits can be removed by selective breeding programs. Groves and Oliver (17) argue that recombinant DNA systems may not work in all cases in strain improvement programs since it may be necessary to transfer many enzymes involved in a particular pathway in order to successfully modify the host cell's physiological profile.

There are reports of the successful cloning of the ribosomal genes (173) from *Yarrowia lipolytica* and preliminary reports of the cloning of URA3 (174,175), the alkaline extracellular protease (176), and genes from the biotin and adenine biosynthetic pathway (173). The ribosomal genes from *Y. lipolytica* were cloned from genomic DNA libraries constructed in the *E. coli* vector pBR322 (173). The libraries were probed with DNA fragments from clones of the ribosomal coding region from *S. cerevisiae*. Of interest is the observation that the pattern of ribosomal RNA gene organization is the same as that of the fission yeast *S. pombe* and suggest that *Y. lipolytica* (as is the case with *S. pombe*) is more closely related to the filamentous fungi than other yeast.

Davidow et al. (175) have been successful in developing an integrative transformation system of *Y. lipolytica*. They successfully cloned the LEU2 gene from *Y. lipolytica* by complementation of a *leu*B6 strain of *E. coli*. The source of the DNA was confirmed to be *Y. lipolytica* by Southern hybridization analysis. The initial plasmid constructed, pLD25, contains a 6.6-kb fragment (Fig. 17) derived from *Y. lipolytica* and contains the complementing activity. A second plasmid was constructed by subcloning a 2.3-kb *Eco*RI fragment into pBR322. This plasmid, pLD40 (Fig. 16), was found to possess the *leu*2 complementing activity as demonstrated by transforming a *leu*B strain of *E. coli*. Variations of the lithium

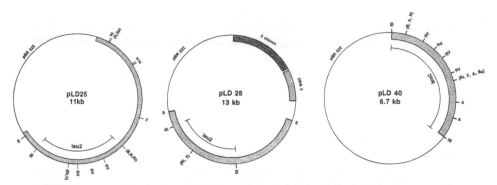

Figure 17. Restriction map of *Y. lipolytica* LEU2-containing plasmids, pLD25. A partial restriction map based on single and some double restriction digests is shown. Sites enclosed in parentheses have not been ordered with respect to each other. All sizes indicated are approximations based on agarose gel analysis. The thin line represents pBR322 DNA (in the standard format with the *Eco*RI site at 12 o'clock), and the thick line represents insert DNA from *Y. lipolytica*. The insert has 3 *Eco*RV (RV), 8 *Ava*I (A—but only one has been mapped) 1 *Sph*I (Sp.), 1 *Kpn*I (K), 2 *Sal*I (S), 4*Eco*RI (RI), 2 *Xho*I (X), and 2 closely spaced *Bgl*II (B) sites. The insert lacks *Hind*III, *Cla*I, *Bam*HI, and *Nru*I sites. The region marked *LEU2* is subcloned in pLD40. The restriction sites in pBR322 are not shown. pLD28: In addition to a 5.3-kb *Y. lipolytica Sal*I piece containing the *LEU2* region, this plasmid contains a 2.2-kb *Eco*RI segment with the origin of replication from the yeast 2-μm plasmid and a 1.1-kb *Hind*III piece with the *S. cerevisiae* URA3 gene inserted into the corresponding sites of pBR322. pLD40: The restriction site abbreviations are as previously indicated. In addition, sites for *Nco*I (N), *Apa*I (Ap), and *Bst*XI (Bs) have been mapped. The *Y. lipolytica* LEU2-containing segment is between 2.3 and 2.4 kb long and was inserted into the *Eco*RI site of pBR322. (From 175, courtesy of Dr. L. S. Davidow.)

acetate transformation procedure were used to transfer the plasmids into a *leu*2–35 *Y. lipolytica* strain. The modifications of the protocol that resulted in higher levels of transformation included (1) using linearized plasmids; (2) the cells were harvested from cultures in later stages of growth than early log; and (3) carrier DNA was added. Transformation efficiencies of greater than 10^4/μg of DNA were achieved using these modifications. All of the resulting transformants were found to be leucine-independent and stable when grown in the absence of selective pressure, suggesting integration. That the entire plasmid had integrated into the host chromosome was confirmed by Southern hybridization analysis and by the observation that plasmids could be rescued only from genomic DNA cleaved with the appropriate restriction enzymes, ligated, and transformed into *E. coli*.

Plasmid pL28 (Fig. 17) consisting of pLD25, the 2-μm plasmid origin of replication, and the URA3 gene from *S. cerevisiae*, was constructed to test complementation of URA3 in a *Y. lipolytica* strain ATCC 20688 (MATa *leu*2–3 *ura*3–11) and LEU2 from *Y. lipolytica* in *S. cerevisiae.* Selection on a uracil-deficient medium yielded no *Y. lipolytica* transformants, whereas the LEU2 gene of *Y. lipolytica* did complement an *S. cerevisiae, leu*1 strain, though weakly. To date attempts to isolate ARSs from this yeast have been unsuccessful (177).

The ability to shuttle the LEU2 plasmids between the transformed host and *E. coli* with recovery of the intact plasmid indicate the efficient insertion of the entire plasmid into the yeast chromosome. With the availability of several cloned genes (174,175), a protocol for efficient transformation, and several characterized host strains, the development of this species should proceed rapidly. Integrating vectors with the capacity to be shuttled into *E. coli* can be used efficiently in complementation studies and for the recovery of specific genes.

VIII.   *ZYGOSACCHAROMYCES* SP.*

Members of this genus are osmotolerant halophilic yeasts and are commonly found in the sediment of soft drinks and fruit juices. They are common spoilage organisms of foods with high sugar content, such as jams and syrups (178). One species, Z. rouxii, is used in Japan for the manufacturing of soy sauce and miso (179).

A. *Z. bailii* and *Z. bisporus*

Toh-e et al. (180) have reported the isolation of two sets of plasmids from *Z. bailii* and *Z. bisporus* that are quite similar in their structure to the 2-μm plasmid of *S. cerevisiae.* The plasmids have been designated isomers pSB1 and pSB2 from *Z. bailii* and isomers pSB3 and pSB4 from *Z. bisporus.* The plasmids have several characteristics in common with the 2-μm plasmid of *S. cerevisiae:* they are each approximately 6 kb in size, possess a pair of inverted repeats, exist in two isomeric forms, and have a similar copy number. Of interest is the observation that these plasmids do not hybridize with 2-μm plasmids or with each other.

The plasmid pSB3 has been sequenced (181). An ARS sequence consisting of 168 bp was located contiguous to the inverted repeat region. This ARS was demonstrated to function in *S. cerevisiae* despite the fact that it bears no homology with the *S. cerevisiae* ARS consensus sequence. The plasmid, pSB3, is capable of stable replication in *Z. rouxii* as well as in *Z. bisporus.* Investigators were able to demonstrate that the plasmid can function in an *S. cerevisiae leu*⁻ strain when carrying LEU2 but owing to inefficient partitioning is not maintained stably. The plasmid requires the presence of two *trans*-acting genes (B and C) in addition to a *cis*-acting site that receives signals from gene products B and C—a situation identical to the *S. cerevisiae* 2-μm plasmid. A direct comparison of the DNA

---

*The yeasts discussed in this section were formerly in the genus *Saccharomyces* and in a recent revision of yeast taxonomy were transfered to *Zygosaccharomyces (185).*

sequences from the two plasmids shows sequence homology at three locations of 12 bp each.

B. *Z. rouxii*

Another 2-μm-like plasmid has been isolated from *Z. rouxii* (179), and its nucleotide sequence has been determined (182). The plasmid, designated isomers pSR1 and pSR2, is 6,251 bp in size, with a pair of identical inverted repeats 959 bp in length that are separated by two unique sequences of 2,654 and 1,679 bp. Each of the inverted repeats contains an ARS. This plasmid hybridizes with pSB4 from *Z. bisporus* but does not hybridize with the 2-μm plasmid from *S. cerevisiae*.despite this difference, a chimeric plasmid composed of pSR1 and carrying the HIS3 gene of *S. cerevisiae* was capable of replication in *S. cerevisiae*(182). These plasmids will provide materials for development of a gene-cloning system in this genus. Members of this genus have characteristics that make them attractive for the development of novel fermentation processes and/or sources of enzymes for novel strain constructions through intergenic fusion (183). These characteristics include growth at temperatures ranging from 35 to 45°C, an ability to utilize cellobiose (184), and osmotolerance.

IX. *PACHYSOLEN TANNOPHILUS*

Pachysolen tannophilus is the first yeast reported to be capable of directly fermenting pentoses (e.g., D-xylose) to ethanol (186,188). Pentoses, primarily D-xylose, are the major products of hemicellulose degradation and are a potential substrate for the large-scale production of ethanol. In batch culture Slininger et al. (187) have shown *P. tannophilus* growing on 50 g/liter D-xylose to produce 0.34 g ethanol per gram of D-xylose consumed, representing a theoretical yield of 66%. Development of this microorganism for industrial purposes will require strain improvement in several areas. Although *P. tannophilus* efficiently converts D-xylose to ethanol, Slininger et al. (189) have shown that ethanol production is adversely affected when D-xylose concentrations exceed 50 g/liter and that D-xylose uptake may be inhibited when ethanol exceeds 20 g/liter. Strain improvement is hampered by the fact that the organism is homothallic, and thus improvement via classical genetic techniques will be very difficult.

New strains have been developed using continuous culture methods by selecting for increased growth rates (190). These strains did not, however, show improvement in ethanol production. Studies concerning the biochemical nature of D-xylose fermentation have shown that the enzymes NADPH linked D-xylose reductase and NAD-linked xylitol dehydrogenase activities in *P. tannophilus* are induced by D-xylose, L-arabinose, and D-galactose (191,192). Kurtzman (193) has suggested that the pathway for xylose assimilation and fermentation should be elucidated with the goal of transferring some of these genes by recombinant DNA techniques to more appropriate hosts—i.e., strains of yeast that are more ethanol-tolerant. He suggests that *C. utilis* would be an appropriate host since this yeast is already able to utilize xylose as a carbon source.

Recently the gene encoding ornithine carbamoyltransferase (OTCase) has been cloned by Maleszka and Skrzypck (194,195) from a *Pachysolen* genomic library by complementation of an *E. coli argB* mutation and *arg3* in *S. cerevisiae*. OTCase is a central enzyme in the biosynthesis of arginine in microorganisms. The goal of this work is to utilize OTCase as a selectable marker in the isolation of genes involved in the assimilation and fermentation of D-xylose and ultimately the transferral of these genes to the more ethanol-tolerant yeast *S. cerevisiae*. An interesting observation in the expression of *P. tannophilus* OTCase in *S. cerevisiae* centers around the subcellular localization of this enzyme. OTCase in the yeast *P. tannophilus* is associated with the mitochondria while in *S. cerevisiae* it is found exclusively in the cytosol. However, when *P. tannophilus* OTCase was expressed in *S. cerevisiae*, the enzyme was localized in the mitochondria. In addition to providing researchers with a selectable marker, OTCase derived from *P. tannophilus* may also prove useful in the study of protein targeting in yeasts. Additional requirements for the utilization of *P. tannophilus* are the same as those for other yeasts—i.e., the development of a transformation system and the isolation of functional ARSs.

X. SCHWANNIOMYCES OCCIDENTALIS

There are many yeasts capable of utilizing starch as a carbon source and, as discussed earlier, have been used in associative fermentations with *S. cerevisiae* to avoid excessive conversion costs. Considerable savings could be realized by the use of an amylolytic yeast that is capable of the efficient fermentation of a wide range of carbon sources. *Schwanniomyces occidentalis* has been shown to be efficient in the conversion of inexpensive carbon compounds such as starch and inulin to ethanol by means of a number of amylolytic and degradative enzymes. The facile conversion of biomass to ethanol as well as other attractive physiological properties has prompted Ingeldew in an exhaustive review of the biochemistry and physiology of the genus *Schwanniomyces* to refer to *Schwanniomyces* as a potential "superyeast" (203). Yeasts of this genus have been shown to produce ethanol at 86% of the theoretical yield in associative fermentations (203 and references therein). Starch to ethanol conversion efficiencies of 95% have also been reported (203 and references therein). *Schwanniomyces* spp. have shown promise in intergenic fusions with nonamylolytic yeasts in a number of strain improvement programs with the goal of producing organisms for the direct conversion of starch to ethanol and in biomass conversion to SCP (203 and references therein). Genetic studies of this organism have been limited to preliminary reports on the isolation of auxotrophic and ethanol tolerant mutants.

Work in our laboratory has focused on *Schwanniomyces occidentalis* as a potentially efficient expression system for the large-scale commercial production of heterologous gene products and as a model for protein secretion and export through the cell wall. The organism has many attractive properties in this regard: It is able to secrete proteins greater than 140,000 Da efficiently into the culture medium (203,204), it does not hyperglycosylate, (204) and it does not produce measurable amounts of secreted proteases (204). Moreover, it is capable of growth on inexpensive substrates (203 and references therein).

The isolation of specific genes from *Schwanniomyces* by complementation in *S. cerevisiae* has been successful. The gene encoding orotidine-5′-phosphate decarboxylase (ODC) was isolated by Klein and Roof (205) from a partial Sau3A genomic library cloned into the *S. cerevisiae* expression vector pYcDE8. The gene was localized to a 1.1-kbp fragment which was found to complement a number of *ura3* strains of *S. cerevisiae* as efficiently as the homologous URA3 gene. These results suggest that the ODC promoter from *Schwanniomyces* is fully functional in *S. cerevisiae*.

An extremely efficient transformation system has also been developed for *Schwanniomyces occidentalis* by Klein and Favreau (206–208) using auxotrophic mutants generated by UV and EMS mutagenesis. One of the systems is based on yeasts deficient in the adenine biosynthesis pathway—namely, phosphoribosyla-minoimidazole-carboxylase. As a selectable marker the gene encoding this enzyme was isolated by complementation in *S. cerevisiae* using a partial Sau3A genomic library in pYcDE8. The resulting plasmid, pADE, was then used to successfully transform *ade* strains of *Schwanniomyces* by modifications of the spheroplasting protocol of Beggs (20). The fact that pADE and other plasmid derivatives were maintained as extrachromosomal elements was demonstrated by Southern hybridization analysis of intact total DNA isolated from the *ade*+ transformants (Fig. 18) and the recovery of plasmids from *E. coli* transformed with total DNA preparations from the same transformants. As was the case for *C. albicans* and *S. pombe*, the extrachromosomal DNA was found to be of high molecular mass, composed of tandem repeats of the introduced vector. The hybridizing material shown in Figure 18 is reduced to a single species, that comigrates with linearized pADE, when treated with restriction enzymes that cleave pADE once in the *S. cerevisiae* vector sequences.

In addition to selectable markers, an ARS sequence was also found, and a number of useful cloning vectors were constructed. These range from vectors for integrating foreign DNA into the host genome to extrachromosomal elements maintained in high copy number. Of interest was the observation that all of these vectors transformed spheroplasted *Schwanniomyces* at an efficiency of >2 × 10³

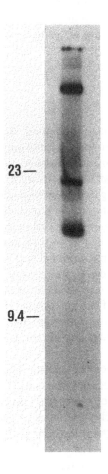

Figure 18. Autoradiogram of Southern hybridization analysis of total DNA isolated from a *Schwanniomyces ade* mutant transformed with the plasmid pADE. The intact DNA was fractionated in a 0.7% agarose gel, electrophoresed for 70 hr at 30 V, transferred to nitrocellulose, and probed with ^{32}P-labeled pYcDE8. Previous work (205) has shown that pYcDE8 does not hybridize to *Schwanniomyces* DNA under the conditions used in this study. The molecular size markers in kilobase pairs (kbp) generated by a *Hind*III digest of bacteriophage λ DNA are shown. Intact genomic DNA as identified by ethidium bromide staining and hybridization with known *Schwanniomyces* sequences migrates just above the 23-kbp marker. The hybridizing extrachromosomal DNAs are reduced to a single homogenous species upon treatment with restriction enzymes that cleave the *S. cerevisiae* cloning vector pYcDE8 only once.

transformants per microgram of plasmid DNA. The availability of a highly efficient cloning system represents a significant step in the commercial and academic exploitation of this organism, ranging from the production of commercially valuable proteins to the enhancement of the fermentation efficiencies of members of this genus.

XI. CONCLUSION

Recent developments in recombinant DNA technology have permitted the exploration of many underutilized yeast species. Interest in these species is not only for an expansion of industrial exploitation, but also as potential models for addressing many questions involving our fundamental understanding of eukaryotic cell biology. Cloning and transformation systems that are being developed in such yeasts as *Candida* and *Schizosaccharomyces* will accelerate progress in our study of pathogenesis, in the case of *C. albicans,* and our understanding of cell cycle regulation and protein kinases, in the case of *S. pombe.* Because of its well-known genetics and physiology, *Saccharomyces cerevisiae* has become a critical tool in the isolation and characterization of genes from other yeasts through complementation. We have cited numerous examples of the use of recombinant DNA technology to expand our knowledge of the fundamentals of the physiology of many industrially important yeasts. A clear example is reflected by the advances that have been made in the genus *Pichia* and methanol catabolism via alcohol oxidase.

We can expect the volume of reports on the development of cloning and transformation systems in the non-*Saccharomyces* yeasts to increase dramatically during the next decade as the tools of molecular biology become more commonplace in the laboratory. This expansion will be quickly followed by an increase in our understanding of the biochemistry of these organisms. Attempts to find viable alternatives to current host/vector gene expression systems for industrial scale production will continue, as will the quest for more representative models for the cell biologist.

The cloning and transformation systems in the non-*Saccharomyces* yeasts in conjunction with classical genetics, and protoplast fusion, provide powerful keys to unlock the secretes of an extremely important portion of the eukaryotic world.

ACKNOWLEDGMENTS

We would like to extend our thanks to our colleagues who contributed materials to this review, to Paula Allred of The Upjohn Company's Corporate Technical Library for her invaluable assistance, and R. Hiebsch for Figures 1, 2, and 3. We would especially like to thank Laurie Tuinstra and Kathy Hiestand for their patience, effort, and skill in the preparation of this manuscript, and Drs. J. Cregg and W. M. Ingledew for their comments and suggestions.

REFERENCES

1. Case, M. E., Transformation of *Neurospora crassa* utilizing plasmid DNA, in *Genetic Engineering of Microorganisms for Chemicals,* A. Hollaender, D. DeMoss, S. Kaplan, J. Konisky, D. Savage, and R. S. Wolfe (Eds.). Plenum Press, New York, pp. 87–100 (1981).
2. Upshall, A., Filamentous fungi in biotechnology, *Biotechniques,* 4:158–166 (1986).
3. Timberlake, W. E. (Ed.), *Molecular Genetics of Filmentous Fungi.* Alan R. Liss, New York (1985).
4. Stewart, G. G., I. Russell, R. D., Klein, and R. R. Hiebsch (Eds.), *Biological Research on Industrial Yeasts,* Vols. I, II, and III. CRC Press, Boca Raton, FL (1987).
5. Korhola, M., and E. Vaisanen (Eds.). *Gene Expression in Yeast.* Foundation for Biotechnical and Industrial Fermentation Research, Helsinki (1985).
6. Phaff, H., Biology of yeasts other than Saccharomyces, in *Biology of Industrial Microorganisms,* A. L. Demain and N. A. Solomon (Eds.). Benjamin/Cummings, Menlo Park, CA, pp. 537–562 (1985).
7. De Mot, R., and H. Verachtert, Biocatalysis and biotechnology with yeasts, *ASM News,* 50:526–531 (1984).
8. Abbott, B. J., and W. E. Gledhill, The extracellular accumulation of metabolic products by hydrocarbon-degrading microorganisms, in *Advances in Applied Microbiology,* D. Perlman (Ed.). Academic Press, New York, pp. 249–389 (1971).
9. Klyosov, A. A., Enzymatic conversion of cellulosic materials to sugars and alcohols, *Appl. Biochem. Biotechnol.,* 12:249–300 (1986).
10. Linko, P., Fuels and industrial chemicals through biotechnology, 2, *Biotechnol. Adv.,* 3:39–63 (1985).
11. Ratledge, C. R., Biotechnology as applied to the oils and fats industry, *Feete Seifen Anstrichmitted,* 86:379–389 (1984).
12. Spencer, J. F. T., and D. M. Spencer, Genetic improvement of industrial yeast, *Annu. Rev. Microbiol.,* 37:121–142 (1983).
13. Elander, R. P., New genetic approaches to industrially important fungi, *Biotechnol. Bioeng.,* 22 (Suppl. 1):49–61 (1980).
14. Perberdy, J. F., and L. Ferenczy (Eds.). *Fungal Protoplasts: Application in Biotechnology and Genetics.* Marcel Dekker, New York (1985).
15. Perberdy, J. F., Fungal protoplasts, isolation, reversion and fusion, *Annu. Rev. Microbiol.,* 33:21–39 (1979).
16. Stewart, G. G., The genetic manipulation of industrial yeast strains, *Can. J. Micriobiol,* 27:973–990 (1981).
17. Groves, D. P., and S. G. Oliver, Formation of intergenic hybrids of yeast by protoplast fusion of *Yarrowia* and *Kluyveromyces* species, *Curr. Genet.,* 8:49–55 (1984).
18. Oppenoorth, W. F. F., Transformation in yeast: Evidence of a real genetic change by the action of DNA. *Nature,* 193:706 (1962).

19. Hinnen, A. B., J. B., Hicks, and G. R. Fink, Transformation of yeast, *Proc. Natl. Acad. Sci. USA,* 75:1929–1933 (1978).

20. Beggs, J. D., Transformation of yeast by a replicating hybrid plasmid, *Nature,*2 75:104–108 (1978).

21. Ferenczy, L., Protoplast fusion in yeast, in *Fungal Protoplasts: Applications in Biochemistry and Genetics,* J. F. Peberdy and L. Ferenczy (Eds.). Marcel Dekker, New York, pp. 279–306 (1985).

22. Yamada, T., and K. Sakaguchi, Polyethylene glycol induced uptake of bacteria into yeast protoplasts, *Agric. Biol. Chem.,* 45:2301–2309 (1981).

23. Ahn, J. S., and M. Y. Pack, Use of liposomes in transforming yeast cells, *Biotechnol. Lett.,* 7:553–556 (1985).

24. Kimura, A., and M. Morita, Fermentative formation of CDP-choline by intact cells of yeast: *Saccharomyces carlsbergensis* (IFO 0641) treated with a detergent. Triton X-100, *Agric. Biol. Chem.,* 39:1469–1474 (1975).

25. Iimura, Y., K. Gotoh, K. Ouchi, and T. Nishiya, Yeast transformation without the spheroplasting process, *Agric. Biol. Chem.,* 47:897–901 (1983).

26. Ito, H., Y. Fukuda, K. Morita, and A. Kimura, Transformation of intact yeast cells with alkali cations, *J. Bacteriol.,* 153:163–168 (1983).

27. Ito, H., K. Murata, and A. Kumra, Transformation of intact yeast cells treated with alkali cations or thiol compounds, *Agric. Biol. Chem.,* 48:341–347 (1984).

28. Klebe, R. J., J. V. Harriss, Z. D. Sharp, and M. G. Douglas, A general method for polyethylene-glycol-induced genetic transformation of bacteria and yeast, *Gene,* 25:333–341 (1983).

29. Broach, J. R., The role of site specific recombination and expression of the yeast plasmid 2 micron circle, in *Yeast Molecular Biology—Recombinant DNA: Recent Advances,* M. Esposito (Ed.). Noyes Publications, Park Ridge, NJ, pp. 93–112 (1982).

30. Williamson, D. H., The yeast ARS element, six years on: A progress report, *Yeast,*1:1–14 (1985).

31. Broach, J. R., and J. B. Hicks, Replication and recombination functions associated with the yeast 2 micron circle, *Cell,* 21:501–508 (1980).

32. Volkert, F. C., and J. R. Broach, The mechanism of propogation of the yeast 2μm circle plasmid, in *Biological Research on Industrial Yeast,* vol. 3, G. G. Stewart, I. Russell, R. D. Klein, and R. R. Hiebsch (Eds.). CRC Press, Boca Raton, FL, pp. 145–170 (1987).

33. Clark, L., and J. Carbon, Isolation of a yeast centromere and construction of functional small circular chromosomes, *Nature,* 287:504–509 (1980).

34. Murray, A. W., and J. W. Szostak, Construction of artificial chromosomes in yeast, *Nature,* 305:189–193 (1983).

35. Kearsey, S., Structural requirements for the function of a yeast chromosomal replicator, *Cell,* 37:299–307 (1984).

36. Chattoo, B., and F. Sherman, Selection of *lys*2 mutants of the yeast *Saccharomyces cerevisiae* by the utilization of α-aminoadipate, *Genetics,* 93:51–65 (1979).

37. Winston, F., D. T. Chaleff, B. Valent, and G. R. Fink, Mutations affecting TY-mediated expression of the HIS4 gene of *Saccharomyces cerevisiae, Genetics,* 107:179–197 (1984).

38. Boeke, J. D., F. la Croute, and G. R. Fink, A positive selection for mutants lacking

orotidine-5′-phosphate decarboxylase activity in yeast: 5-fluoro-orotic acid re-sistance, *Mol. Gen. Genet.,* 197:345–346 (1984).

39. Roggenkamp, R., H. Hansen, M. Eckart, Z. Janowicz, and C. P. Hollenberg, Transformation of methyltrophic yeast *Hansenula polymorpha* by autonomous replication and integration vectors, *Mol. Gen. Genet.,* 202:302–308 (1986).

40. McKnight, G. L., and B. L. McConaughy, Selection of functional cDNAs by complementation in yeast, *Proc. Natl. Acad. Sci. USA,* 80:4412–4416 (1983).

41. McKnight, G. L., H. Kato, A. Upshall, M. D. Parker, G. Saari, and P. J. O'Hara, Identification and molecular analysis of a third *Aspergillus nidulans* alcohol dehydrogenase gene, *EMBO J.,* 4:2093–2099 (1985).

42. Struhl, K., Direct selection for gene replacement events in yeast, *Gene,* 26:231–242 (1983).

43. Gritz, L., and J. Davis, Plasmid encoded hygromycin-B resistance: The sequence of the hygromycin B phosphotransferase gene and its expression in *E. coli* and *S. cerevisiae, Gene,* 25:179–188 (1983).

44. Jaminez, A., and J. Davies, Expression of a transposable antibiotic resistance element in *Saccharomyces, Nature,* 287:869–871 (1979).

45. Henderson, R. C., B. S. Cox, and R. Tubb, The transformation of brewing yeasts with a plasmid containing the gene for copper resistance, *Curr. Genet.,* 9:133–138 (1985).

46. Urdea, M. S., J. P. Merryweather, G. T. Mullenbach, et al., Chemical synthesis of a gene for human epidermal growth factor urogastrone and it expression in yeast, *Proc. Natl. Acad. Sci. USA,* 80:7461–7465 (1983).

47. Valenzuela, P., A. Medina, W. J. Rutter, G. Ammerer, and B. D. Hall, Synthesis and assembly of hepatitis B virus surface antigen particles in yeast, *Nature,* 298:347–350 (1982).

48. Miyanohara, A., A. Toh-e, C. Nozaki, F. Hamada, N. Othono, and K. Matsubara, Expression of hepatitis B surface antigen gene in yeast, *Proc. Natl. Acad. Sci. USA,,* 80:1–5 (1983).

49. Hitzeman, R. A., C. Y. Chen, F. E. Hagie, et al., Expression of hepatitis B virus surface antigen in yeast, *Nucleic Acids Res.,* 11:2745–2762 (1983).

50. Murray, K., S. A. Bruce, A. Hinnen, et al., Hepatitis B virus antigens made in microbial cells immunise against viral infection, *EMBO J.,* 3:645–650 (1984).

51. Mellor, J., M. J. Dobson, N. A. Roberts, et al., Efficient synthesis of enzymatically active calf chymosin in *Saccharomyces cerevisiae, Gene,* 24:1–14 (1983).

52. Goff, C. G., D. T. Moir, T. Kohno, et al., Expression of calf prochymosin in *Saccharomyces cerevisiae, Gene,* 27:35–46 (1984).

53. Bitter, G. A., and K. M. Egan, Expression of heterologous genes in *Saccharomyces cerevisiae* from vectors utilizing the glyceraldehyde-3-phosphate dehydrogenase gene promoter, *Gene,* 32:263–274 (1984).

54. Tuite, M. F., M. J. Dobson, N. A. Roberts, et al., Regulated high efficiency expression of human interferon-alpha in *Saccharomyces cerevisiae, EMBO, J.,* 1:603–608 (1982).

55. Dobson, M. J., M. F. Tuite, J. Mellor, et al., Expression in *Saccharomyces cerevisiae* of human interferon-alpha directed by the TRP1 5′ region, *Nucl. Acids Res.,* 11:2287–2302 (1983).

56. Gill, G. G., P. G. Zaworski, K. R. Marotti, et al., Expression of human tissue plasminogen activator (t-PA) in *Saccharomyces cerevisiae*, in *Biological Research on Industrial Yeasts, vol. 1.*, Stewart, G. G., I. Russell, R. D. Klein, and R. R. Hiebsch (Eds.), CRC Press, Boca Raton, FL, pp. 165–176 (1987).

57. MacKay, V. L., Secretion of heterologous proteins in yeast, in *Biological Research on Industrial Yeasts, Vol 2.*, Stewart, G. G., I.Russell, R. D. Klein, and R. R. Hiebsch (Eds.), CRC Press, Boca Raton, FL, pp. 27–36 (1987).

58. Zsebo, K. M., H.-S. Lu, J. C. Fieschko, et al., Protein secretion from *Saccharomyces cerevisiae* directed by the prepro-α-factor leader region, *J. Biol. Chem.*, 261:5858–5865 (1986).

59. Stepien, P. P., R. Brousseau, R. Wu, S. Narang, and D. Y. Thomas, Synthesis of a human insulin gene VI. Expression of the synthetic proinsulin gene in yeast, *Gene,* 24:289–297 (1983).

60. Vlasuk, G. P., G. H. Bencen, R. M. Scarborough, et al., Expressions and secretion of biologically active human atrial natriuretic peptide in *Saccharomyces cerevisiae, J. Biol. Chem.*, 261:4789–4796 (1986).

61. Derynck, R., A. Singh, and D. V. Goeddel, Expression of the human interferon-α cDNA in yeast, *Nucl. Acids Res.*, 11:1819–1837 (1983).

62. Barr, P. J., R. A., Hallewell, S. Rosenberg, and A. J. Brake, Active site modified protease α-1-antitrypsin inhibitors and their production, Eur. Pat. Appl. 85107126.6 (1985).

63. Struhl, K., W. Chen, D. E. Hill, I. A. Hope, and M. A. Oettinger, Constitutive and coordinately regulated transcription of yeast genes: Promoter elements, positive and negative regulatory sites, and DNA binding proteins, *Cold Spring Harbor Symp.*,50:489–504 (1985).

64. Broach, J. R., Y. Y. Li, L. C. C. Wu, and M. Jayaram, Vectors for high-level inducible expression of cloned genes in yeast, in *Experimental Manipulation of Gene Expression,* M. Inouye (Ed.). Academic Press, New York, pp. 83–117 (1983).

65. Guarente, L., Use of gene fusions to study biological problems, in *Genetic Engineering: Principles and Methods,* Vol. 6, J. K. Setlow and A. Hollander (Eds.). Plenum Press, New York, pp. 233–251 (1985).

66. Beardsmore, A. J., S. H. Collins, K. A. Powell, and P. J. Senior, Production and use of genetically modified microorganisms, Eur. Pat. Appl. 82302608.3 (1982).

67. Windlass, J. D., M. J. Worsey, E. M. Pioli, et al., Improved conversion of methanol to single-cell protein by *Methylophilus methylotrophus, Nature,* 287:396–401 (1980).

68. Hanley, S., and R. Yocum, Expression of the lactose utilization genes of *K. lactis* in *S. cerevisiae, Yeast,* 2:5148 (1986).

69. Santos, M. G. G., H. Aboutboul, J. B. Faria, et al., Genetic improvement of *Saccharomyces* for ethanol production from starch, *Yeast,* 5:511–516 (1989).

70. Barnett, J. A., The stability of biological nomenclature: Yeasts, *Nature,* 322:599 (1986).

71. Barnett, J. A., R. W. Payne, and D. Yarrow (Eds.), *Yeasts: Characteristics and Identification.* Cambridge University Press, London (1983).

72. Moreton, R. S., Growth of *Candida utilis* on enzymatically hydrolyzed potato waste,

J. Appl. Bacteriol, 44:373–382 (1978).

73. J. Ledere, M. B. Blondin, R. Ratomahenina, A. Arnaud, and P. Galzy, Selection and study of mutants of *Dekkera intermedia* and *Candida wickerhamii* derepressed for β-glucosidase production — useful for ethanol production from glucose and cellobiose, *FEMS Microbiol. Lett.*, 30:389–392 (1985).

74. Kohchi, C., and A. Toh-e, Nucleotide sequence of *Candida pelliculosa* β-glucosidase gene, *Nucl. Acids Res.*, 13:6273–6282 (1985).

75. Tubb, R. S., Amylolytic yeasts for commercial applications, *TIBTECH*, 4:98–104 (1986).

76. Skogman, H., The symba process, *Starch*, 28:278–282 (1976).

77. Solomons, G. L., Single cell protein, *CRC Crit. Rev. Biotechnol.*, 1:21–58 (1983).

78. Lucas, C., and N. van Uden, Transport of hemicellulose monomers in the xylose-fermenting yeast *Candida shehatae* — glucose, mannose, galatose and arabinose uptake facilitated diffusion system; potential ethanol producer, *Appl. Microbiol. Biotechnol.*, 23:491–495 (1986).

79. Johannsen, E., L. Eagle, and G. Bedenhann, Protoplast fusion used in the construction of presumptive polyploids of the D-xylose fermenting yeast *Candida shehatae*, *Curr. Genet.*, 9:313–319 (1985).

80. Guiraud, J. P., J. M. Caillaud, and P. Galzy, Optimization of alcohol production from Jerusalem artichokes, *Eur. J. Appl. Microbiol. Biotechnol.*, 14:81–85 (1982).

81. Moulin, G., and P. Galzy, Alcohol production from whey, in *Proceedings of the Sixth International Fermentation Symposium*, M. Moo-Young and C. W. Robbins (Eds.). Pergamon Press, Toronto, pp. 181–187 (1981).

82. Vandarnme, E. J., Penicillin acylases and β-lactamases, in *Microbial Enzymes and Bioconversions*, Vol. V. *Economic Microbiology*. Academic Press, New York, pp. 467–522 (1980).

83. Yonehara, T., and Y. Tani, Comparative studies on ATP production from adenosine by a methanol yeast, *Candida biodinii*, *Agric. Biol. Chem.*, 50:899–905 (1986).

84. Liu, Z. T., Z. H. Yi, K. Y. Li, H. Z. Hau, Y. M. Hia, and H. F. Fang, Studies on thermotolerant yeast producing alpha, omega dibasic acid from n-alkanes—*Candida tropicalis* mutagenesis, *Eur. Cong. Biotechnol.*, 3:71–76 (1984).

85. Saltzman, B., Fungal infections in immunocompromised patients, *Mediguide Oncol.*, 6:1–5 (1986).

86. Odds, R. C., Candida and *Candidiosis*, University Park Press, Baltimore (1979).

87. Hughes, J., D. Culver, J. White, et al., Nosocomial infection surveillance, 1980–82, *MMWR*, 32:1–16 (1983).

88. Fraser, D. W., J. I. Ward, L. Ajello, and B. D. Plikaytis, Aspergillosis and other systemic mycoses: The growing problem, *JAMA*, 242:1631–1635 (1979).

89. Hawkins, C., and D. Armstrong, Fungal infections in the immunocompromised host, *Clin. Haematol.*, 13:599–630 (1984).

90. Shepherd, M. G., R. T. M. Poulter, and P. A. Sullivan, *Candida albicans:* Biology, genetics and pathogencity, *Annu. Rev. Microbiol.*, 39:579–614 (1985).

91. Olaiya, A. F., and S. J. Sogin, Ploidy determination of *Candida albicans*, *J. Bacteriol*, 140:1043–1049 (1979).

92. Rigsby, W. S., L. J. Torres-Bauza, J. W. Wills, and T. M. Townes, DNA content, kinetic complexity and the ploidy question in *Candida albicans*, *Mol. Cell.*

*Biol.,*2:853–862 (1982).

93. Sarachek, A., D. D. Rhoads, and R. H. Schwarzhoff, Hybridization of *Candida albicans* through fusion of protoplasts, *Arch. Microbiol.,* 129:1–8 (1981).

94. Poulter, R., K. Jeffrey, M. H. Hubbard, M. G. Shepherd, and P. A. Sullivan, Parasexual genetic analysis of *Candida albicans, J. Bacteriol.,* 146:833–840 (1981).

95. Rosenbluh, A., M. Mevarech, Y. Kottin, and J. A. Gorman, Isolation of genes from *Candida albicans* by complementation in *Saccharomyces cerevisiae, Mol. Gen. Genet.,* 200:500–502 (1985).

96. Kurtz, M. B., M. W. Cortelyau, and D. R. Kirsch, Integrative transformation of *Candida albicans,* using a cloned *Candida* ADE2 gene, *Mol. Cell. Biol.,* 6:142–149 (1986).

97. Montes, L. F., and Wilborn, W. H., Fungus-host relationship in candidiosis, *Arch. Dematol.,* 121:119–124 (1985).

98. Meyer, R. D., Cutaneous and mucosal manifestations of the deep seated mycotic infections, *Acta Derm. Venerol. (Stockh.),* 121(Suppl.):57–72 (1986).

99. Kunze, G., C. Petzoldt, R. Bode, I. A. Samsonover, F. Bottcher, and D. Birnbaum, Transformation of the industrially important yeasts *Candida maltosa* and *Pichia guilliermondii, J. Basic Microbiol.,* 25:141–144 (1985).

100. Kunze, G., C. Petzoldt, R. Bode, I. A. Samsonova, M. Hecker, and D. Birnbaum, Transformation of *Candida maltosa* and *Pichia guilliermondii* by a plasmid containing *Saccharomyces cerevisiae* ARG4 DNA, *Curr. Genet.,* 9:205–209 (1985).

101. Kawamura, M., M. Takagi, and K. Yano, Cloning of a LEU gene and an ARS site of *Candida maltosa, Gene,* 24:157–162 (1983).

102. Takagi, M., S. Kawai, M. C. Chang, I. Shibuya, and K. Yano, Construction of a host-vector system in *Candida maltosa* by using an ARS site isolated from its genome, *J. Bacteriol.,* 167:51–555 (1986).

103. Kohchi, C., and A. Toh-e, Cloning of *Candida pelliculosa* β-glucosidase gene and its expression in *Saccharomyces cerevisiae, Mol. Gen. Genet.,* 203:89–99 (1986).

104. Freer, S. N., and R. W. Detroy, Characterization of cellobiose fermentations to ethanol by yeasts, *Biotechnol. Bioeng.,* 25:541–557 (1983).

105. Kawamoto, S., C. Nozaki, A. Tanaka, and S. Fukui, Fatty acid β-oxidation system in microbodies of n-alkane grown *Candida tropicalis, Eur. J. Biochem.,* 83:609–613 (1978).

106. Osumi, M., F. Fukuzumi, Y. Teranishi, A. Tanaka, and S. Fukui, Development of microbodies in *Candida tropicalis* during incubation in a n-alkane rich medium, *Arch. Microbiol.,* 103:1–11 (1975).

107. Tanaka, A., M. Osumi, and S. Fukui, Peroxisomes of alkane-grown yeast: Fundamental and applied aspects, *Ann. N.Y. Acad. Sci.,* 386:183–199 (1982).

108. Harder, W., and M. Veenhuis, Physiological significance and biogenesis of yeast microbodies, in *Biological Studies of Industrial Yeast, vol. 3,* G. G. Stewart, I. Russell, R. D. Klein, and R. R. Hiebsch (Eds.). CRC Press, Boca Raton, FL, pp. 125–158 (1987).

109. Kamiryo, T., and K. Okazaki, High level expression and molecular cloning of genes encoding *Candida tropicalis* peroxisomal proteins, *Mol. Cell. Biol.,* 4:2131–2141 (1984).

110. Lazarow, P. W., R. A. Rachubinski, and Y. Fujiki, Cloning of cDNAs cloding for

catalase and other peroxisomal proteins of *Candida tropicalis, J. Cell. Biochem.,*9e(Suppl.):120 (1985).

111. Ho, N. W. Y., H. C. Gao, J. J. Huang, P. E. Stevis, S. F. Chang, and G. T. Tsao, Development of a cloning system for *Candida* species, *Biotechnol. Bioeng. Symp.,*14:295–391 (1984).

112. Zheng, Y. Z., and C. A. Reddy, Cloning of a *Candida utilis* gene which complements *leu2* mutation in *S. cerevisiae, Curr. Genet.,* 10:573–578 (1986).

113. Hsu, W. H., P. T. Magee, B. B. Magee, and C. A. Reddy, Construction of a new yeast cloning vector containing autonomous replication sequences from *Candida utilis, J. Bacteriol.,* 154:1033–1039 (1983).

114. Hsu, W. H., and C. A. Reddy, Development of a simple generalized transformation system for yeasts utilizing Kan^r/G418 selection procedure,*Abstr. Annu. Mtg. Am. Soc. Microbiol.,* 143-H210 (1986).

115. Dickerson, R. C., and J. S. Markin, Molecular cloning and expression in *E. coli* of a yeast gene coding for β-galactosidase, *Cell,* 15:123–130 (1978).

116. Breuning, K. D., U. Dahlens, S. Das, and C. P. Hollenberg, Analysis of a eucaryotic β-galactosidase gene: The N-terminal end of the yeast *Kluyveromyces lactis* protein shows homology to the *Escherichia coli lacZ* gene product, *Nucl. Acids Res.,*12:2327–2341 (1984).

117. Chen, X. J., M. Saliola, C. Falcone, M. M. Bianchi, and H. Fukuhara, Sequence organization of the circular plasmid pKD1 from the yeast *Kluyveromyces drosophilarum, Nucl. Acids Res.,* 14:14471–14481 (1986).

118. Falcone, C., M. Saliola, X. J. Chen, L. Frontali, and H. Fukuhara, Analysis of a 1.6-μm circular plasmid from the yeast *Kluyveromyces drosophilarum:* Structural and molecular dimorphism, *Plasmid,* 15:248–252 (1986).

119. Das, S., and C. P. Hollenberg, A high-frequency transformation system for the yeast *Kluyveromyces lactis, Curr. Genet.,* 6:123–128 (1982).

120. Gunge, N., Linear DNA killer plasmids from the yeast *Kluyveromyces, Yeast,*2: 153–162 (1986).

121. Thomas, D. Y., T. Vernet, C. Boone, D. Greene, S. Lollie, and H. Bussey, The yeast killer toxin expression systems, functional analysis and biotechnological application, in *Biological Studies of Industrial Yeast, vol. 1,* G. G. Stewart, I. Russell, R. D. Klein, and R. R. Hiebsch (Eds.). CRC Press, Boca Raton, FL pp. 155–164 (1987).

122. Wesolowski, M., A. Algeri, P. Goffrini, and H. Fukuhara, Killer plasmids of the yeast *Kluyveromyces lactis.* I. Mutations affecting the killer phenotype, *Curr. Genet.,*5: 191–197 (1982).

123. Wesolowski, M., P. Dumazert, and H. Fukuhara, Killer plasmids of the yeast *Kluyveromyces lactis.* II. Restriction endonucleas maps, *Curr. Genet.,* 5:199–203 (1982).

124. Wesolowski, M., A. Algeri, and H. Fukuhara, Killer plasmids of the yeast *Kluyveromyces lactis.* III. Plasmid recombination, *Curr. Genet.,* 5:205–206 (1982).

125. De Louvencourt, H. Fukuhara, H. Heslot, and M. Wesolowski, Transformation of *Kluyveromyces lactis* by killer plasmid DNA, *J. Bacteriol.,* 154:737–742 (1983).

126. Thompson, A., and S. G. Oliver, Physical separation and functional interaction of *Kluyveromyces lactis* and *Sacharromyces cerevisiae* ARS elements derived from killer plasmid DNA, *Yeast,* 2:179–191 (1986).

127. Van der Walt, J. P., Genus *Kluyveromyces*, in *The Yeasts*, J. Lodder (Ed.). North Holland, Amsterdam, pp. 316–378 (1970).
128. Negoro, H., Inulase from *Kluyveromyces fragilis, J. Ferment. Technol.,* 56:102–107 (1978).
129. Farahnak, F., T. Seki, D. D. Y. Ryu, and D. Ogrydziak, Construction of lactose-assimilating and high-ethanol-producing yeasts by protoplast fusion, *J. Bacteriol.,* 51:362–367 (1986).
130. Das, S., E. Kellerman, and C. P. Hollenberg, Transformation of *Kluyveromyces fragilis, J. Bacteriol.,* 158:1165–1167 (1984).
131. Sugisaki, Y., N. Gunge, K. Sakaguchi, M. Yamasaki, and G. Tamura, Transfer of DNA killer plasmids from *Kluyveromyces* to *Kluyveromyces fragilis* and *Candida pseudotropicalis, J. Bacteriol.,* 164:1373–1375 (1985).
132. Swift, H. E., Fuels and chemicals from single carbon sources, *Am. Sci.,* 71:616–620 (1983).
133. Ushio, K., K. Inouye, K. Nakamura, S. Oka, and A. Ohno, Stereochemical control in microbiol reduction 4. Effect of cultivation conditions on the reduction of β-keto esters by methylotrophic yeast, *Tetrahedron Lett.,* 27:2657–2660 (1986).
134. Dijkhuizen, L., T. A. Hansen, and W. Harder, Methanol, a potential feedstock for biotechnological processes, *Trends Biol. Sci.,* 3:262–267 (1985).
135. Van Dijken, J. P., R. Otto, and W. Harder, Growth of *Hansenula polymorphia* in a methanol limited chemostat: Physiological response due to the involvement of methanol oxidase as a key enzyme in methanol metabolism, *Arch. Microbiol.,* 111:137–144 (1976).
136. Van Dijken, J. P., M. Veenhuis, N. J. W. van Rij Kreger, and W. Harder, Microbodies in methanol-assimulating yeasts, *Arch. Microbiol.,* 102:41–44 (1975).
137. Van Dijken, J. P., M. Veenhuis, and W. Harder, Peroxisomes of methanol-grown yeasts, *Ann. N.Y. Acad. Sci.,* 386:200–216 (1982).
138. Hazeu, W., S. C. deBruyn, and P. Bos, Methanol assimilation by yeasts, *Arch. Microbiol.,* 87:185–188 (1972).
139. Roggenkamp, R., Z. Janowicz, B. Stanikowski, and C. P. Hollenberg, Biosynthesis and regulation of peroxisomal methanol oxidase from the methylotrophic yeast *Hansenula polymorpha, Mol. Gen. Genet.,* 144:489–193 (1984).
140. Roa, M., and G. Blobel, Biosynthesis of peroxisomal enzymes in the methylotrophic yeast *Hansenula polymorpha, Proc. Natl. Acad. Sci. USA,* 80:6872–6876 (1983).
141. Ledeboer, A. M., L. Edens, J. Maat, et al., Molecular cloning and characterization of a gene coding for methanol oxidase in *Hasenula polymorpha, Nucl. Acids Res.,*13:3063–3082 (1985).
142. Janowicz, Z. A., M. R. Eckart, C. Drewke, et al., Cloning and characterization of the DAS gene encoding the major methanol assimilating enzyme from the methylotrophic yeast *Hansenula polymorpha, Nucl. Acids Res.,* 13:3043–3062 (1985).
143. Tikhomirova, L. P., R. N. Ikonomova, and E. N. Kuznetsova, Evidence for autonomous replication and stabilization of recombinant plasmids in the transformation of yeast *Hansenula polymorpha, Curr. Gen.,* 10:741–747 (1986).
144. Sakai, K., J. Sakaguchi, and M. Yamamoto, High-frequency cotransformation by copolymerization of plasmids in the fission yeast *Schizosaccharomyces pombe, Mol.*

Cell. Biol., 4:651–656 (1984).

145. Roggenkamp, R., H. Hansen, M. Eckart, Z. Janowicz, and C. P. Hollenberg, Transformation of methylotrophic yeast *Hansenula polymorpha* by autonomous replication and integration vectors, *Mol. Gen. Genet.,* 202:302–308 (1986).

146. Ellis, S. B., P. F. Brust, P. S. Koutz, A. F. Waters, M. M. Harpold, and T. R. Gingeras, The isolation of alcohol oxidase and two other methanol regulatable genes from the yeast, *Pichia pastoris, Mol. Cell. Biol.,* 5:1111–1121 (1985).

147. Tschopp, J. F., P. F. Burst, J. M. Cregg, C. A. Stillman, and T. R. Gingers, Expression of the *LacZ* gene from two methanol-regulated promoters in *Pichia pastoris, Nucl. Acids Res.* 15:3859–3876 (1987).

148. Chen, C. Y., H. Oppermann, and R. A. Hitzeman, Homologous versus heterologous gene expression in yeast, *Saccharomyces cerevisiae, Nucl. Acids Res.,* 12:8951–8970 (1984).

149. Cregg, J. M., K. J. Barringer, A. Y. Hessler, and K. R. Madden, *Pichia pastoris* as a host system for transformations, *Mol. Cell. Biol.,* 5:3376–3385 (1985).

150. Yamada, S., K. Kobe, N. Iuzo, K. Nakamichi, and I. Chibata, Production of L-phenylalanine from *trans*-cinnamic acid with *Rhodotorula glutinis* containing L-phenylalanine ammonia lyase activity, *Appl. Environ. Microbiol.,* 42:773–778 (1981).

151. Ambrus, C. M., J. L. Ambrus, C. Horvath, et al., Phenylalanine depletion for management of phenylketonuria: Use of enzyme reactors with immobilized enzymes, *Science,* 201:837–839 (1978).

152. Gilbert, H. J., J. R. Stephenson, and M. Tully, Control of synthesis of functional mRNA coding for phenylalanine ammonia lyase from *Rhodosporidium toruloides, J. Bacteriol.,* 153:1147–1154 (1983).

153. Gilbert, H. J., I. N. Clarke, R. K. Gibson, J. R. Stephenson, and M. Tully, Molecular cloning of the phenylalanine ammonia lyase gene from *Rhodosporidium toruloides* in *E. coli* K-12, *J. Bacteriol.,* 161:314–320 (1985).

154. Tully, M., and H. J. Gilbert, Transformation of *Rhodosporidium toruloides, Gene,* 36:235–240 (1985).

155. Gyax, A., and P. Thuriaux. A revised chromosome map of the fission yeast *Schizosaccharomyces pombe, Curr. Genet.,* 8:85–92 (1984).

156. Sipiczki, M., W. D. Heyer, and J. Kohli, Preparation and regeneration of protoplasts and spheroplasts for fusion and transformation of *Schizosaccharomyces pombe, Curr. Microbiol.,* 12:169–174 (1985).

157. Beach, D., and P. Nurse, High-frequency transformation of the fission yeast *Schizosaccharomyces pombe, Nature,* 290:140–142 (1981).

158. Kaufer, N. F., V. Simanis, and P. Nurse, Fission yeast *Schizosaccharomyces pombe* correctly excises a mammalian RNA transcript intervening sequence, *Nature,* 318:78–80 (1985).

159. Watts, F., C. Castle, and J. D. Beggs, Aberrant splicing of *Drosophila* alcohol dehydrogenase transcripts in *Saccharomyces cerevisiae, EMBO J.,* 2:2085–2091 (1983).

160. Langford, C. J., J. Nellen, J. Niessing, and D. Gallwitz, Yeast is unable to excise foreign intervening sequences from hybrid gene transcripts, *Proc. Natl. Acad. Sci. USA,* 80:1496–1500 (1983).

161. Umesono, K., Y. Hiraoka, T. Toda, and Yanogida, Visualization of chromosomes in mitotically arrested cells of the fission yeast *Schizosaccharomyces pombe, Curr. Genet.*, 7:123–128 (1983).

162. Toda, T., I. Uno, T. Ishikawa, et al., In yeast, RAS proteins are controlling elements of adenylate cyclase, *Cell*, 40:27–36 (1985).

163. Fukui, Y., T. Kozasa, Y. Kaziro, T. Takeda, and M. Yamamoto, Role of ras homology in the life cycle of *Schizosaccharomyces pombe, Cell*, 44:329–336 (1986).

164. Yamamoto, M., A. Yoshihiro, and M. Yanagida, Cloning of a gene from the fission yeast *S. pombe* which complements *E. coli* pyrB, the gene for aspartate transcarbamylase, *Mol. Gen. Genet.*, 182:426–429 (1981).

165. Sakaguchi, J., and M. Yamamoto, Cloned *ura*1 locus of *Schizosaccharomyces pombe* propagates autonomously in this yeast assuming a polymeric form, *Proc. Natl. Sci. USA*, 79:7819–7823 (1982).

166. Guerrini, A. M., F. Ascenzioni, C. Tribioli, and P. Donini, Transformation of *Saccharomyces cerevisiae* and *Schizosaccharomyces pombe* with linear plasmids containing 2μ sequences, *EMBO J.*, 4:1569–1573 (1985).

167. Wright, A., K. Maundrell, W-D. Heyer, D. Beach, and P. Nurse, Vectors for the construction of gene banks and the integration of cloned genes in *Schizosaccharomyces pombe* and *Saccharomyces cerevisiae, Plasmid*, 15:156–158 (1986).

168. Wright, A., K. Maundrell, and S. Shall, Transformation of *Schizosaccharomyces pombe* by non-homologous, unstable integration of plasmids into the genome, *Curr. Genet.*, 10:503–508 (1986).

169. Toda, T., Y. Nakaseko, O. Niwa, and M. Yanagida, Mapping of rNA genes by integration of hybrid plasmids in *Schizosaccharomyces pombe, Curr. Genet.*, 8:93–97 (1984).

170. Ueng, P. P., K. J. Volpp, J. V. Tecker, C. S. Gong, and L. F. Chen, Molecular cloning of the *Escherichia coli* gene encoding xylose isomerase, *Biotechnol. Lett.*, 7:153–158 (1985).

171. Pearson, B. M., D. A. McKenzie, and M. H. J. Keenan, Production of biotin by yeasts—e.g. *Sporobalomyces roseus, Rhodotorula rubrum* and *Yarrowia lipolytica:* screening using biotin free culture media, *Lett. Appl. Microbiol.*, 2:25–28 (1986).

172. Bassel, J. B., and R. K. Mortimer, Identification of mutants preventing n-hexadecane uptake among 26 alkane non-utilizing mutants of *Yarrowis (Saccharomycopsis) lipolytica, J. Am. Oil. Chem. Soc.*, 63:462 (1986).

173. Clare, J. J., L. S. Davidow, D. C. J. Gardner, and S. G. Oliver, Cloning and characterization of the ribosomal RNA gene of the dimorphic yeast, *Yarrowia lipolytica, Curr. Genet.*, 10:449–452 (1986).

174. Davidow, L. S., D. Apostolakos, I. Stasko, and J. R. DeZeeuw, Cloning genes by complementation from an integrating vector library in the yeast *Yarrowia lipolytica* using the plasmid pLD40, *Abstr. 13th Int. Conf. Yeast Genetics and Molecular Biology*, 63 (1986).

175. Davidow, L. S., D. Apostolakos, M. M. O-Donnell, et al., Integrative transformation of the yeast *Yarrowia lipolytica, Curr. Genet.*, 10:39–48 (1985).

176. Davidow, L. S., M. M. O'Donnell, F. S. Kaczmarck, et al. Cloning of the alkaline

extracellular protease gene from *Yarrowia lipolytica, J. Bacteriol, 169*:4621–4629 (1987).

177. Wing, R. A., and D. M. Ogrydziak, Development of the genetics of the dimorphic yeast *Yarrowia lipolytica* — autonomous replicating sequence attempted isolation, *UCLA Symp. Mol. Cell. Biol.,* 34:367–381 (1985).

178. Pitt, J. I., Resistance of some food spoilage yeasts to preservation, *Food Technol. Aust.,* 26:238–241 (1974).

179. Toh-e, A., S. Tada, and Y. Oshima, 2-μm DNA-like plasmids in the osmophilic haploid yeast *Saccharomyces rouxii, J. Bacteriol.,* 151:1380–1390 (1982).

180. Toh-e, A., H. Araki, I. Utatsu, and Y. Oshima, Plasmids resembling 2-μm DNA in osmotolerant yeasts *Saccharomyces bailii* and *Saccharomyces bisporus, J. Gen. Microbiol.,* 130:2527–2534.

181. Toh-e, A., and I. Utatsu, Physical and functional structure of a yeast plasmid, pSB3, isolated from *Zygosaccharomyces bisporus, Nucl. Acids Res.,* 13:4267–4283 (1985).

182. Araki, H., A. Jearnpipatkul, H. Tatsumi, et al., Molecular and functional organization of yeast plasmid pSR1, *J. Mol. Biol.,* 182:191–203 (1985).

183. Pina, A., I. L. Calderin, and T. Benitez, Intergenic hybrids of *Saccharomyces cerevisiae* and *Zygosaccharomyces fermentati,* obtained by protoplasts fusion, *Appl. Environ. Microbiol.,* 51:995–1003 (1986).

184. Lodder, J. (Eds.), *The Yeasts: A Taxonomic Study,* North-Holland, Amsterdam, pp. 555–1083 (1970).

185. Yarrows, D., Genus 33. *Zygosaccharomyces* Barker, in *The Yeasts: A Taxonomic Study,* N. J. W. Krueger–van Rij (Ed.), third edition, Elsevier, Groningen, Holland, pp. 449–465 (1984).

186. Schneider, H., P. Y. Wang, Y. K. Chang, and R. Malezka, Conversion of D-xylose into ethanol by the yeast *Pachysolen tannophilus, Biotechnol. Lett.,* 3:89–92 (1981).

187. Slininger, P. J., R. J. Bothast, J. E. Cauwenberge, and C. P. Kurtzman, Conversion of D-xylose to ethanol by the yeast *Pachysolen tannophilus, Biotechnol. Bioeng.,* 24:371–384 (1982).

188. Slininger, P. J., R. J. Bothast, M. R. Okos, and M. R. Ladisch, Comparative evaluation of ethanol production by xylose-fermenting yeast presented high xylose concentrations, *Biotechnol. Lett.,* 7:431–436 (1985).

189. Slininger, P. J., R. J. Bothast, L. T. Black, and J. E. McGhee, Continuous conversion of D-xylose to ethanol by immobolized *Pachysolen tannophilus, Biotechnol. Bioeng.,* 24:2241–2251 (1982).

190. Bolen, P. L., and P. J. Slininger, Continuous culture selection of mutant strains of *Pachysolen tannophilus* capable of rapid aerobic growth on D-xylose, *Dev. Indust. Microbiol.,* 25:449–457 (1984).

191. Smiley, K. L., and P. L. Bolen, Demonstration of D-xylose reductase and D-xylitol dehydrogenase in *Pachysolen tannophilus, Biotechnol. Lett.,* 4:607–610 (1982).

192. Bolen, P. L., and R. W. Detroy, Induction of NADPH-linked D-xylose reductase and NAD-linked Xylitol dehydrogenase activities in *Pachysolen tannophilus* by D-xylose, L-arbinose or D-galactose, *Biotech. Bioeng.,* 27:302–307 (1985).

193. Kurtzman, C. P., Biology and physiology of the D-xylose fermenting yeast *Pachysolen tannophilus,* in *Advances in Biochemical Engineering/Biotechnology,* Vol. 27, A. Fiechter (Ed.). Springer-Verlag, Berlin, pp. 73–83 (1983).

194. Malesyka, R., and M. Skrzypek, Cloning of the ornithine carbonyl-transferase gene from *Pachysolen tannophilus* and its expression in *S. cerevisiae, Abstr. 13th Int. Conf. on Yeast Genetics and Molecular Biology,* 5224 (1986).

195. Malesyka, R., Personal communication.

196. Gillum, A. M., E. Y. H. Tsay, and D. R. Kirsch, Isolation of *Candida albicans* gene for orotidine-5'-phosphate decarboxylase by complementation in *S. cerevisiae ura3* and *E. coli* pyrf mutations, *Mol. Gen. Genet.,* 198:179–182 (1984).

197. Kelly, R., S. M. Miller, M. B. Kurtz, and D. R. Kirsch, Directed mutagenesis in *Candida albicans:* One-step gene disruption to isolate *ura3* mutants, *Mol. Cell. Biol.,* 7:199–207 (1987).

198. Kurtz, M. B., M. W. Cortelyou, S. M. Miller, M. Lai, and D. R. Kirsch, Development of autonomously replicating plasmids for *Candida albicans, Mol. Cell. Biol.,* 7:209–217 (1987).

199. Gleeson, M. A., and P. E. Sudbery, The methylotrophic yeasts, *Yeast* 4:1–15 (1988).

200. Tschopp, J. F., G. Sverlow, R. Kosson, W. Craig, and L. Grinna, High-level secretion of glycosylated invertase in the methylotrophic yeast, *Pichia pastoris, Bio/Tech.,* 5:1305–1308 (1987).

201. Cregg, J. M., J. F. Tschopp, C. Stillman, et al., High-level expression and efficient assembly of hepatitis B surface antigen in the methylotrophic yeast, *Pichia pastoria, Bio/Tech.,* 5:479–485 (1987).

202. A. C. Douma, M. Veenhuis, W. deKoning, et al., Dihydroxyacctone synthesis is localized in the peroxisomal matix of methanol grown *Hansenula polymorpha, Arch. Microbiol, 143*:237–243 (1985).

203. Ingledew, W. M., *Schwanniomyces:* A potential superyeast?, *CRC Rev. Biotechnol.,* 5:159–176 (1987).

204. Deibel, M. R. Jr., R. R. Hiebsch, and R. D. Klein, Secreted amylolytic enzymes from *Schwanniomyces occidentalis:* Purification by fast protein liquid chromatography (FPLC) and preliminary characterization, *Prep. Biochem.,* 18(1):77–120 (1988).

205. Klein, R. D., and L. L. Roof, Cloning of the orotidine 5'-phosphate decarboxylase (ODC) gene of *Schwanniomyces occidentalis* by complementation of the *ura3* mutation in *S. cerevisiae, Curr. Genet.,* 13:29–35 (1988).

206. Favreau, M. A., and R. D. Klein, Transformation of *Schwanniomyces occidentalis* with an ADE gene cloned from *Schwanniomyces, XVIth International Congress of Genetics,* Toronto, Canada (1988).

207. Klein, R. D., and M. A. Favreau, Plasmid vectors for transformation of *Schwanniomyces occidentalis, XIVth International Conf. Yeast Genetics and Molecular Biology,*Espoo, Finland (1988).

208. Klein, R. D., and M. A. Favreau, Transformation of *Schwanniomyces occidentalis* with an ADE gene cloned from *Schwanniomyces* (submitted, 1988).

209. Rathner, M., Protein expression in yeasts, *Bio/Tech.,* 7:1129–1133 (1989).

10

Host Cell Control of Heterologous Protein Production in *Saccharomyces cerevisiae*

Rathin C. Das* and Douglas A. Campbell

Miles Inc.
Elkhart, Indiana

I. INTRODUCTION

Efforts to achieve heterologous protein production in yeasts have largely centered on manipulations of plasmids carrying foreign genes of interest. Genetic alterations of host cells to enhance productivity have received far less attention. This imbalance may reflect the very real challenges of addressing by classical genetic means and molecular methods what is likely to be a complex, multigene situation. Nonetheless, a start has been made toward refining the role of yeast hosts in plasmid-directed heterologous protein synthesis.

Laboratory strains of the yeast *Saccharomyces cerevisiae* amenable to transformation by plasmid DNAs often carry properties that render them unsuitable as production hosts. These can include high frequencies of *petite* segregation, poor metabolic vigor in both synthetic and complex media, and limited carbohydrate utilization repertoires. Some of these deficiencies can be eliminated by judicious choice of initial strains and by the application of standard genetic procedures. We wish here, however, to address aspects of heterologous protein expression in which the host cell might well play a significant role. These are host cell control of plasmid maintenance (copy number and stability), and host cell control of

---

**Present affiliation*: Miles Research Center, West Haven, Connecticut.

heterologous protein synthesis and secretion. The first half of the chapter deals with host cell modifications. Indeed, host cell modifications that enhance either or both copy numbers and stability of heterologous genes could well lead to increased production of heterologous gene product. The several properties within these two categories may be interrelated, and practical production strategies may need to embody compromises among them. The later part of this review deals with a brief summary of intracellular expression of foreign genes. We also discuss the recent status of heterologous protein secretion in yeast.

II. HOST CELL MODIFICATIONS AFFECTING PLASMID MAINTENANCE

We can illustrate the role of host cells in plasmid maintenance by considering the mitotic behavior of the principal vector classes in current use (Parent et al., 1985). These include (1) self-replicating plasmids carrying either chromosomally derived replication initiation sequences (autonomous replication sequences, or ARSs) or replication initiation sequences derived from the native yeast 2-μm circle episome (Beggs, 1978; Botstein and Davis, 1982); (2) self-replicating plasmids that in addition carry elements of chromosomally derived centromeric sequences (Clarke and Carbon, 1980); and (3) nonreplicating plasmids that integrate by recombination into the continuity of host chromosomes at genetically specific locations (Hinnen et al., 1978).

With the exception of the third group of vectors—those that integrate into host chromosomes and so exhibit mitotic stabilities characteristic of chromosomes themselves—none of the self-replicating plasmids are wholly stable. Plasmids carrying most or all of the self-replicating elements of the native yeast 2-μm circle episome, however, appear decidedly more mitotically stable, on average, than plasmids carrying ARS elements (Beggs, 1978). Moreover, the mitotic stability of self-replicating plasmids appears to reflect, at least in part, plasmid copy number—that is, the greater the copy number, the greater the mitotic stability (Futcher and Cox, 1984). Plasmids carrying centromeric elements derived from yeast chromosomes are maintained at stabilities that approach those of native yeast chromosomes (Clarke and Carbon, 1980), but their copy numbers are perforce limited to one or a few copies per cell.

From the standpoint of industrial production, then, self-replicating plasmids embodying replication elements of the yeast 2-μm circle episome are potentially the vectors of choice (Broach, 1983). If the apparent high copy numbers exhibited by these plasmids (and hence high numbers of copies of the heterologous gene of interest) lead proportionately to increased levels of gene product, genetic manipulations of yeast hosts harboring such plasmids that might augment plasmid copy numbers and/or plasmid stability are surely warranted. On the other hand, the extreme stability of integrating plasmids provides a useful basis for genetic explorations of the host that might increase heterologous gene copy number and, perhaps, product yield. We discuss below some recent efforts in these areas.

A. Cir° Hosts

Self-replicating plasmids embodying replication elements of the native yeast 2-μm circle episome are among the most mitotically stable vectors yet constructed (Beggs, 1978), but these plasmids are still notably less stable than the native episome. Futcher and Cox (1983) have determined that the 2-μm circle episome is lost from growing cell populations at a rate of 7.6×10^{-5} per cell per generation, despite the fact that cells harboring the episome (cir⁺) exhibit a growth-selective disadvantage of about 1% per generation with respect to cells that have lost the plasmid (cir°). Futcher (1986) has proposed that the high intrinsic stability of the 2-μm circle episome is attributable to overreplication of the plasmid mediated by intraplasmid recombinational events that permit the two divergent replication forks, initiated from the plasmid's single replication origin, to advance in tandem, thereby generating multimeric plasmid forms in a single cell generation.

It might be anticipated, therefore, that hybrid plasmids carrying appropriate 2-μm circle elements could also partake of this copy amplification mechanism. In addition, replicative competition between 2-μm circle-based plasmids and the resident episome might be obviated if the yeast host were free of the native episome (cir°). Accordingly, Futcher and Cox (1984) have compared the rates of mitotic loss of a set of 2-μm circle-based hybrid plasmids in both cir⁺ and cir° hosts. In nearly all cases, rates of plasmid loss were constant over as many as 100 generations, and ranged from 8×10^{-4} to 5×10^{-1} plasmid-free cells per plasmid-containing cell per generation. Comparison of plasmid stabilities in closely related cir⁺ and cir° hosts, however, revealed an inconsistent pattern of stabilization. In two cases, the anticipated greater stability in cir° hosts was borne out, with plasmid loss rates reduced roughly twofold. Of two other plasmids examined, one was equally stable in both hosts, and one was about twice as stable in the cir⁺ host as in the cir° host. A further group of three plasmids exhibited comparatively low stabilities in the cir⁺ host, ranging from 3 to 5×10^{-2}, but were rapidly and nearly completely lost from cir° cells. The mitotic behavior of this latter group is surely due to insertional disruption or loss of essential 2-μm circle functions in the hybrid plasmids.

In an unrelated study employing a plasmid carrying the entire 2-μm circle episome, Blanc et al. (1979) found a marked contrast in plasmid stability between the two hosts; in the cir° host the plasmid exhibited a stability of about 1×10^{-2}; in the cir⁺ host some 90% of cells had lost the plasmid after only 30 generations.

The extreme instability of this plasmid, in contrast to those examined by Futcher and Cox (1984), remains unexplained.

It is clear from these observations that cir° hosts can provide a measure of enhanced stability for some hybrid plasmids, but not all. It is also clear that the replicative and distributional elements of 2-μm circle-based plasmids interact with host replicative capacities in ways not yet fully understood (Kikuchi, 1983; Jayaram et al., 1985; Wu et al., 1987).

Since most laboratory and industrial strains of *Saccharomyces cerevisiae* carry the 2-μm circle episome, methods for deriving cir° variants are of interest. The most common curing method is based on the observation that some 2-μm circle-based hybrid plasmids can successfully outcompete the endogenous 2-μm circle episome upon prolonged passage of transformed cells under plasmid-selective conditions (Dobson et al., 1980; Erhart and Hollenberg, 1981,1983). Cir° derivatives are found among cells that have lost the selectable hybrid plasmid. Harford and Peeters (1987) have suggested that hybrid plasmids embodying the entire 2-μm circle episome are superior vectors for competitive curing of the endogenous episome; some partial 2-μm circle-based vectors appear not to effect episome curing (Erhart and Hollenberg, 1983). In a study comparing such induced cir° strains with cir° strains arising spontaneously in glucose-limited continuous cultures, Mead et al. (1987) have observed that induced (but not spontaneous) cir° derivatives exhibit a significant reduction in specific growth rate relative to their cir+ progenitors. The phenomenon obtains for both haploid cells (8–12% reduction in specific growth rates) and diploid cells (25% reduction). This cautionary result, as yet unexplained, suggests that the method employed to obtain cir° hosts may significantly influence later production performance.

B. Host Mutations Affecting Plasmid Maintenance

The complexity of host-plasmid replicative interactions can be appreciated by examining a body of related findings on host genetic controls of self-replicating plasmids. Kikuchi and Toh-e (1986) have isolated mutations in yeast in which 2-μm circle-based plasmids, as well as the native 2-μm circle episome, are poorly maintained. The mutations collectively define a single chromosomal gene, designated MAP1 (*ma*intenance of *p*lasmid). Other self-replicating plasmids carrying chromosomal ARS replication origins and/or centromeric sequences are also destabilized in the presence of *map*1 mutations. Since actual *rates* of plasmid loss were not determined in this study, the degree of plasmid destabilization in the mutant host remains quantitatively uncertain.

In an analogous investigation, Maine et al. (1984) have described the isolation and partial characterization of host mutations that destabilize centromere-bearing, self-replicating plasmids. The 40 mutant isolates define 16 genetic loci and can be further grouped into two general phenotypic classes. One class of mutations effects loss of all centromere-containing plasmids tested; the other class, comprising 4 of the 16 genes, effects loss of only some of the plasmids tested and appears specific for particular autonomous replication sequences (ARSs). Whether the mutations act to reduce plasmid replicative efficiency, or to increase plasmid mis-segregation, is as yet unknown, though it would appear that both kinds of defects might be represented among these functional classes. Whether the *MAP1* locus identified by Kikuchi and Toh-e (1984) or a phenotypically similar locus identified by Larionov et al. (1984) can be included among the 16 genes identified by Maine et al. (1984) remains to be seen.

It is clear from the foregoing results that host cell genetic functions significantly influence hybrid plasmid stabilities, and that the interactions between host and plasmid in this regard are complex. These studies suggest that, given the existence of host mutations that lead to plasmid instability, there may also exist genetic alterations in the host that could lead to increased plasmid stability. We are unaware of any published information on this latter, potentially useful possibility.

C. Alterations in Host Cell Ploidy

Many industrial yeast cultivars are polyploid. That polyploidy per se can result in improved yields of metabolic products is well established. It is reasonable to examine, therefore, whether ploidy increases might also lead to improvements in heterologous protein production.

In *Saccharomyces,* augmented ploidy, at least from haploid (N) to tetraploid (4N), correlates directly with increases in per-cell mass and global protein content; that is, tetraploids are four times as massive as haploids (DiIorio et al., 1987). The per-cell activities of at least some native enzymes are also proportional to gene dosage (Reichert and Winter, 1975; DiIorio et al., 1987). The larger size of cells of greater ploidy suggests the possibility that self-replicating plasmids could be maintained at higher copy numbers per cell, with a concomitant improvement in plasmid stability. Larger cell sizes might also promote more equitable segregation of plasmids to daughter cells at cell division.

Evidence in support of these ideas has been adduced by comparing the mitotic stability and copy number of a 2-μm circle-based plasmid in closely related haploid and diploid hosts (Mead et al., 1986). Copy numbers in the diploid of both the hybrid plasmid and the resident 2-μm circle episome were roughly twice the haploid average; the frequency of plasmid loss in the diploid was more than an order of magnitude less than those in the two haploids comprising the diploid, and remained nearly constant through 100 generations in continuous culture.

These results are encouraging and might profitably be extended to cells of ploidy greater than diploid and to vectors expressing a heterologous protein. Takagi et al. (1985a) have addressed this issue directly by constructing three isogenic ploidy series (N, 2N, 4N), by transformation-associated cell fusion (Harashima et al., 1984; Takagi et al., 1985b), from haploids carrying one of three types of self-replicating plasmids: 2-μm circle-based, ARS-based, and ARS centromere-based. In all three cases the heterologous gene was *Escherichia coli lacZ*, the β-galactosidase structural gene, fused to a yeast HIS5 promoter.

Measurements of heterologous enzyme activity taken in this system by Takagi et al. (1985a) fail to fulfill predictions suggested by the results of Mead et al. (1986). Indeed, the relative specific activity of intracellular β-galactosidase *declined* with increased cell ploidy, independent of plasmid structure. The extent of the decline, relative to haploid values, was about twofold in the diploid and three- to fivefold in the tetraploid. Surprisingly, the cells in each ploidy series (N, 2N, 4N) harbored near-identical plasmid copy numbers, independent of cell ploidy.

Given the counterintuitive import of these results, it is appropriate here to examine features of the system of Takagi et al. (1985a) that might have produced this outcome. First, *E. coli* β-galactosidase may not be the appropriate heterologous enzyme for quantitative expression studies of this kind. It is a tetramer in its active form, and the results did not show that all of the yeast-produced β-galactosidase protein was functionally active. Hence, enzymatic determinations alone may not fully reflect the degree of heterologous gene activity. Second, the method used here to construct diploid and tetraploid cells—transformation-associated cell fusion (Takagi et al., 1985b)—may attenuate in unknown ways the extent of subsequent plasmid replications and, hence, plasmid copy numbers. We suggest, at minimum, that this method will also produce cells of genetically unknowable ploidy, if, as is likely in most applications, progenitor cells are normal, heterothallic haploids. We suggest that other methods for generating yeast polyploids, ones less fraught with potential complications and resulting in strains whose ploidy can be directly confirmed genetically, are much to be preferred (Campbell et al., 1981; Takagi et al., 1983; DiIorio et al., 1987). Transformation of such strains by the intact cell method of Ito et al. (1983)—and/or recent modifications of this method (Bruschi et al., 1987)—would not be expected to alter cell ploidy, which, in any event, could again by genetically checked. Finally, Takagi et al. (1985a) report enzyme measurements on 24- and 48-hr stationary-phase batch cultures. We note that levels of intracellular proteases reach a maximum in stationary phase yeast cells (Pringle, 1975); per-cell protease activities might even be augmented in cells of higher ploidy. These technical considerations lead us to suggest that the potential of augmented host cell ploidy to enhance heterologous protein production remains an open option, and that further examinations of this potential are warranted.

D. Effects of Selective Marker on Plasmid Maintenance

Transformation of yeast cells by plasmids is inefficient, and strong selection is necessary to detect the rare minority of cells that have incorporated the plasmid

(Hinnen et al., 1978). Hence a yeast-active gene (normally of yeast origin) in the transforming plasmid that complements a homologous, usually nutritional defect in the recipient host is normally employed, whether the plasmid be self-replicating or integrative. Alternatively, systems that make use of drug resistance genes (Sakai and Yamamoto, 1986) or metal ion (e.g., copper) resistance (Henderson et al., 1985) have found application in the transformation of industrial cultivars in which introduction of nutritional defects for selection purposes would be undesirable.

The choice of nutritional selective markers in yeast transformations has, to date, been largely influenced by historical accident. For example, Beggs (1978) has cloned copies of the yeast LEU2 gene into a series of 2-μm circle-based plasmids that exhibit, in some cases, unusually high copy numbers and stabilities (Futcher and Cox, 1984). That these features might be attributable to a particular LEU2 gene copy is suggested by the observation that other LEU2-bearing plasmids of structure similar to those of Beggs (1978) do not exhibit high plasmid copy numbers or stabilities (Dobson et al., 1981; Takagi et al., 1985a). Erhart and Hollenberg (1983) have provided an answer. A portion of the promoter of one of the LEU2 alleles of Beggs is deleted, and the per-gene expression level of the LEU2 gene product, β-isopropylmalate dehydrogenase, is less than 5% that of the normal LEU2 gene. Erhart and Hollenberg (1983) have suggested that the selection procedure for detecting cells transformed by plasmids carrying the defective LEU2 allele is itself responsible for establishing high copy numbers: only those transformed cells able rapidly to establish plasmid copies sufficient for survival on selective (leucine-deficient) medium will be recovered. Since plasmid copy number is, in part, a self-perpetuating characteristic in individual cell lineages (Gerbaud and Guerineau, 1980; Futcher and Cox, 1984), the high copy numbers established initially are likely to persist.

Loison et al. (1986) and Marquet et al. (1986) have reported another apparent approach to plasmid stabilization. Here the selective marker is the yeast URA3 gene. In addition, a second mutation in the host gene FUR1, specifying uridine phosphoribosylpyrophosphate transferase, was introduced into the plasmid-bearing cells by selection for 5-fluorouracil resistance. The fur1 mutation blocks entry of exogenous uracil into the phosphorylation and methylation (to cytosine) steps of the yeast pyrimidine biosynthetic pathway (Jones and Fink, 1982). Double mutant strains (ura3 fur1) are thus unable to grow in the presence of exogenous uracil; only the plasmid-borne URA3 gene product permits cell growth. Plasmid loss, then, is effectively lethal to the cell. Marquet et al. (1986) report that in such a host, under glucose-limited continuous culture in uracil-containing medium, levels of self-replicating vector-specified human α1-antitrypsin remained constant over at least 150 generations. Marquet et al. (1986) have concluded from this result that plasmid loss has been abolished in the double mutant host. We submit that this interpretation is misplaced. The double mutant host does not *prevent* plasmid loss; it renders plasmid loss *undetectable*. The constant production level of heterologous gene product observed in this case would be expected even if a fraction of cells lost the plasmid in each cell generation. Since, by design, plasmid loss is lethal, plasmidless cells must eventually cease to proliferate and to synthesize the heterologous protein, but there is no reason to believe that the added *fur*1 mutation impairs those cells that have lost the plasmid from continuing to proliferate for several, indeed many, generations (Futcher and Cox, 1984).

In short, while such approaches may appear to improve the efficiency of heterologous protein production in mass cultures, the reality of self-replicating plasmid instability is not ameliorated by this experimental approach.

III. EXPRESSION OF HETEROLOGOUS GENES

In recent years a considerable number of heterologous genes has been expressed and their gene products secreted in *Saccharomyces cerevisiae*. Successful expression of foreign genes in yeast has been achieved by cloning them into plasmid vectors that replicate at high copy number. As noted earlier, two types of self-replicating vectors are available. The first type contains ARS (autonomously replicating sequence) sequences which are presumably chromosomal origins of replication (Campbell, 1983); the second type of vector is based on sequences derived from endogenous yeast 2-μm circle DNA (Broach, 1982). These vectors also contain one or more of several other yeast genes including URA3 (Bach et al., 1979), TRP1 (Tschumper and Carbon, 1980), and LEU2 (Beggs, 1978), which serve as the selection marker for yeast transformation. Genes that confer resistance to various drugs have also been used as selectable markers, especially if the host yeast strains lack appropriate auxotrophic mutations. These drug-resistance genes include those specifying chloramphenicol acetyltransferase (Cohen et al., 1980), aminoglycoside 3' phosphotransferase (G418 resistance) (Jimenez and Davies, 1980; Webster and Dickson, 1983), thymidine kinase (McNiel and Friesen, 1981), hygromycin B phosphotransferase (Gritz and Davies, 1983), and dihydrofolate reductase (tetrahydrofolate dehydrogenase) (Miyajima et al., 1984). For efficient expression of foreign genes in yeast, the vectors are equipped with regulatory elements for transcription initiation and termination as well as regulatory sequences for translation initiation.

A. Yeast Promoters

The first yeast promoter used for foreign gene expression in yeast was the promoter of the gene encoding alcohol dehydrogenase I (ADHI) (Hitzeman et al., 1981), which is a very strong constitutive promoter. Constitutive promoters of other highly expressed genes commonly used for production of foreign proteins are those of phosphoglycerate kinase (PGK) (Dobson et al., 1982a; Tuite et al., 1982),

N(5'-phosphoribosyl)-anthranilate isomerase (TRP1) (Dobson et al., 1982b), glyceraldehyde 3-phosphate dehydrogenase (GPD) (Holland and Holland, 1980;Edens et al., 1984; Urdea et al., 1983; Bitter and Egan, 1984), and α factor (MFα1) (Brake et al., 1984; Bitter et al., 1984). Regulated promoters that have been successfully used for foreign gene expression are galactokinase (GAL1) (Broach et al., 1983), UDP-glucose-4-epimerase (GAL 10) (Kornbluth et al., 1987), acid phosphatase (PHO5) (Kramer et al., 1984), alcohol dehydrogenase II (ADH II) (Price et al., 1987), invertase (SUC2) (Smith et al., 1985), and as explained later, hybrid GPD(G) promoter (Bitter, 1987).

Table 1 lists examples of heterologous genes that have been expressed intracellularly in yeast by the use of different yeast promoters. There is wide variation in the level of production of heterologous proteins in yeast even with the use of the same promoters. For example, the yields of hepatitis B surface antigen (HBsA) and human interferon (IFN-α2) directed by the PGK promoter are 1–2% and up to 5% of total yeast protein, respectively (Hitzeman et al., 1983; Mellor et al., 1985), while activable prochymosin is produced as 5% of the total cellular protein (Mellor et al., 1983). Likewise, the ADHI promoter generates bacterial β-lactamase at 10% of total yeast protein (Roggenkamp et al., 1983), while the same promoter produces about 5% IFN-α2 (Hitzman et al., 1983) and only 1–2% HBsA (Valenzuela et al., 1982).

Recently, Hallewell et al. (1987) reported the production of human superoxide dismutase (SOD) in yeast to a level of 70% of total cell protein using the yeast GPD promoter. In contrast, the yield of human epidermal growth factor (EGF), when directed by the same GPD promoter, was only 0.1% of total cell protein (Urdea et al., 1983). Some heterologous proteins (rat growth hormone and human proinsulin) are very unstable when produced intracellularly in yeast. Successful expression of proinsulin in yeast was achieved only when the gene was fused with the promoter and the coding sequence of the yeast galactokinase (GAL1) gene (Stepien et al., 1983). Human EGF was produced only at very low levels, although EGF-specific mRNA could be detected (Urdca et al., 1983).

In a recent study, Bitter and Egan (1984) have reported the construction of a yeast expression plasmid containing a hybrid GPD promoter. This hybrid promoter utilizes a 365-bp DNA fragment from the yeast GAL1–GAL10 intergenic region (UAS$_G$) inserted 240 bp 5' to the TATA sequence of the GPD promoter sequence. A constitutive GPD promoter was thus converted to a galactose-inducible GPD(G) promoter. Expression of human immune interferon (IFN-γ) in yeast was attempted using the GPD(G) promoter-based vector. Indeed, after achieving high cell density in the presence of glucose, the level of IFN-γ production in yeast was induced by galactose 1,000-fold in less than one cell generation, to a yield of more than 2 g IFN-γ per liter of culture (Fieschko et al., 1987).

It has been demonstrated that the yeast PGK1 gene when present on a 2-μm replicative plasmid generates PGK protein at up to 30% of the total yeast protein

Table 1. Yeast Promoters Utilized for Intracellular Expression of Heterologous Genes

Promoter	Gene	Original source of the gene	Yield of the expressed protein[a]	Reference
ADHI	Interferon-α1	Human leucocyte	2–5	Hitzeman et al., 1981
	Surface antigen	Hepatitis B virus	1–2	Valenzuela et al., 1982
	β-Lactamase	*Escherichia coli*	2–10	Roggenkamp et al., 1983
	Hemagglutinin	WSN influenza virus	N/A[b]	Jabbar et al., 1985
	Fe protein subunit of nitrogenase	*Klebsiella pneumoniae*	N/A	Berman et al., 1985
ADC1	Acetylcholine receptor-α	*Torpedo californica*	1	Fujita et al., 1986
ARG3	α1-Antitrypsin	Human	1	Cabezon et al., 1984
CUP1	Metallothionein-I	Monkey	N/A	Thiele et al., 1986
	Metallothionein-II	Monkey	N/A	Thiele et al., 1986
Chelatin	Serum albumin	Human	1	Etcheverry et al., 1986
CYC1	Zein	Maize	5	Coraggio et al., 1986
	α-Gliadin	Wheat	0.1	Neill et al., 1987
GAL1	Capsid protein and glycoproteins E1 and E2	Sindbis virus	2.5–3	Wen and Schlesinger, 1986
	G protein	Vesicular stomatitis virus	2.5–3	Wen and Schlesinger, 1986
	G glycoprotein	Respiratory syncytial virus	0.05	Ding et al., 1987
	Proinsulin	Human	0.01	Stepien et al., 1983
	Prochymosin	Calf		Goff et al., 1984
GAL10	p60[v-src]	Rous Sarcoma virus	N/A	Kornbluth et al., 1987
	p60[c-src]			
GAL1–10/CYC1	β-Galactosidase	*Kluyveromyces lactis*	15	Velati-Bellini et al., 1986
GPD	Reverse transcriptase domain of *pol* gene	Human immunodeficiency virus	N/A	Barr et al., 1987

Promoter	Protein	Source	molecules/cell	Reference
	Superoxide dismutase	Human	30–70	Hallewell et al., 1987
	Interferon-αCon$_1$	Synthetic	1	Bitter and Egan, 1984
	Surface antigen	Hepatitis B virus	2–4	Bitter and Egan, 1984
	Thaumatin	Thamatococcus danielli Benth.	3 × 10	Edens et al., 1986
GPD(G)	Core antigen	Hepatitis B virus	40	Kniskern et al., 1986
	Epidermal growth factor	Human	0.1	Urdea et al., 1983
	α1-Proteinase inhibitor	Human	5–6	Hoylaerts et al., 1986
	Interferon-γ	Human	2 g/liter	Fieschko et al., 1986
	Thimidine kinase	Herpes simplex virus	N/A	Edens et al., 1984
PGK	Prochymosin	Calf	5	Mellor et al., 1983
	Middle-T antigen	Polyoma virus	0.1	Belsham et al., 1986
	Legumin	Pea (Pisum sativum L.)	1–2	Yarwood et al., 1987
	Lipase	Human gastric tissue	0.55	Bodmer et al., 1987
	Interferon-α2	Human	1–5	Mellor et al., 1985
	Surface antigen	Hepatitis B virus	1–2	Hitzeman et al., 1983a
PHO5	Reverse transcriptase	Cauliflower mosaic virus	N/A	Takatsuji et al., 1986
	Tissue inhibitor of metalloproteinases	Human	0.4	Kaczorek et al., 1987
	Oxidation-resistant mutant of α1-antitrypsin	Human	N/A	Rosenberg et al., 1984
	Lysozyme	Human	19.2 mg/liter	Hayakawa et al., 1987
	Large envelope protein	Hepatitis B virus	4	Dehoux et al., 1986
	gag Protein	HTLV-III virus	N/A	Kramer et al., 1986

Continued

Promoter	Gene	Original source of the gene	Yield of the expressed protein[a]	Reference
	Human growth hormone	Synthetic	10 molecules/cell	Tokunaga et al., 1985
	c-*myc* Protein	Human	N/A	Miyamoto et al., 1985
	β-*neo*-Endorphin	Synthetic	10 molecules/cell	Oshima et al., 1986
TRP1	Interferon-α	Human	1.2	Dobson et al., 1982b
URA3	β-Galactosidase	*Escherichia coli*	N/A	Rose et al., 1981

[a]Unless otherwise stated, the numbers in this column represent the percent of total protein.
[b]Not available.

(Chen et al., 1984). However, when the PGK1 coding sequence was replaced by a heterologous gene (Hitzeman et al., 1983), the level of expression of the heterologous gene product decreased by 1 or 2 orders of magnitude. Such dramatic differences in the level of expression have been shown to be due to a reduced steady-state mRNA level with a concomitant decrease in the level of protein production (Chen et al., 1984). Our current experience strongly suggests that in order to overcome limitations on the overall level of gene expression, each individual gene of interest has to be adapted to its regulatory signals either through vector modifications or host strain improvement. Also, proteins that are otherwise unstable in yeast may be expressed at very high levels by employing regulated expression systems, as demonstrated by Fieschko et al. (1987).

More recent work has focused on developing vectors that direct secretion of heterologous proteins into the yeast growth medium. Such developments have facilitated the production at high yields of a number of heterologous proteins that otherwise were unstable when produced intracellularly.

B. Secretion of Heterologous Proteins

Saccharomyces cerevisiae has a well-defined secretion system. Analyses of secretion-defective (*sec*) mutants generated by Schekman and his colleagues (Novick et al., 1980, 1981) and of N-linked glycosylation-defective (*alg*) mutants developed by Huffaker and Robbins (1982, 1983) have demonstrated that some of the early steps in the biosynthesis, glycosylation, and organelle movement in the yeast secretory pathway are similar in their general characteristics to those of mammalian cells. Yeast cells, however, secrete only a limited number of their own proteins into the growth medium. Most of these proteins are larger than 50,000 Da (Bitter, 1986). In addition, native proteins secreted into the growth medium collectively constitute only 0.5% of total yeast proteins. Secretion of such low levels of native yeast proteins is of distinct advantage for product recovery and purification of a secreted heterologous protein when produced at a very high level. Some heterologous proteins are found to be insoluble and inactive when produced intracellularly in yeast (Kingsman et al., 1985). This often appears to be due to the lack of formation of disulfide bonds or attachment of oligosaccharides. The formation of disulfide bonds and the addition of oligosaccharides in a protein take place in the yeast secretory organelles, and are often necessary for the correct protein conformation and hence protein activity. More recent work on the production of stable and active heterologous proteins in yeast has generally focused on the secretion of recombinant proteins rather than on their intracellular accumulation in yeast.

Secretion of calf chymosin in yeast is one of the earliest examples of the application of molecular genetics to the high-yield production of a functionally active heterologous protein. As mentioned earlier, the first report of the intracellular production of prochymosin in yeast was published by Mellor et al. in 1983. They incorporated the preprochymosin gene between the 5' and 3' control regions of the yeast PGK1 gene and were able to direct the expression of chymosin in the form of preprochymosin and prochymosin. The chymosin polypeptides were localized intracellularly; when activated by acidification, active chymosin that clotted milk could be recovered. Moir et al. (1985) subsequently achieved secretion of prochymosin in yeast cells harboring a plasmid in which the prochymosin gene was fused in the promoter, signal sequence, and part of the structural gene of yeast invertase (SUC2). Approximately 10% of the invertase prochymosin fusion protein was secreted. Acidic growth conditions converted all of the extracellular chymosin to its active form, but most of the expressed chymosin remained internal and resistant to acid activation (Smith et al., 1985).

In later experiments, Goff et al. (1984) fused the prochymosin gene with the highly regulated yeast GAL1 promoter, and made plasmid constructions with and without the SUC2 transcription terminator. The two constructions yielded equal amounts of prochymosin-specific mRNA and acid-activable prochymosin. However, approximately 80% of the prochymosin in these yeast strains remained insoluble, even after cell breakage. Even the yeast triose phosphate isomerase promoter bearing an invertase signal sequence (which increases the level of prochymosin levels to 0.25% of the total cell protein) did not result in the secretion of more than one tenth of the total prochymosin (Moir et al., 1985; Smith et al., 1985). Likewise, a PHO5 signal sequence (Arima et al., 1983) and a MFα1 secretion signal resulted in less efficient secretion of prochymosin. The natural calf secretion signal proved to be totally ineffective in directing the transport of the protein outside the yeast cell (Smith et al., 1985).

Successful secretion of prochymosin to a considerable level into the yeast growth medium was finally achieved by the development of "supersecreting" mutant (*ssc*) strains of *S. cerevisiae* (Smith and Gill, 1985; Smith et al., 1985). The isolation of such supersecreting mutants of yeast was made possible by the development of a rapid "footprinting" screening assay that detects high prochymosin secretor colonies on nutrient agar plates. Of 120,000 mutagenized colonies examined by this assay, 39 strains bearing partially dominant or recessive mutations were identified as supersecretors. Some of these mutants produce 8–10 times more extracellular prochymosin than the parent strain. Complementation analysis of the recessive mutations indicates that mutations in two genes, SSC1 and SSC2, are the strongest and most easily manipulated. Haploid mutant strains with both the *ssc*1 and *ssc*2 mutations were more efficient secretors than either single mutant. Thus, secretion into the yeast growth medium of 80–85% of the prochymosin was achievable in such mutant strains, both in laboratory-scale shake flask cultures and in fermenters having cell densities of at least 25 g (dry weight) per liter. These supersecretor mutants are expected to be of general value in secretion of a wide variety of heterologous proteins. In fact, it has already been shown that the *ssc*1 mutant strains secrete 10-fold more bovine growth hormone (BGH) precursor protein, and that *ssc*2 mutants increase the yield also, though to a lesser extent (Smith et al., 1985).

The exact mechanisms of action of the *ssc* mutations are not clear. Smith et al. (1985) speculate that the mutations result in the generation of new bypass routes around a single validating step in the secretory pathway, thus facilitating supersecretion of prochymosin. Such bypass routes may still be protein-specific. This may explain why pre-BGH secretion is improved dramatically in the *ssc*1 mutants but only marginally in *ssc*2 mutants. Development of additional alternative routes may then alleviate the problem of marginal BGH secretion.

To date, a large number of foreign proteins have been secreted into the yeast growth medium, and their numbers are ever increasing. Initial success on the secretion of a foreign protein in yeast was accomplished by using the protein's own signal sequence in association with a yeast promoter. Some of the noteworthy examples are: human IFN-α1 and IFN-α2 (Hitzeman et al., 1983), *E. coli* β-lactamase (Roggenkamp et al., 1981), wheat α-amylase (Rothstein et al., 1984), plant thaumatin (Eden et al., 1984), and mouse immunoglobulin light and heavy chains (Wood et al., 1985). In most cases the overall yield of the secreted protein was low, and the recovered proteins exhibited evidence of aberrant processing of signal sequences. An alternative approach was soon developed in which the heterologous gene of interest was fused to a yeast signal as leader sequence under the control of a yeast promoter. Signal sequences from four secretory yeast proteins, e.g., acid phosphatase (PHO5) (Hinnen et al., 1983), invertase (SUC2) (Smith et al., 1985; Chang et al., 1986), killer toxin (K1) (Skipper et al., 1987), and α-factor (MFα1) (reviewed in Bigelis and Das, 1988) have been successfully utilized so far. Native yeast acid phosphatase and invertase are normally localized in the periplasmic space, and less than 5% of these enzymes is secreted into the growth medium. Indeed, secretion of significant amounts of foreign proteins into the yeast growth medium using the signal sequences of acid phosphatase or invertase has not yet been successful. For example, only 10% of the expressed human IFN-α is secreted when the acid phosphatase promoter and the first 14 amino acids of the acid phosphatase signal sequence are used. Similarly, as noted earlier, by using the signal sequence and part of the structural gene of invertase, secretion of calf prochymosin was achieved only to a level of about 10% of expressed protein unless a hypersecreting mutant strain was utilized (Smith et al., 1985). More recently, Chang et al. (1986) have demonstrated that by utilizing the invertase signal sequence under the control of the invertase promoter, as much as 30% of the expressed human IFN-α2 could be secreted into the growth medium.

Secretion and processing sequences of the yeast mating pheromone, α factor, have been used most extensively to promote secretion of a wide variety of heterologous proteins into the yeast culture medium. The α factor gene MFα1 has been cloned (Kurjan and Herskowitz, 1982), and the biosynthesis and processing of the MFα1 gene product have been characterized (Julius et al., 1983, 1984a).

The MFα1 gene product is composed of 165 amino acids and is the precursor of α factor, a 13-amino acid peptide. A stretch of approximately 20 hydrophobic N-terminal amino acids of the precursor protein serves as the signal sequence, which is followed by a 61-amino acid prosequence having three N-glycosylation sites. The precursor also contains four tandem repeats of the mature α factor sequence. The four α factor structural sequences are separated by spacer peptides of the sequence Lys-Arg-Glu-Ala-Glu-Ala or Lys-Arg-Glu-Ala-Asp-Ala. Three processing steps are involved in the release of mature α factor from the precursor molecule: (1) cathepsin B or trypsinlike cleavage by the yeast KEX2 gene product (Julius et al., 1984) at the carboxyterminal ends of the Lys-Arg dibasic residues; (2) dipeptidylaminopeptidase A (product of the yeast STE13 gene) cleavage of the N-terminal dipeptides—e.g., Glu-Ala, Asp-Ala, or Val-Ala (Julius et al., 1983); and (3) cleavage of C-terminal basic residues, by a carboxypeptidase B-like enzyme which has been shown recently to be the product of the KEX1 gene (Dmochowska et al., 1987).

For further details of the cloning, biosynthesis, processing, and secretion of α factor, readers should refer to Kurjan and Herskowitz (1982), Julius et al (1983), and Julius et al. (1984). An up-to-date list of foreign proteins or peptides that have been secreted into the yeast growth medium is shown in Table 2.

Brake and his co-workers (1984) were the first to demonstrate the efficacy of the prepro α factor sequence in successful secretion of heterologous proteins into the culture medium of yeast. As much as 25% of the total human epidermal growth factor (hEGF) synthesized in yeast with the help of the α factor leader sequence was found in the yeast culture medium. Expression of hEGF in yeast was achieved constitutively using MFα1 promoter. Also, regulated expression of hEGF was obtained from a strain having a temperature-sensitive SIR mutation. The products of the four known SIR genes are required for α factor expression, and a mutation in any one of the SIR genes leads to the expression of a nonmating phenotype. Indeed, in a sir3 mutant hEGF was produced at less than 10 ng/liter at 37°C, but was increased to 4 mg/liter within several hours after shifting to 24°C (Brake et al., 1984). Such sir3 mutation-dependent regulation of heterologous protein production has also been observed in the case of prepro α factor-directed secretion of E. coli β-galactosidase in yeast (Das et al., 1989).

Although yeast-derived hEGF retained its biological activity, amino acid sequence analysis indicates that only the Lys-Arg segment of the spacer peptide of prepro α factor has been accurately processed from the amino terminal end of hEGF (Brake et al., 1984). Similar incomplete processing of the spacer sequence was observed in the case of secretion of β-endorphin (Bitter et al., 1984), synthetic IFN-α (Bitter et al., 1984), and human IFN-α1 (Singh et al., 1984). Interestingly, overproduction of α factor from multicopy vectors in wild-type yeast cells leads

Table 2. Heterologous Proteins Secreted into the Growth Medium by *Saccharomyces cerevisiae*

Proteins/peptides	Natural occurrence	Source of the DNA	Promoter utilized	Signal sequence used	Amount secreted	Reference
Mammalian Proteins						
α-Amylase	Mouse	Mouse salivary	ADC1	Own	NK[b]	Thomsen, 1983
	Mouse	Mouse pancreas	MFα1	Prepro α factor	153 μg/ml	Filho et al., 1986
Antithrombin III	Human	Human	ADH1	Own	30 ng/ml	Broker et al., 1987
			CYC1	Own	14–31 ng/ml	
			GAL1	Own	100 ng/ml	
Atrial natriuretic peptide	Human	Synthetic[a]	MFα1	Prepro α factor	570 μg/liter	Vlasuk et al., 1986
Calcitonin	Human	Synthetic[a]	MFα1	Prepro α factor	12 mg/liter	Zsebo et al., 1986
Consensus α–interferon	Human	Synthetic[a]	MFα1	Prepro α factor	100 mg/liter	Zsebo et al., 1986
Connective tissue activating peptide III	Human	Synthetic[a]	MFα1	Prepro α factor	2.5 × 10 units/liter	Mullenbach et al., 1986
β-Endorphin	Human	Synthetic[a]	MFα1	Prepro α factor	7 mg/liter	Zsebo et al., 1986
Epidermal growth factor	Human	Synthetic[a]	MFα1	Prepro α factor	4.6 mg/liter	Brake et al., 1984
Granulocyte macrophage colony stimulating factor	Human	Human	MFα1	Prepro α factor	NK[b]	Cantrell et al., 1985

(Continued)

Proteins/peptides	Natural occurrence	Source of the DNA	Promoter utilized	Signal sequence used	Amount secreted	Reference
Granulocyte macrophage colony stimulating factor	Rat	Rat	MFα1	Prepro α factor	1×10^8 units/liter	Miyajima et al., 1986
	Mouse	Mouse	ADH2	Prepro α factor	50–60 mg/liter	Price et al., 1987
Growth hormone releasing factor	Human	Human	MFα1	Prepro α factor	NK[b]	Brake et al., 1984
HI-30 protein	Human	Human	MFα1	Prepro α factor	500 μg/liter	Lehman et al., 1987
Interferon-α1	Human	Human	MFα1	Prepro α factor	1×10 units/liter	Singh et al., 1984
Interferon-α2	Human	Human	ADHI	Own	5[c]	Hitzeman et al., 1983b
Insulin	Human	Human	PGK	Invertase	2×10 units/liter	Chang et al., 1986
	Human	Human	MFα1	Prepro α factor	0.18–2 mM	Thim et al., 1985
Insulinlike growth factors	Human	Human	MFα1	Prepro α factor	NK[b]	Brake et al., 1984
Interleukin-2	Cow	Cow	ADH2	Prepro α factor	50–60 mg/liter	Price et al., 1987
	Mouse	Mouse	MFα1	Prepro α factor	10 μg/liter	Miyajima et al., 1985
	Human	Human	MFα1	Prepro α factor	NK[b]	Brake et al., 1984

Protein	Source		Gene	Signal	Yield	Reference
Lysozyme	Human	Synthetic	GAL10	Chicken lysozyme	0.96 mg/liter	Jigami et al., 1986
Gp350	Epstein-Barr virus	Viral	MFα1	Prepro α factor	N/A	Schultz et al., 1987
Prochymosin	Calf	Calf	GAL10	Prepro α factor	N/A	Schultz et al., 1987
			GAL1	Invertase	7.9[c]	Smith et al., 1985
				Own	11.0[c]	Smith et al., 1985
				Alkaline phosphatase	13.0[c]	Smith et al., 1985
			SUC2	Invertase	51.0[c]	Smith et al., 1985
			TRP1	Invertase	27–40[c]	Smith et al., 1985
			MFα1	Prepro α factor	3.4[c]	Smith et al., 1985
Plant proteins						
α-Amylase	Wheat	Wheat	PGK	Own	30–60[c]	Rothstein et al., 1984
Fungal proteins						
Glycoamylase	Aspergillus awamori	Aspergillus awamori	ENO1	Own	5–10 mg/liter	Innis et al., 1985
	Rhizopus sp.	Rhizopus sp.	GPD	Own	100–300 mg/liter	Ashikari et al., 1986
			PGK	Own	50–100 mg/liter	(unpublished)
			PHO5	Own	50–100 mg/liter	(unpublished)
Cellobiohydrolase I	Trichoderma reesei	Trichoderma reesei	ENO1	Own	10–50 mg/liter	Shoemaker et al., 1984

(Continued)

Proteins/peptides	Natural occurrence	Source of the DNA	Promoter utilized	Signal sequence used	Amount secreted	Reference
Cellobiohydrolase II	*Trichoderma reesei*	Trichoderma reesei	PGK	Own	20–100 mg/liter	Knowles et al., 1985
Endoglucanase I	*Trichoderma reesei*	Trichoderma reesei	PGK	Own	20–100 mg/liter	Knowles et al., 1985
Protease	*Mucor pucillus*	Mucor pucillus	GAL7	Own	100–200 mg/liter	Yoshizumi and Ashikari, 1987
Bacterial proteins Glucanase	*Cellulomonas fimi*	Cellulomonas fimi		K1 toxin		Skipper et al., 1985
Others						
Somatomedin C	Angler fish	Synthetic	MFα1	Prepro α factor	12 mg/liter	Ernst, 1986
Somatostatin	Angler fish	Angler fish	MFα1	Prepro α factor	200 ng/10^7 cells	Green et al., 1986

[a]The DNA sequence was based on the amino acid sequence of the native peptide.
[b]Not known.
[c]Percent of total secreted protein.

to incompletely processed α factor having the Glu-Ala-Glu-Ala sequence at the N terminus, indicating that the dipeptidylaminopeptidase A cleavage is the rate-limiting step (Julius et al., 1983). Moreover, the efficiency of cleavage by this endoproteinase has been demonstrated to be dependent on the amino acid sequence context around the Lys-Arg processing site. Subsequent studies on the secretion of heterologous proteins have circumvented such problems of incomplete processing by deleting the Glu-Ala processing sites and retaining only the Lys-Arg sequence at the fusion junction of prepro α factor and the heterologous protein(s) (Brake et al., 1984; Bitter et al., 1984).

One of the problems associated with the secretion of heterologous proteins using the α factor system is the internal cleavage of nonspecific sites within the mature protein. Bitter et al. (1984) have demonstrated such undesirable processing of β-endorphin and Con$_1$ α-interferon. The protease(s) effecting such internal cleavage are neither the KEX2 gene product nor protease B (Zsebo et al., 1986). Such internal cleavage of β-endorphin was demonstrated in two different yeast strains, one having normal vacuolar protease activity (PEP4) and the other (pep4-3) having only 5% of that activity due to a mutation in the PEP4 gene (Zubenko et al., 1982). Similarly, IFN-α Con$_1$ was processed at internal sites during its secretion from a pep4-3 yeast strain. However, the frequency of such processing was of low abundance, and no more than 5% of the secreted IFN-α Con$_1$ was internally processed. Singh et al. (1984) have obtained secretion of IFN-α1 using a pep4-3 yeast mutant similar to that used by Bitter et al. (1984) for β-endorphin secretion. No internal cleavage of secreted IFN-α1 was reported, however. In fact, very little information on internal cleavage of any other heterologous proteins secreted from yeast is available. It is possible that in reality such cleavage is not strain- but conformation-dependent. Interestingly, pep4-3 strains are sometimes useful for increasing the level of production of some heterologous proteins—e.g., β-endorphin (Bitter et al., 1984) or α1-antitrypsin (Rosenberg et al., 1984)—although this host mutation has very little effect on the level of production of other heterologous proteins (e.g., IFN (Kingsman et al., 1985) and HI-30 (R. C. Das and D. J. Lehman, unpublished observation).

Recent studies have indicated that yeast is capable of carrying out in vivo posttranslational modifications of mammalian proteins. Some of these modifications, which include acetylation, phosphorylation, and glycosylation, are essential for the stability of the expressed heterologous proteins, while others may be necessary for solubility or the expression of biological activity. Some notable examples are the N-acetylation of superoxide dismutase (Hallewell et al., 1987), fatty acid acetylation of RAS proteins (Fujiyama and Tamanoi, 1986), and phosphorylation of mouse FOS protein (Sambucetti et al., 1986) and c-myc protein (Miyamoto et al., 1985). Also, influenza hemagglutinin (Jabbar et al., 1985), mouse immunoglobulin (Wood et al., 1985), human α1-antitrypsin (Hoylaerts et al., 1986), and fungal glucoamylase (Innis et al., 1985) are glycosylated when expressed in yeast. However, such glycosylation is heterogeneous in nature and is of only high-mannose type. While core glycosylation in yeast and other eucaryotic systems is similar (Huffaker and Robbins, 1983), yeast lacks the ability to generate oligosaccharide side chains with complex carbohydrates—e.g., sugar residues other than mannose. The role of glycosylation in eukaryotic systems is not fully understood. Thus, if complex carbohydrates are essential for the biological activity of a eukaryotic protein, it is likely that yeast will not be the host of choice for the commercial production of that protein.

IV. CONCLUSION

The field of heterologous protein production in yeast has grown into prominence only during the past few years, as evidenced by the examples cited in this review. Indeed, there have been significant advances in our understanding of the basic aspects of the expression of heterologous genes and secretion of gene products in yeast, some of which have already been utilized for the industrial production of recombinant proteins. For example, yeast-produced recombinant hepatitis B vaccine has already been marketed. And a number of other yeast-produced eukaryotic proteins of pharmaceutical value are in clinical trials at present. Future research on heterologous protein production in yeast will continue to concentrate on the development of high-yielding yeast strains, as well as on expression vectors capable of self-amplification and stable maintenance at high copy number. Better understanding of yeast chromosomal structure and the control and regulation of transcription in yeast will facilitate construction of such high-copy expression vectors. The complexity of the secretion pathway in yeast is being slowly unraveled, and the mechanisms of protein translocation, glycosylation, folding, processing, and localization are being increasingly better understood. It is conceivable that improved knowledge of these processes will facilitate the design of artificial sequences capable of copious secretion of specifically modified, biologically active, novel, heterologous proteins. Also, continuing advances in medium optimization, fermentation engineering, and process development will broaden the scope of economically viable future production of novel polypeptides of heterologous origin in yeast.

ACKNOWLEDGMENTS

We thank Dr. David Saari and Ms. Cheryl Kobold of the Miles Library for their generous assistance in assembling the references, and Dr. James G. Yarger and Dr. Ramunas Bigelis for helpful comments on the manuscript.

NOTE ADDED IN PROOF

Recently, it has been demonstrated that the SSC1 protein is very similar to Ca^{2+} ATPases (Rudolph et al., 1989). The protein is localized in the secretory pathway and functions as a Ca^{2+} pump affecting intracellular transport and protein secretion.

Also, since submission of this manuscript quite a number of publications have appeared on the production of various heterologous proteins in yeast. Readers are referred to those cited at the end of the References section.

REFERENCES

Arima, K., Oshima, T., Kubota, I., Nakamura, N., Mizunaga, T., and Toh-e, A. (1983) The nucleotide sequence of the yeast *PHO5* gene: A putative precursor of repressible acid phosphatase contains a signal peptide. *Nucl. Acids Res.* 11:1657–1672.

Ashikari, T., Nakamura, T., Tanaka, Y., et al. (1986) *Rhizopus* raw-starch-degrading glucoamylase: Its cloning and expression in yeast (*Saccharomyces cerevisiae*). *Agric. Biol. Chem.* 50:957–964.

Bach, M. L., Lacroute, F., and Bostian, D. (1979) Evidence for transcriptional regulation of orotidine 5'-phosphate in yeast by hybridization of mRNA to the yeast structural gene cloned in *E. coli. Proc. Natl. Acad. Sci. USA.*76:386–390.

Barr, P. J., Porrer, M. D., Lee-Ng, C. T., Gibson, H. L., and Lucir, P. A. (1987) Expression of active human immunodeficiency virus reverse transcriptase in *Saccharomyces cerevisiae. Biotechnology* 5:486–489.

Beggs, J. D. (1978) Transformation of yeast by a replicating hybrid plasmid. *Nature* 275:104–109.

Beggs, J. D., Guerineau, M., and Atkins, J. F. (1978) A map of the restriction targets in yeast 2 micron plasmid DNA cloned on bacteriphage lambda. *Mol. Gen. Genet.* 148:287–294.

Belsham, G. J., Barker, D. G., and Smith, A. E. (1986) Expression of polyoma virus middle-T antigen in *Saccharomyces cerevisiae. Eur. J. Biochem.*156:413–421.

Berman, J., Zilberstein A., Solomon, D., and Zamio, A. (1985) Expression of a nitrogen-fixation gene encoding a nitrogenase subunit in yeast. *Gene* 35:1–9.

Bigelis, R., and Das, R. C. (1988) Secretion research in industrial mycology. In *Protein Transfer and Organelle Biogenesis*, R. C. Das and P. W. Robbins (Eds.). Academic Press, New York, pp. 773–810.

Bitter, G. A. (1986) Engineering *Saccharomyces cerevisiae* for the efficient secretion of heterologous proteins. In *Microbiology—1986*. American Society for Microbiology Washington, DC, pp. 330–334.

Bitter, G. A., and Egan, K. M. (1984) Expression of heterologous genes in *Saccharomyces cerevisiae* from vectors utilizing the glyceraldehyde-3-phosphate dehydrogenase gene promoter. *Gene* 32:263–274.

Bitter, G. A., Chen, K. K., Banks, A. R., and Lai, P. H. (1984) Secretion of foreign proteins from *Saccharomyces cerevisiae* directed by α-factor gene fusions. *Proc. Natl. Acad. Sci. USA* 81:5330–5334.

Bitter, G. A., Egan, K. M., Koski, R. A., Jones, M. D., Elliott, S. G., and Griffin, J. C. (1987) Expression and secretion vectors for yeast. In *Methods in Enzymology,* Vol. 153: *Recombinant DNA,* Part D, R. Wu and L. Grossman (Eds.). Academic Press, New York, pp. 516–544.

Blanc, H., Gerbaud, C., Slonimski, P. P., and Guerineau, M. (1979) Stable yeast transformation with chimeric plasmids using a 2μm-circular DNA-less strain as a recipient. *Mol. Gen. Genet.* 176:335–342.

Bodmer, N. W., Angal, S., Yarranton, G. T., et al. (1987) Molecular cloning of a human gastric lipase and expression of the enzyme in yeast. *Biochem. Biophys. Acta* 909:237–244.

Boel, E., Hjort, I., Svensson, B., et al. (1984) Glucoamylase G1 and glucoamylase G2 from *Aspergillus niger* are synthesized from two different but closely related messenger RNA species. *EMBO J.* 3:1097–1102.

Botstein, D., and Davis, R. W. (1982) Principles and practice of recombinant DNA research with yeast. In *The Molecular Biology of the Yeast Saccharomyces—Metabolism and Gene Expression.* J. N. Strathern, E. W. Jones, and J. R. Broach (Eds.). Cold Spring Harbor Laboratory, Cold Spring Harbor, pp. 607–636.

Brake, A. J., Merryweather, J. P., Coit, D. G., et al. (1984) α-Factor directed synthesis and secretion of mature foreign proteins in *Saccharomyces cerevisiae. Proc. Natl. Acad. Sci. USA* 81:4642–4646.

Broach, J. R. (1982) The yeast plasmid 2μ circle. *Cell* 28:203–204.

Broach, J. R. (1983) Construction of high copy yeast vectors using 2-μm circle sequences. In *Methods in Enzymology,* Vol. 101. R. Wu, L. Grossman, and K. Moldave (Eds.). Academic Press, New York, pp. 307–325.

Broker, M., Ragg, H., and Karges, H. E. (1987) Expression of human antithrombin III in *Saccharomyces cerevisiae* and *Schizosaccharomyces pombe. Biochim. Biophys. Acta* 908:203–213.

Bruschi, V. C., Comer, A. R., and Howe, G. A. (1987) Specificity of DNA uptake during whole cell transformation of *S. cerevisiae. Yeast* 3:131–137.

Cabezon, T., DeWilde, M., Herion, P., Loriau, R., and Bollen, A. (1984) Expression of human α₁-antitrypsin cDNA in the yeast *Saccharomyces cerevisiae. Proc. Natl. Acad. Sci. USA* 81:6594–6598.

Campbell, D., Doctor, J. S., Feuersanger, J. H., and Doolittle, M. M. (1981) Differential mitotic stability of yeast disomes derived from triploid meiosis. *Genetics* 98:239–255.

Campbell, J. L. (1983) Yeast DNA replication. In *Genetic Engineering Principles and Methods,* Vol. 5. J. K. Setlow and A. Hollaender (Eds.). Plenum, New York, pp. 109–146.

Cantrell, M. A., Anderson, D., Cerretti, D. P., et al. (1985) Cloning, sequence and expression of a human granulocyte/macrophage colony-stimulating factor. *Proc. Natl. Acad. Sci. USA* 82:6250–6254.

Chang, C. N., Matteucci, M., Perry, L. J., Wulf, J. J., Chen, C. Y., and Hitzeman, R. A. (1986) *Saccharomyces cerevisiae* secretes and correctly processes human interferon hybrid proteins containing yeast invertase signal peptides. *Mol. Cell. Biol.* 6:1812–1819.

Chen, C. Y., Oppermann, H., and Hitzeman, R. A. (1984) Homologous versus heterologous gene expression in the yeast, *S. cerevisiae. Nucl. Acids Res.* 12:8951–8970.

Clarke, L., and Carbon, J. (1980) Isolation of a yeast centromere and construction of functional small circular chromosomes. *Nature* 287:504–509.

Cohen, J. D., Eccleshall, T. R., Needleman, R. B., Federoff, H., Buchferer, B. A., and Marmur, J. (1980) Functional expression of the *Escherichia coli* plasmid gene coding for chloramphenicol acetyltransferase. *Proc. Natl. Acad. Sci. USA* 77:1078–1082.

Coraggio, I., Compagno, C., Martegano, E., et al. (1986) Transcription and expression of zein sequences in yeast under natural plant or yeast promoters. *EMBO J.* 5:459– 465.

Das, R. C., Shultz, J. L., and Lehman, D. J. (1986) α-Factor leader sequence-directed transport of *Escherichia coli* β-galactosidase in the secretory pathway of *Saccharomyces cerevisiae*. *Mol. Gen. Genet.* 218:240–248 .

Dehoux, P., Ribes, V., Sobezak, E., and Streeck, R. E. (1986) Expression of the hepatitis B virus large envelope protein in *Saccharomyces cerevisiae*. *Gene* 48:155–163.

DiIorio, A. A., Weathers, P. J., and Campbell, D. A. (1987) Comparative enzyme and ethanol production in an isogenic yeast ploid series. *Curr. Genet.* 12:9–14.

Ding, M., Wen, D., Schlesinger, M. J., Wertz, G. W., and Ball, L. A. (1987) Expression and glycosylation of the respiratory syncytial virus G protein in *Saccharomyces cerevisiae*. *Virology* 159:450–453.

Dmochowska, A., Dignard, D., Henning, D., Thomas, D. Y., and Bussey, H. (1987) Yeast *KEX1* gene encodes a putative protease with a carboxypeptidase B-like function involved in killer toxin and α-factor precursor processing. *Cell* 50:573–584.

Dobson, M. J., Futcher, A. B., and Cox, B. S. (1980) Loss of 2 μm DNA from *Saccharomyces cerevisiae* transformed with the chimeric plasmid pJDB219. *Curr. Genet.* 2:201–205.

Dobson, M. J., Kingsman, S. M., and Kingsman, A. J. (1981) Sequence variation in the *LEU2* region of the *Saccharomyces cerevisiae* genome. *Gene* 16:133–139.

Dobson, M. J., Tuite, M. F., Roberts, N. A., et al. (1982a) Conservation of high efficiency promoter sequences in *Saccharomyces cerevisiae*. *Nucl. Acids. Res.*10:2625–2637.

Dobson, M. J., Tuite, M. F., Mellor, J., et al. (1982b) Expression in *Saccharomyces cerevisiae* of human interferon-alpha directed by the *TRP1* 5' region. *Nucl. Acids. Res.*11:2287–2302.

Edens, L., Bom, I., Lederboer, A. M., et al. (1984) Synthesis and processing of the plant protein thaumatin in yeast. *Cell* 37:629–633.

Erhart, E., and Hollenberg, C. P. (1981) Curing of *Saccharomyces cerevisiae* 2-μm DNA by transformation. *Curr. Genet.* 3:83–89.

Erhart, E., and Hollenberg, C. P. (1983) The presence of a defective *LEU2* gene on 2μ DNA recombinant plasmids of *Saccharomyces cerevisiae* is responsible for curing and high copy number. *J. Bacteriol.* 156:625–635.

Etcheverry, T., Forrester, W., and Hitzeman, R. (1986) Regulation of the chelatin promoter during the expression of human serum albumin or yeast phosphoglycerate kinase in yeast. *Biotechnology* 4:726–730.

Fieschko, J. C., Egan, K. M., Ritch, T., Koski, R. A., Jones, M., and Bitter, G. A. (1986) Controlled expression and purification of human immune interferon from high cell density fermentations of *Saccharomyces cerevisiae*. *Biotechnol. Bioeng.* 29:1113–1121.

Filho, S. A., Galembeck, E. V., Faria, J. B., and Schenberg Frascino, A. C. (1986) Stable yeast transformants that secrete functional α-amylase encoded by cloned mouse pancreatic cDNA. *Biotechnology* 4:311–315.

Fujita, N., Nelson, N., Fox, T. D., et al. (1986) Biosynthesis of the *Torpedo californica* acetylcholine receptor α subunit in yeast. *Science* 231:1284–1287.

Fujiyama, A., and Tamanoi, F. (1986) Processing and fatty acid acylation of RAS1 and RAS2 proteins in *Saccharomyces cerevisiae. Proc. Natl. Acad. Sci. USA* 83:1266–1270.

Futcher, A. B. (1986) Copy number amplification of the 2 μm circle plasmid of *Saccharomyces cerevisiae. J. Theor. Biol.* 119:197–204.

Futcher, A. B., and Cox, B. S. (1983) Maintenance of the 2 μm circle plasmid in populations of *Saccharomyces cerevisiae. J. Bacteriol.* 154:612–622.

Futcher, A. B., and Cox, B. S. (1984) Copy number and stability of 2-μm circle-based artificial plasmids of *Saccharomyces cerevisiae. J. Bacteriol.* 157:283–290.

Gerbaud, C., and Guerineau, M. (1980) 2μm Plasmid copy number in different yeast strains and repartition of endogenous and 2μm chimeric plasmids in transformed strains. *Curr. Genet.* 1:219–228.

Goff, C. G., Moir, D. T., Kohno, T., et al. (1984) The expression of calf prochymosin in *Saccharomyces cerevisiae. Gene* 27:35–46.

Gritz, L., and Davies, J. (1983) Plasmid encoded hygromycin B resistance: The sequence of hygromycin B phosphotransferase and its expression in *Escherichia coli* and *Saccharomyces cerevisiae. Gene* 25:179–188.

Hallewell, R. A., Mills, R., Takamp-Olson, P., et al. (1987) Amino terminal acetylation of authentic human Cu, Zn superoxide dismutase produced in yeast. *Biotechnology* 5:363–366.

Harashima, S., Takagi, A., and Oshima, Y. (1984) Transformation of protoplasted yeast cells is directly associated with cell fusion. *Mol. Cell. Biol.* 4:771–778.

Harford, M. N., and Peeters, M. (1987) Curing of endogenous 2 micron DNA in yeast by recombinant vectors. *Curr. Genet.* 11:315–319.

Hayakawa, T., Toibana, G., Marumoto, R., et al. (1987) Expression of human lysozyme in an insoluble form in yeast. *Gene* 56:53–59.

Henderson, R. C. A., Cox, B. S., and Tubb, R. (1985) The transformation of brewing yeasts with a plasmid containing the gene for copper resistance. *Curr. Genet.* 9:133–138.

Hinnen, A., Hicks, J. B., and Fink, G. R. (1978) Transformation of yeast. *Proc. Natl. Acad. Sci. USA* 75:1929–1933.

Hinnen, A., Meyhack, B., and Tsapis, R. (1983) High expression and secretion of foreign proteins in yeast. In *Gene Expression in Yeast, Foundation for Biotechnical and Industrial Fermentation Research*, Vol. 1. M. Korhola and E. Vaisanen (Eds.). Kauppakirjapaino Oy, Helsinki, pp. 157–166.

Hitzeman, R. A., Hagie, F. F., Levine, H. L., Geoddel, D. W., Ammerer, G., and Hall, B. D. (1981) Expression of human gene for interferon in yeast. *Nature* 293:717–722.

Hitzeman, R. A., Chen, C. Y., Hagie, F. E., et al. (1983a) Expression of Hepatitis B virus surface antigens in yeast. *Nucl. Acids Res.* 11:2745–2763.

Hitzeman, R. A., Leung, D. W., Perry, L. J., Kohr, W. J., Levine, H. L., and Goeddel, D. V. (1983b) Secretion of human interferons by yeast. *Science* 219:620–625.

Holland, J. P., and Holland, M. J. (1980) Structural comparison of two non-tandemly repeated yeast glyceraldehyde-3-phosphate dehydrogenase genes. *J. Biol. Chem.* 255:2596–2605.

Hoylaerts, M., Weyers, A., Boller, A., Harford, N., and Cabezon, T. (1986) High-level production and isolation of human recombinant α1-proteinase inhibitor in yeast. *FEBS Lett.* 204:83–87.

Huffaker, T., and Robbins, P. (1982) Temperature-sensitive yeast mutants deficient in asparagine-linked glycosylation. *J. Biol. Chem.* 257:3203–3210.

Huffaker, T. C., and Robbins, P. W. (1983) Yeast mutants deficient in protein glycosylation. *Proc. Natl. Acad. Sci. USA* 80:7466–7470.

Innis, M. A., Holland, M. J., McCabe, P. C., et al. (1985) Expression, glycosylation and secretion of an *Aspergillus* glucoamylase by *Saccharomyces cerevisiae. Science* 228 :21–26.

Ito, H., Fukuda, Y., Murata, K., and Kimura, A. (1983) Transformation of intact yeast cells treated with alkali cations. *J. Bacteriol.* 153:163–168.

Jabbar, M. A., Sivasubramanian, N., and Nayak, D. P. (1985) Influenza viral (A/WSN/33) hemagglutinin is expressed and glycosylated in the yeast *Saccharomyces cerevisiae. Proc. Natl. Acad. Sci. USA* 82:2019–2023.

Jarayam, M., Sutton, A., and Broach, J. R. (1985) Properties of *REP3*: A *cis*-acting locus required for stable propagation of the *Saccharomyces cerevisiae* plasmid 2µm circle. *Mol. Cell. Biol.* 5:2466–2475.

Jigami, Y., Muraki, M., Harada, N., and Tanaka, H. (1986) Expression of synthetic human-lysozyme gene in *Saccharomyces cerevisiae:* Use of a synthetic chicken-lysozyme signal sequence for secretion and processing. *Gene* 43:273–279.

Jimenez, A., and Davies, J. (1980) Expression of a transposable antibiotic resistance element in *Saccharomyces. Nature* 287:869–871.

Jones, E. W., and Fink, G. R. (1982) Regulations of amino acid and nucleotide biosynthesis in yeast. In *The Molecular Biology of the Yeast Saccharomyces—Gene Expression.* J. N. Strathern, E. W. Jones, and J. R. Broach (Eds.). Cold Spring Harbor Laboratory, Cold Spring Harbor, pp. 181–299.

Julius, D., Blair, L., Brake, A., Sprague, G., and Thorner, J. (1983) Yeast α-factor is processed from a larger precursor polypeptide: The essential role of a membrane bound dipeptidyl aminopeptidase. *Cell* 32:839–852.

Julius, G., Brake, A., Bliar, L., Kunbawa, R., and Thorner, J. (1984a) Isolation of the putative structural gene for the lysine-arginine-cleaving endopeptidase required for processing of yeast prepro-α-factor. *Cell* 37:1075–1089.

Julius, D., Schekman, R., and Thorner, J. (1984b) Glycosylation and processing of prepro-α-factor through the yeast secretory pathway. *Cell* 36:309–318.

Kaczorek, M., Honore, N., Ribes, V., et al. (1987) Molecular cloning and synthesis of biologically active human tissue inhibitor of metalloproteinases in yeast. *Biotechnology* 5:595–598.

Kikuchi, Y. (1983) Yeast plasmid requires a *cis*-acting locus and two plasmid proteins for its stable maintenance. *Cell* 35.487–493.

Kikuchi, Y., and Toh-e, A. (1986) A nuclear gene of *Saccharomyces cerevisiae* needed for stable maintenance of plasmids. *Mol. Cell. Biol.* 6:4053–4059.

Kingsman, S. M., Kingsman, A. J., Dobson, M. J., Mellar, J., and Roberts, N. A. (1985) Heterologous gene expression in *Saccharomyces cerevisiae. Biotechnol. Genet. Eng. Rev.* 3:377–415.

Knowles, J., Penttile, M., Teeri, T., Nevalainen, H., and Salvueri, I. (1985) *International*

Patent Publication No. WO 85/04672.

Kornbluth, S., Jove, R., and Hanafusa, H. (1987) Characterization of avian and viral p60^{svc}proteins expressed in yeast. *Proc. Natl. Acad. Sci. USA* 84:4455–4459.

Kramer, R. A., Dechara, T. M., Schaber, M. D., and Hilliker, S. (1984) Regulated expression of a human interferon gene in yeast: Control by phosphate concentration or temperature. *Proc. Natl. Acad. Sci. USA* 81:367–370.

Kramer, R. A., Schaber, M. D., Skalka, A. M., Ganzuly, K., Wong-Staal, F., and Reddy, E. P. (1986) HTLV-III *gag* protein is processed in yeast cells by the virus *pol*-protease. *Science* 231:1580–1584.

Kriskern, P. J., Hagopian, A., Montgomery, D. L., et al. (1986) Unusually high level expression of a foreign gene (hepatitis B virus case antigen) in *Saccharomyces cerevisiae. Gene* 46:135–141.

Kurjan, J., and Herskowitz, I. (1982) Structure of a yeast pheromone gene (MFα): A putative α-factor precursor contains four tandem copies of mature α-factor. *Cell*30:933–943.

Larionov, V., Kouprina, N., and Karpova, T. (1984) Stability of recombinant plasmids containing the *ars* sequence of yeast extrachromosomal rDNA in several stains of *Saccharomyces cerevisiae. Gene* 28:229–235.

Larionov, V. L., Karpova, T. S., Kouprina, N. Y., and Jouravleva, G. A. (1985) A mutant of *Saccharomyces cerevisiae* with impaired maintenance of centromeric plasmids. *Curr. Genet.* 10:15–20.

Lehman, D. J., Das, R. C., and Kaumeyer, J. F. (1987) Expression and secretion of human inter-α-trypsin inhibitor-related polypeptides in *Saccharomyces cerevisiae.* In *Abstracts of Yeast Cell Biology Meeting.* Cold Spring Harbor Laboratory, Cold Spring Harbor, NY, p. 62.

Loison, G., Nguyen-Juilleret, M., Alouani, S., and Marquet, M. (1986) Plasmid-transformed *ura3 fur1* double-mutants of *S. cerevisiae*: An autoselection system applicable to the production of foreign proteins. *Biotechnology* 4:433–437.

Maine, G. T., Sinha, P., and Tye, B. K. (1984) Mutants of *S. cerevisiae* defective in the maintenance of minichromosomes. *Genetics* 106:365–385.

Marquet, M., Alouani, S., and Brown, S. W. (1986) Plasmid stability during continuous culture in a *Saccharomyces cerevisiae* double mutant transformed by a plasmid carrying a eukaryotic gene. *Biotechnol. Lett.* 8:535–540.

McNeil, J. B., and Friesen, J. (1981). Expression of the herpes simplex virus thymidine kinase gene in *Saccharomyces cerevisiae. Mol. Gen. Genet.* 184:386–393.

Mead, D. J., Gardner, D. C. J., and Oliver, S. G. (1986) Enhanced stability of a 2μ-based recombinant plasmid in diploid yeast. *Biotechnol. Lett.* 8:381–396.

Mead, D. J., Gardner, D. C. J., and Oliver, S. G. (1987) Phenotypic differences between induced and spontaneous 2μm-plasmid-free segregants of *Saccharomyces cerevisiae. Curr. Genet.* 11:415–418.

Mellor, J., Dobson, M. J., Roberts, N. A., et al. (1983) Efficient synthesis of enzymatically active calf chymosin in *Saccharomyces cerevisiae. Gene* 24:1–14.

Mellor, J., Dobson, M. J., Roberts, N. A., Kingsman, A. J., and Kingsman, S. M. (1985) Factors affecting heterologous gene expression in *Saccharomyces cerevisiae. Gene* 33:215–226.

Miyajima, A., Bond, M. W., Otsu, K., Arai, K.-I., and Arai, N. (1985) Secretion of mature mouse interleukin-2 by *Saccharomyces cerevisiae:* Use of a general secretion vector containing promoter and leader sequences of the mating pheromone α-factor. *Gene* 37:155–161.

Miyajima, A., Miyajima, I., Arai, K.-I., and Arai, N. (1984) Expression of plasmid R388 encoded type II dihydrofolate reductase as a dominant selective marker in *Saccharomyces cerevisiae. Mol. Cell. Biol.* 4:407–414.

Miyajima, A., Otsu, K., Schreurs, J., Bond, N. W., Abrams, J. S. and Arai, K. (1986) Expression of murine and human granulocyte-macrophage colony stimulating factors in *S. cerevisiae:* Mutagenesis of the potential glycosylation sites. *EMBO J.* 5:1193–1197.

Miyamoto, C., Chizzonite, R., Crowl, R., et al. (1985) Molecular cloning and regulated expression of the human c-*myc* gene in *Escherichia coli* and *Saccharomyces cerevisiae:* Comparison of the protein products. *Proc. Natl. Acad. Sci. USA* 82: 7232–7236.

Moir, D. T., Mao, J., Duncan, M. J., Smith, R. A., and Kohno, T. (1985) Production of calf chymosin by the yeast *S. cerevisiae. Dev. Indust. Microbiol.* 26:75–85.

Mullenbach, T. G., Tabrizi, A., Blacher, R. W., and Steimer, K. S. (1986) Chemical synthesis and expression in yeast of a gene encoding connective tissue activating peptide III. *J. Biol. Chem.* 261:719–722.

Nakamura, Y., Sato, T., Emi, M., Miyanohara, A., Nishide, T., and Matsubara, K. (1986) Expression of human salivary α-amylase gene in *Saccharomyces cerevisiae* and its secretion using the mammalian signal sequence. *Gene* 50:239–245.

Neill, J. D., Litts, J. C., Anderson, O. D., Greene, F. C., and Stiles, J. I. (1987) Expression of a wheat α-gliadin gene in *Saccharomyces cerevisiae. Gene* 55:303–317.

Novick, P., Field, C., and Schekman, R. (1980) Identification of 23 complementation groups required for posttranslational events in the yeast secretory pathway. *Cell* 21: 205–215.

Novick, P., Ferro, S., and Schekman, R. (1981) Order events in the yeast secretory pathway. *Cell* 25:460–469.

Nunberg, J. H., Meade, J. H., Cole, G., et al. (1984) Molecular cloning and characterization of the glucoamylase gene of *Aspergillus awamori. Mol. Cell. Biol.* 4:2306–2315.

Oshima, T., Arima, K., Matsubara, K., Tanaka, S., and Nakazato, H. (1986) Expression of chemically synthesized α-Neo-endorphin genes under the control of the *PHO5* gene of *Saccharomyces cerevisiae. Agric. Biol. Chem.* 50:1161–1167.

Parent, S. A., Fenimore, C. M., and Bostian, K. A. (1985) Vector systems for the expression, analysis and cloning of DNA sequences in *S. cerevisiae. Yeast* 1:83–138.

Price, V., Mochizuki, D., March, C. J., et al. (1987) Expression, purification and chacterization of recombinant murine granulocyte-macrophage colony-stimulating factor and bovine interleukin-2 from yeast. *Gene* 55:287–293.

Pringle, J. R. (1975) Methods for avoiding proteolytic artefacts in studies of enzymes and other proteins from yeast. In *Methods in Cell Biology,* Vol. 12. D. M. Prescott (Ed.). Academic Press, New York, pp. 149–184.

Reichert, U., and Winter, M. (1975) Gene dosage effects in polyploid strains of *Saccharomyces cerevisiae* containing *gua-1* wild-type and mutant alleles. *J. Bacteriol.* 124:1041–1045.

Roggenkamp, R., Kustermann-Kuhn, B., and Hollenberg, C. P. (1981) Expression and processing of bacterial β-lactamase in the yeast *Saccharomyces cerevisiae*. *Proc. Natl. Acad. Sci. USA* 78:4466–4470.

Roggenkamp, R., Hoppe, J., Hollenberg, C. P. (1983) Specific processing of the bacterial beta-lactamase precursor in *S. cerevisiae*. *J. Cell. Biochem.* 22:141–149.

Rose, M., Casadaban, M. J., and Botstein, D. (1981) Yeast genes fused to β-galactosidase in *Escherichia coli* can be expressed normally in yeast. *Proc. Natl. Acad. Sci. USA* 78:2460–2464.

Rothstein, S. J., Lazarus, C. M., Smith, W. E., Baulcombe, D. C., and Gatenby, A. A. (1984) Secretion of a wheat α-amylase expressed in yeast. *Nature* 308:662–665.

Rudolph, H. K., Antebi, A., Fink, G. R., et al. (1989) The yeast secretory pathway is perturbed by mutations in *PMR1*, a member of Ca^{2+} ATPase family. *Cell* 58:133–145.

Sakai, K., and Yamamoto, M. (1986) Transformation of the yeast, *Saccharomyces calsbergensis*, using an antibiotic resistance marker. *Agric. Biol. Chem.* 50:1177–1182.

Sambucetti, L. C., Schaber, M., Kramer, R., Crowl, R., and Curran, T. (1986) The *fos* gene product undergoes extensive post-translational modification in eukaryotic but not in prokaryotic cells. *Gene* 43:69–77.

Schultz, L. D., Tanner, J., Hofman, K. J., et al. (1987) Expression and secretion in yeast of a 400 kDa envelope glycoprotein derived from Epstein-Barr virus. *Gene* 54:113–123.

Shoemaker, S., Schweickart, V., Ladner, M., et al. (1983) Molecular cloning of exo-cellobiohydrolase I derived from *Trichoderma reesei* strain L27. *Biotechnology* 1:691–696.

Shoemaker, S. P., Gelfand, D. H., Innis, M. A., Kwok, S. U., Landmer, M. B., and Schweickart, V. (1984) European Patent Application 84110305.4.

Singh, A., Lugovoy, J. M., Kohr, W. J., and Perry, L. J. (1984) Synthesis, secretion and processing of α-factor-interferon fusion proteins in yeast. *Nucl. Acids Res.* 12:8927–8938.

Skipper, N., Sutherland, M., Davies, R. W., et al. (1987) Secretion of a bacterial cellulase by yeast. *Science* 230:958–960.

Smith, R. A., and Gill, T. (1985) Yeast mutants that have improved secretion efficiency for calf prochymosin. *J. Cell. Biochem.* 9C(Suppl.):157.

Smith, R. A., Duncan, M. J., and Moir, D. T. (1985) Heterologous protein secretion from yeast. *Science* 229:1219–1223.

Stepien, P. P., Brousseau, R., Wu, R., Narang, S., and Thomas, D. Y. (1983) Synthesis of a human insulin gene. VI. Expression of the synthetic proinsulin gene in yeast. *Gene* 24:289–297.

Takagi, A., Harashima, S., and Oshima, Y. (1983) Construction and characterization of isogenic series of *Saccharomyces cerevisiae* polyploid strains. *Appl. Environ. Microbiol.* 45:1034–1040.

Takagi, A., Chua, E. N., Boonchird, C., Harashima, S., and Oshima, Y. (1985a) Constant copy numbers of plasmids in *Saccharomyces cerevisiae* hosts with different ploidies. *Appl. Microbiol. Biotechnol.* 23:123–129.

Takagi, A., Harashima, S., and Oshima, Y. (1985b) Hybridization and polyploidization of *Saccharomyces cerevisiae* strains by transformation-associated cell fusion. *Appl. Environ. Microbiol.* 49:244–246.

Takatsuji, H., Hirochika, H., Fukushi, T., and Ikeda, J.-E. (1986) Expression of cauliflower mosaic virus reverse transcriptase in yeast. *Nature* 319:240–243.

Teeri, T., Salovuori, I., and Knowles, J. (1983) The molecular cloning of the major cellulase gene from *Trichoderma reesei*. *Biotechnology* 1:696–699.

Thiele, D. J., Walling, M. J., and Hamer, D. H. (1986) Mammalian metallothionein is functional in yeast. *Science* 231:854–856.

Thim, L., Hansen, M. T., Norris, K., et al. (1986) Secretion and processing of insulin precursors in yeast. *Proc. Natl. Acad. Sci. USA* 83:6766–6770.

Thomsen, K. K. (1983) Mouse α-amylase synthesized by *Saccharomyces cerevisiae* is released into culture medium. *Carlsberg Res. Commun.* 48:545–555.

Tokunaga, T., Iwai, S., Gomi, H., et al. (1985) Expression of a synthetic human growth hormone gene in yeast. *Gene* 39:117–120.

Tonouchi, N., Shoun, H., Uozumi, T., and Beppu, T. (1986) Cloning and sequencing of a gene for mucor rennin, an aspartate protease from *Mucro pusillus*. *Nucl. Acids Res.* 14:7557–7568.

Tschumper, G., and Carbon, J. (1980) Sequence of a yeast DNA fragment containing a chromosomal replicator and the *TRP1* gene. *Gene* 10:157–166.

Tuite, M. F., Dobson, M. J., Roberts, N. A., et al. (1982) Regulated high efficiency expression of human interferon-alpha in *Saccharomyces cerevisiae*. *EMBO J.* 1:603–608.

Urdea, M. S., Merryweather, J. P., Mullenbach, D. C., et al. (1983) Chemical synthesis of a gene for human epidermal growth factor urogastrone and its expression in yeast. *Proc. Natl. Acad. Sci. USA* 80:7461–7465.

Valenzuela, P., Medina, A., Rutter, W. J., Ammerer, G., and Hall, B. D. (1982) Synthesis and assembly of hepatitis B virus surface antigen particles in yeast. *Nature* 298:347–350.

Velati-Bellini, A., Pedroni, P., Martegani, E., and Alberghina, L. (1986) High levels of inducible expression of cloned β-galactosidase of *Kluyveromyces lactis* in *Saccharomyces cerevisiae*. *Appl. Microbiol. Biotechnol.* 25:124–131.

Vlasuk, G. P., Bencen, G. H., Scarborough, R. M., et al. (1986) Expression and secretion of biologically active human atrial natriuretic peptide in *Saccharomyces cerevisiae*. *J. Biol. Chem.* 261:4789–4796.

Webster, T. D., and Dickson, R. C. (1983) Direct selection of *Saccharomyces cerevisiae* resistant to the antibiotic G418 following transformation with a DNA vector carrying the kanamycin resistance gene of Tn903. *Gene* 26:243–252.

Wen, D., and Schlesinger, M. J. (1986) Regulated expression of sindbis and vesicular stomatitis virus glycoproteins in *Saccharomyces cerevisiae*. *Proc. Natl. Acad. Sci. USA* 83:3639–3643.

Wood, C. R., Boss, M. A., Kenton, J. M., Calvert, J. E., Roberts, N. A., and Emtage, J. S. (1985) The synthesis and *in vivo* assembly of functional antibodies in yeast. *Nature* 314:446–449.

Wu, L. C., Fisher, P. A., and Broach, J. R. (1987) A yeast plasmid partitioning protein is a karyoskeletal component. *J. Biol. Chem.* 262:883–891.

Yarwood, J. N., Harris, N., Delauncy, A., et al. (1987) Construction of a hybrid cDNA encoding a major legumin precursor polypeptide and its expression and localization in *Saccharomyces cerevisiae*. *FEBS Lett.* 222:175–180.

Yoshizumi, H., and Ashikari, T. (1987) Expression, glycosylation and secretion of fungal hydrolysases in yeast. *Trends Biotechnol.* 5:277–281.

Zsebo, K. M., Lu, H.-S., Fieschko, J. C., et al. (1986) Protein secretion from *Saccharomyces cerevisiae* directed by the prepro-α-factor leader region. *J. Biol. Chem.* 262:5858–5865.

Zubenko, G. S., Park, F. J., and Jones, E. W. (1982) Genetic properties of mutations at the *PEP-4* locus in *Saccharomyces cerevisiae. Genetics* 102:679–690.

RECENT REFERENCES ADDED IN PROOF

Kikuchi, I., Yamamoto, Y., Taniyami, Y., et al. (1988) Secretion in yeast of human lysozymes with different specific activities created by replacing valine-110 with proline by site-directed mutagenesis. *Proc. Natl. Acad. Sci. USA* 85:9411–9415.

Mak, P., McDonnell, D. P., Weigel, N. L., Schrader, W. T., and O'Malley, B. W. (1989) Expression of functional chicken oviduct progesterone receptors in yeast (*Saccharomyces cerevisiae*). *J. Biol. Chem.* 264:21613–21618.

Pichuantes, S., Babe, L. M., Barr, P. J., and Craik, C. S. (1989) Recombinant HIV1 proteases secreted by *Saccharomyces cerevisiae* correctly processes myristylated *gag* polyprotein. *Proteins* 6:324–337.

Sabin, E. A., Lee-Ng, C. T., Shuster, J. R., and Barr, P. J. (1989) High-level expression and *in vivo* processing of chimeric ubiquit in fusion proteins in *Saccharomyces cerevisiae. Bio/technology* 7:705–709.

Sakai, A., Shimizu, Y., and Hishinuma, F. (1988) Isolation and characterization of mutants which show an oversecretion phenotype in *Saccharomyces cerevisiae. Genetics* 119:499–506.

Stetler, G. L., Forsyth, C., Gleason, T., Wilson, J., and Thompson, R. C. (1989) Secretion of active, full- and half-length human secretory leukocyte protease inhibitor by *Saccharomyces cerevisiae. Biotechnology* 7:55–60.

11
Future Prospects

Chandra J. Panchal
VetroGen Corporation
London, Ontario, Canada

There is no question about the enormous contribution yeasts have made to the welfare and betterment of humanity since the very beginnings of civilization as we know it. It is unthinkable what the world would be like without bread, beer, and wine made through the contributions of yeast. It is no wonder, then, that the study of modern biochemistry began with yeast and that to this day yeast is by far the microorganism most intimately associated with our economic as well as social well-being. With the yeast biotechnology base already in existence, it is safe to predict that yeast will continue to play a major role in industry for many more years to come. With the knowledge of genetic manipulation of yeast, however, the emphasis on industrial use of yeast will, I believe, gradually shift from traditional biotechnology to the "new" biotechnology. As a result, more and more attention will undoubtedly be placed on the appropriateness of particular yeast strains for the required usage. As the knowledge base for yeast molecular biology increases, the choices for strain selection will be increased accordingly, permitting the yeast biologist to effectively combine the selection process with the knowledge of large-scale industrial process to achieve the targeted goals.

Although the most commonly used yeast *Saccharomyces cerevisiae* is a simple unicellular organism, it has many of the characteristics of a complex, highly sophisticated eukaryotic system. Over the last few years it has become quite clear that many of the features found in this yeast have elements of commonality in higher eukaryotes. Most of these have been alluded to in the preceding chapters. It is not surprising, then, that *S. cerevisiae* has been considered a model eukaryote for biochemical and genetic studies. These include:

1. Study of cell specialization and cell-cell interactions: *S. cerevisiae* has specialized cells, *MATa* and *MATα* haploids, and *MATa/α* diploids, each behaving in a distinct and predictive manner. The specialized cells play a unique role during the life cycle of the organism, in a somewhat similar fashion to specialized cells in a multicellular organism which play distinct roles within the organism.

2. Study of differential gene expression: Cell differentiation in yeast requires inductive signals that trigger specific events such as gene transcription and cell-cycle arrest. The mating-type locus acts as the "master regulatory" element determining the fate of the cell.

3. Study of cell-cycle events: Many of the genes controlling cell-cycle events in yeast have similar homologues in higher eukaryotes.

4. Study of secretion: *S. cerevisiae* has a well-defined secretory pathway through which glycosylated or unglycosylated proteins can be secreted into either the periplasmic space or the extracellular medium. Similarities to secretion in higher eukaryotes are striking, and yeasts have extensively been used to secrete heterologous proteins of various origins.

5. Study of oncogenes: The yeast *ras* genes have been found to have properties similar to those of some of the oncogenes in mammalian cells and can, in fact, be replaceable by the latter genes for functionality in yeast.

6. Study of retroviruses: The yeast transposable element *Ty* has genetic similarities to mammalian retroviruses, except for the lack of the *env* gene necessary for propagation of the virus. The *Ty* element, however, has been a subject of intensive study for elucidation of the behavior of retroviruses and for the possibility of developing a vaccine (e.g., against AIDS virus).

7. Study of site-specific recombination: The existence of site-specific homologous recombination in yeast has attracted attention from scientists attempting to do the same in mammalian cells for developing gene therapy protocols.

8. Study of heat-shock proteins: Many of the heat-shock proteins produced by yeast in response to not only high temperature, but also ethanol and osmolarity-induced stress, have similarities to heat-shock proteins produced by higher organisms.

9. Study of catabolite repression: Catabolite repression still remains a mystery, yet the impact of this phenomenon is far reaching in both yeasts as well as higher organisms, including mammalian systems.

10. Study of viral or viruslike systems: The killer system in *Saccharomyces* yeasts is a very sophisticated, well-defined system and has been a good model for the study of secretion as well as virus (or virus-like)-host interactions.

The above represent just some of the attributes of *S. cerevisiae* that make it an extremely important eukaryote for the study of biochemistry and genetics of eukaryotes in general. In time, there is little doubt that more similarities will be discovered between *Saccharomyces* as well as other yeasts and higher organisms.

While the above attributes together with the simplicity of genetic manipulation of *Saccharomyces* yeasts have attracted basic research scientists to yeast, the impetus to strive for appropriate yeast strain selection has come primarily from the applied scientists endeavoring to use yeast for performing a specific function. Thus the use of yeast in producing a variety of microbial, plant, and mammalian proteins, which has been increasing at an accelerated rate in the last few years, has led to the search for "appropriate" strains that are stable, do not degrade the proteins, can be propagated easily, and so forth. It has become quite clear that chosing such an appropriate strain can be crucial to the economic viability of some bioprocesses. Plasmid propagation and maintenance can be highly strain specific. Use of a diploid or a polyploid, rather than a haploid, could be critical for a large-scale fermentation process.

As revealed in the preceding chapters, the appropriateness of the chosen yeast can only be judged by the use that it is being put through. For traditional processes such as baking, brewing, and wine making, the yeasts have in most instances been selected over centuries of usage. While modern strain selection in these cases may be limited to the improvement in efficiencies of the processes, it is generally not crucial to the end product, with some exceptions (e.g., off-flavor removal). In usage of yeast for heterologous gene expression, however, strain selection can be very critical, hence the surge in interest among applied scientists, particularly, either to improve host yeast strains or to look at alternate yeasts.

There has been a growing interest among yeast biologists to study non-*Saccharomyces* yeasts in the last few decades. This, I believe, will continue but at a much faster rate than before. Plasmid expression systems have already been developed in the methylotrophic yeast, *Pichia pastoris*, the lactose-utilizing yeasts, *Klyveromyces*, the fission yeast, *Schizosaccharomyces*, and the amylolytic yeast, *Schwanniomyces*. Other yeasts, such as *Ustilago, Hansenula, Rhodotorula, Candida shehatae*, and *Pichia stipitis*, among many others, are currently being actively studied in many laboratories around the world. The coming years will undoubtedly see the emergence of "novel" yeasts and yeast strains for production of pharmaceutical and/or food proteins. Such yeasts will also be used for fermentations to isolate novel compounds, such as flavor compounds or more efficient enzymes or novel carbohydrates.

Perhaps it may even be possible to have one of these yeasts produce antibiotics!

Index